Systems Engineering in Research and Industrial Practice

Josip Stjepandić · Nel Wognum ·
Wim J. C. Verhagen
Editors

Systems Engineering in Research and Industrial Practice

Foundations, Developments and Challenges

Springer

Editors
Josip Stjepandić
PROSTEP AG
Darmstadt, Germany

Croatian Academy of Sciences and Arts in Diaspora and Homeland (HAZUDD)
Gossau, Switzerland

Wim J. C. Verhagen
Aerospace Engineering, Air Transport and Operations
TU Delft
Delft, The Netherlands

Nel Wognum
Aerospace Engineering, Air Transport and Operations
TU Delft
Delft, The Netherlands

ISBN 978-3-030-33311-9 ISBN 978-3-030-33312-6 (eBook)
https://doi.org/10.1007/978-3-030-33312-6

This Springer imprint is published by the registered company Springer Nature Switzerland AG
The registered company address is: Gewerbestrasse 11, 6330 Cham, Switzerland

The editors of this book wish to dedicate its revenues to foundations that support high-potentials in achieving goals that would have been not possible otherwise. Referring to biographies of great scholars, with their achievements and contributions to humanity and science, we often do not realize that many of them experienced financial problems during their studies, which only could be solved with scholarships and grants. One of such foundations is the Foundation Fra Bonifacije Ivan Pavletić (www.zaklada.biskupija-sisak.hr) from Sisak, Croatia, which is dedicated to financially aiding high-potentials with heavy financial needs.

FOUNDATION
fra Bonifacije Ivan Pavletić

Acknowledgements

The editors, Josip Stjepandić, Nel Wognum, and Wim Verhagen, sincerely wish to express their gratitude to the colleagues of the Croatian Academy of Sciences and Arts in Diaspora and Homeland (HAZUDD), in particular Tomislava Majić, who have supported the editing of the book by adapting phrasing and English and improving graphics and charts.

Used Trademarks

Cimatron	Cimatron
EdgeMaster	Alicona
InsightTec, InsightNet, InsightData	CGEE
Mastercam	CNC Software

Contents

Part III Applications

Part IV Current Challenges

Editors and Contributors

About the Editors

Dr. Josip Stjepandić is Head of business unit 3D Product Creation at PROSTEP AG, the leading product data integration company worldwide. After receiving his grade as a M.Eng. from the University of Zagreb and his Ph.D. from the Graz University of Technology, he has worked for two automotive suppliers in the areas of engineering simulation and design methodology. From 1994 to 1996, he was Associated Professor for Applied Informatics in Mechanical Engineering at the University of Applied Sciences Dortmund. Since 1996, he has been in charge for consultancy and solution development at PROSTEP AG in the areas of design methodology, supplier integration, systems engineering, knowledge-based engineering, product data validation and visualization, configuration management and CAD data exchange for many industries (automotive, aerospace, shipbuilding, machinery). One of his most important contributions is the multiple awarded global supplier portal OpenDESC.com. His current research is focused on systems engineering, modular design, and Digital Twin. He is Vice President of ISTE, the International Society of Transdisciplinary Engineering (www.intsoctransde.org).

Dr. Nel Wognum has received her master's degree in Medical Informatics at Leiden University and her Ph.D. in knowledge-based systems at University of Twente, the Netherlands. She has been Assistant Professor in University of Twente, first in knowledge-based systems, later in organization and management studies. Her field of interest has since long been the (systemic) interaction between various disciplines as in concurrent engineering. She was the President of ISPE from 2004 till 2006. Since 2007, she has worked as a researcher in Wageningen University in the Management Studies Group in the field of supply chain management in the food area. She has authored and co-authored various publications in this field. Since the end of 2014, she is enjoying her retirement, but is still active in the field of transdisciplinary engineering. She is General Secretary of ISTE, the International Society of Transdisciplinary Engineering (www.intsoctransde.org).

Dr. Wim J. C. Verhagen is Assistant Professor at the section of Air Transport and Operations, Faculty of Aerospace Engineering, Delft University of Technology. He obtained his M.Sc. and Ph.D. at the same university. The focus of his early research has been on knowledge-based engineering systems in aircraft lifecycle processes. In recent years, his research has focused on aircraft maintenance research, focusing on the development models, methods, and applications for predictive maintenance and decision support. He has authored a range of publications in the fields of knowledge-based engineering systems and aircraft maintenance.

Contributors

Ronald C. Beckett Swinburne University of Technology, Melbourne, Australia

Alain Biahmou EDAG Engineering GmbH, Fulda, Germany

David Bradley Abertay University, Dundee, UK

Luís E. V. Loures da Costa Aeronautics Institute of Technology (ITA), São José dos Campos, SP, Brazil

Abbas Dehghani School of Mechanical Engineering, Institute of Design, Robotics and Optimisation (iDRO), University of Leeds, Leeds, UK

Songlin Ding Manufacturing Engineering, RMIT University, Melbourne, Australia

Fredrik Elgh School of Engineering, Jönköping University, Jönköping, Sweden

Alain-Jérôme Fougères ECAM Rennes, Rennes, France

Bin He Shanghai Key Laboratory of Intelligent Manufacturing and Robotics, School of Mechatronic Engineering and Automation, Shanghai University, Shanghai, People's Republic of China

Peter Hehenberger School of Engineering, University of Applied Sciences Upper Austria, Wels, Austria

John C. Hsu AIAA, Reston, USA;
INCOSE ESEP, San Diego, USA;
The Boeing Company, Long Beach, USA;
California State University Long Beach, Long Beach, USA;
System Management and Engineering Consulting Services, Cypress, USA;
UK Royal Academy of Engineering, London, UK;
Queens University, Belfast, UK

Katarzyna Jezierska-Krupa Silesian University of Technology, Gliwice, Poland

Joel Johansson School of Engineering, Jönköping University, Jönköping, Sweden

John P. T. Mo Manufacturing Engineering, RMIT University, Melbourne, Australia

Egon Ostrosi Pôle ERgonomie et COnception des Systèmes ERCOS/ELLIADD EA4661, Pôle Industrie 4.0, Université de Bourgogne Franche-Comté, UTBM, Belfort Cedex, France

Margherita Peruzzini University of Modena and Reggio Emilia, Modena, Italy

Fabien Pfaender UTSEUS, Shanghai University, Shanghai, People's Republic of China;
Costech EA2223, Université de Technologie de Compiègne, Compiègne, France

Wojciech Skarka Silesian University of Technology, Gliwice, Poland

Ryszard Skoberla Silesian University of Technology, Gliwice, Poland

Josip Stjepandić PROSTEP AG, Darmstadt, Germany

Luís Gonzaga Trabasso Aeronautics Institute of Technology (ITA), São José dos Campos, SP, Brazil

Patrick Traxler Software Competence Center Hagenberg, Hagenberg im Mühlkreis, Austria

Wim J. C. Verhagen Aerospace Engineering, Air Transport and Operations, TU Delft, Delft, The Netherlands

Timo Wekerle Aeronautics Institute of Technology (ITA), São José dos Campos, SP, Brazil;
Airbus Operations GmbH, Hamburg, Germany

Stefan Wiesner BIBA University, Bremen, Germany

Nel Wognum Aerospace Engineering, Air Transport and Operations, TU Delft, Delft, The Netherlands

Part I
Introduction

Chapter 1
Introduction to the Book

Josip Stjepandić, Nel Wognum and Wim J. C. Verhagen

Abstract The system concept has existed for several decades now, but is still a viable concept to be used to denote a problem area and to adopt a holistic view. The essence of a system is that it consists of elements and relationships between these elements, and that it exerts a function in its environment, provided it is an open system. A system can be defined at different layers of abstraction consisting of subsystems, which themselves may consist of subsystems again. The most complex level includes human beings. The system concept is adopted in Systems Engineering (SE) in which not only the engineering system under development is modeled, but also the development process itself in which many different disciplines need to be involved depending on the (lifecycle) requirements in focus. In this introductory chapter we draw the way we have paved to provide this book from the first idea on. The system concept, the origins, the goals and the expected audience of this book are roughly described. Finally, we give the first insight in the structure of this book and the mutual interdependence of the chapters. This book contains many different contributions in the area of SE, categorized into 4 parts: an introduction to the concept, methods and tools, applications, and challenges.

Keywords Systems · System thinking · Systems engineering

J. Stjepandić (✉)
PROSTEP AG, Dolivostr. 11, 64293 Darmstadt, Germany
e-mail: josip.stjepandic@prostep.com

N. Wognum · W. J. C. Verhagen
Aerospace Engineering, Air Transport and Operations, TU Delft, 2600 AC Delft, The Netherlands
e-mail: p.m.wognum@tudelft.nl

W. J. C. Verhagen
e-mail: w.j.c.verhagen@tudelft.nl

J. Stjepandić et al. (eds.), *Systems Engineering in Research and Industrial Practice*,
https://doi.org/10.1007/978-3-030-33312-6_1

1.1 The System Concept

The system concept has been defined in the 50s by von Bertalanffy [1]. It is used in General System Theory (GST) in a search for a general construct to describe the empirical world and to compare theories in different domains [2]. The ambition to find gaps in theories by comparing them in a systematic way has not been achieved, however.

The merit of system theory can be found in specifically framing and defining the focus of attention. This can be disciplinary, like a waste treatment model, but also interdisciplinary, combining two or more different disciplinary systems, like the waste treatment model and the eco system [3]. Of course, such an integrated model is less acceptable to each of the disciplines, but is a compromise to support communication and the search for an inter-disciplinary solution. Trans-disciplinary systems add a level of analysis, which does not exist on the level of each of the disciplines [4, 5].

With the system concept every object in the real world can be denoted and characterized. In principle, a system consists of elements and relationships between elements [6]. Moreover, a system performs a function in its environment, intentionally or not, provided it is an open system. For example, a stone is a system that can be described by characterizing its elements and the relationships between the elements, which are predominantly chemical. A stone can exert a function in its environment, for example, putting a weight on a pile of papers to prevent them from blowing away in the wind. Systems can be identified on many levels of complexity, with subsystems as elements. For example, a car is a system in which many different interacting subsystems can be identified, consisting of subsystems themselves [7].

On the most complex level, humans are involved. Examples are organizational systems, in which people perform processes with the help of methods, tools and knowledge to achieve the functions of the system in its environment. An example of such a function is satisfying the needs of consumers or customers. An organizational system is called complex, because the goals of the system are not static, due to interactions between goals of people and the intended goals of the organization [3]. The system concept is useful for framing problem areas, because content is separated from context by clearly defining system boundaries. Studying a system, however, cannot be fully successful when context is left out of scope. Interaction between a system and its context needs to be part of the study.

Many different system approaches have been developed and used in the past decades, like GST, system thinking soft-system thinking [8], as well as systems engineering, the subject of this book. Wang has extended this concept to the systems intuition and the collective systems intuition [9]. The goals of these different methods differ, as well as their approach and application domains. However, the system concept is the central concept with a clear boundary and context.

Systems engineering is an approach in which the development of an engineering system itself is also considered as a system, an organizational system. From the start of a systems engineering process, all relevant product and process requirements are taken into account that concern the whole product lifecycle. In the organizational system all relevant disciplines need to be involved in the development process depending on the requirements at focus. For example, when manufacturing requirements need to be considered, manufacturing engineers need to be part of the development team. When market or customer requirements are considered, social-science disciplines need to be involved.

A systems engineering approach, inherently, puts a heavy load on the management of a systems engineering process. Not only the different teams, but also the engineering system itself requires the management of different types of knowledge and information, including the necessary communication. A system engineering project often consists of several, interacting, teams and sub-teams, necessitating extensive communication and information exchange. We refer to Chap. 14 for more detail on the achievements and challenges of SE.

In the book, several systems engineering approaches are presented (see Sect. 1.6).

1.2 Origins of the Book

This book is the result of various discussions during and after the 23rd and 24th ISTE international conference on Transdisciplinary Engineering in Curitiba (Brazil), in October 2016 resp. Singapore, in July 2017 [10, 11]. A number of valuable submissions were selected as well as new submissions invited by the editors Josip Stjepandić, Nel Wognum, and Wim Verhagen. The contributors were primarily recruited from the ISTE community (www.intsoctransde.org). To achieve higher practical relevance, several industry experts were invited to contribute to the book.

1.3 Goals of the Book

This book is an attempt to present the latest developments and best practices of the principles of Systems Engineering (SE). The presentation includes not only current SE processes and methods, but also, very importantly, complex real-life applications and experiences. These applications and experiences are aimed to show that SE is still an indispensable part of business nowadays. The term SE covers a variety of approaches that can be classified as SE approaches. Each such approach must be connected to an innovation or product and process development. Each approach must also consist of methods and tools to enable and support extensive collaboration and information exchange between people from different disciplines, functions, departments or companies.

The first goal of the book is to characterize the SE concept. A second goal of the book is to illustrate the choices that exist in organizing information. These choices encompass selection of methods and tools, technical as well as organisational. The methods and tools show the variety of problems that need to be tackled in practice [12]. They should support trade-offs and finding (near-)optimal solutions [13]. The third goal of this book is to demonstrate that Systems Engineering has become indispensable, used widely in many industries and that the same basic engineering principles can be applied to new, emerging fields like city design. The final goal of the book is to provide sufficient examples that thoroughly illustrate achievements and practices of SE. In addition, many remaining challenges in research and practice are listed.

1.4 Audience

The authors intend this book to be useful for several audiences: industry experts, managers, students, researchers, and software developers. The content is intended to serve both as an introduction to development and assessment of novel approaches and techniques of SE and as a compact reference for more experienced experts. In this role practitioners can use the content to improve their core competencies and use it as a reference during their daily work. Graduate and undergraduate students who have already mastered several basic areas of engineering may find it useful instruction material to practices in modern industrial product creation processes. Researchers can find recent achievements and challenges in various fields of SE.

Engineers in various design domains, such as mechanical, electrical, computer science, and environmental and logistics engineering may find this book helpful to understand the fundamental background as captured in modern systems engineering. It may help them to understand the multi-disciplinary, multi-dimensional and multi-level nature of SE. It may help them to request information they need from and to supply information needed for engineering processes to the relevant stakeholders. It will help stakeholders from various domains to understand how SE works and to participate in different SE teams.

Managers need to understand information representing numerous facets of SE for developing a comprehensive strategy and establishing suitable engineering structures and organization. The decisions they make must advance business competition by meeting quality, cost and time targets. Management and engineering need to exchange information rapidly and seamlessly so that the processes will be adjusted to support the business strategy and so that management can understand and track product issues and maturity. This book presents several methods for organising, transferring, tracking and tracing of information.

Students and researchers in the wide area of engineering need comprehensive information on recent achievements and on directions for future research. The book fulfills this need. For this purpose valuable information can be found in the closing part of this book.

Finally, a further audience may consist of developers of tools and development platforms who usually have a strong software engineering background and are not experienced in applications and process development. In particular, for those who define and implement integration scenarios, this book could be a useful reference.

1.5 Structure of the Book

The present edited book is a collection of 14 chapters written upon invitation from the editors by internationally recognized experts from academia and industry. Singular chapters contribute to various aspects of basic concepts, methods, technologies, industrial applications, and current challenges of SE. The volume is organized in four parts according to the main subjects: Background, New developments and methods, Applications and Current challenges. The structure of the book is illustrated in Fig. 1.1.

The first part of the book presents a brief introduction to Systems Engineering.

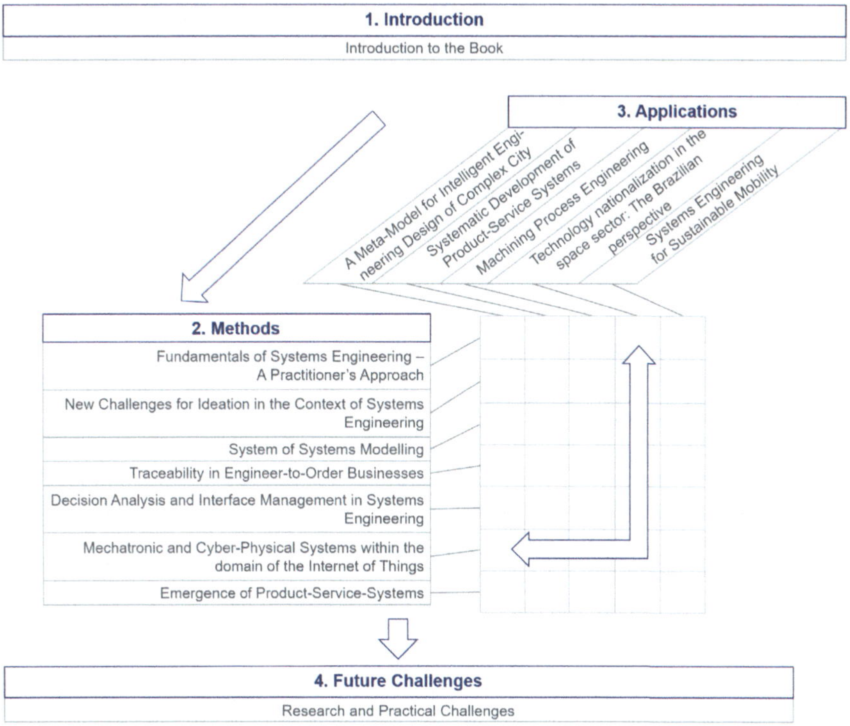

Fig. 1.1 The structure of the book

The second part of the book contains methods for Systems Engineering. It presents fundamentals of SE, ideation methods, system of systems modeling, decision analysis and interface management, interaction between mechatronic and cyber-physical systems, and product-service systems.

The third part addresses applications of SE in industry. It comprises achievements and experiences in city design, product-service systems, machining, the space sector and sustainable mobility.

The fourth part presents current challenges that have been identified by the editors and authors of the book.

1.6 Content of the Book

The book has been structured into four consecutive parts, which will be introduced below.

Part 1: Introduction

Chapter 1 Introduction to the Book (this chapter).

Part 2: Methods

In Chap. 2, John C. Hsu introduces the fundamentals of systems engineering as a practitioner's approach. In today's globalized environment, companies compete fiercely for business. They need world class product quality, no cost overrun and schedule slippage. Customer satisfaction is number 1 and cannot afford system development failures. Practicing systems engineering is the right answer. It is an old subject but has been revitalized since the mid 90s. Systems engineering is now a major theme in this century, has led to reduction in time-to-market, improving quality and reducing costs. However, systems engineering has not been sufficiently understood by the majority of workers (technical and non-technical, professional and non-professional, and financial, etc.), evidenced by many failures been reported, specifically in US Government Accountability Office (GAO) reports. Therefore, continuous systems engineering education is still needed, which is the major theme of this chapter.

In this chapter a history of systems engineering is introduced; including why it is needed, its evolution and revitalization, the fundamentals are presented. The requirement management process needs to be followed to analyze, derive, allocate and trace the requirements. Functional analysis can assist requirement hierarchy developed, vice versa; requirements can also feed function architecture development. This chapter is concluded by overview of the functional allocation and the system synthesis.

In Chap. 3, Wojciech Skarka, Katarzyna Jezierska-Krupa and Ryszard Skoberla present the usage of ideation methods in a Systems Engineering context. The first part presents the classification of ideation methods. This classification is based on different criteria to allow designers to take a look from different points of view. There is also a discussion on the division of ideation methods and how to choose the right method for a given case. The second part deals with background and a review of the most popular methods with special attention for the analysis of ideation methods in complex and multidisciplinary projects over the last few years. The definitions of discussed methods are given in this part. This classification can be a quick way to get to know the issue of ideation methods. This approach allows to elaborate guidelines concerning the usefulness of the analyzed ideation methods in Systems Engineering. Special attention has been given to methods that have been commonly used in recent case studies. The presented two case studies describe the usage of selected methods in examples from the field of design and development of medical devices as well as the automotive industry. Reading this chapter is meant to result in the evaluation of the value of ideation methods depending on the application. The conclusions sum up guidelines concerning the usage of ideation methods and indicate the direction of changes and improvements in these methods.

In Chap. 4, John P. T. Mo and Ronald Beckett illustrate the System of Systems modelling. The design, manufacturing and through-life support of modern engineering systems such as an aircraft or a frigate are complex, multifaceted and may change over time. These engineering systems are working in an environment that has multiple individual users, complicated supply chains, many government and socially affected stakeholders. In essence, these systems are working as a system of interacting semi-autonomous systems each of which are governed by their individual set of rules and could operate with different enterprise structures. Engineers trying to apply the theory of systems engineering to “design” a system of systems find the outcome often unpredictable and uncontrollable, as the linked systems operate with high degree of independence. System operations are embedded in business networks that are evolving and changing all the time. Individuals and organisations participate voluntarily in the networks. They can come and go at any time without warning. These highly uncertain relationships require different approaches. This chapter addresses the modelling requirements for design, development, implementation and operation of a complex system that interacts with many socio-technical systems. The methodology is illustrated by two case studies.

In Chap. 5, Fredrik Elgh and Joel Johansson discuss the topic of traceability in engineer-to-order businesses. A rapidly growing strategy in product design and manufacture, with great potential to improve customer value, is mass-customization. The main idea is to divide the product into modules that can be shared among different product variants. This will support a wide range of options for the end customer to select from, while an internal efficiency, similar to mass-production, can be achieved. This has been a success for many companies acting on the consumer market. However, many manufacturing companies are engineer-to-order (ETO) companies, such as original equipment suppliers (OES). They design a unique solution, often in close collaboration with other companies. The solution can then be manufactured in different quantities depending on the client’s need. For these companies, there is a strategic need for developing high quality engineering support to further utilize and exploit the information and knowledge produced during product development and to succeed with a strategy influenced by the principles of mass-customization. This has to include the implementation and management of systems enabling highly custom-engineered products to be efficiently designed and manufactured. One challenge when introducing such flexible support is to enable traceability of decisions taken, tasks executed, knowledge used and artefacts developed throughout the whole life-cycle of an individual product.

In this chapter, it is shown that traceability can be achieved by introducing support for capturing, structuring and mapping between decisions and resulting outputs, such as geometrical building blocks, knowledge implemented as rules, and the argumentation for the selection, design and specification of these. Three examples are presented where the concept Design Description has been modelled based on an item-oriented, a task-oriented, and a decision-oriented perspective which show the generality of the Design Description concept. The three examples demonstrate how to use the Design Description to enable traceability in platform design, product design, and manufacturing development processes.

In Chap. 6, John C. Hsu concludes the fundamentals of systems engineering by discussing decision analysis and interface management. The cross-cutting technical management process facilitates both the systems design and product realization processes. The eight sub-processes within the cross-cutting technical management process are: technical planning, requirements management, interface management, technical risk management, configuration management, technical data management, technical assessment, and decision analysis. The technical management processes link project management with the technical team. Subsequently, individual members and tasks are integrated into a functioning system that meets cost and schedule pre-requisites. The cross-cutting functions serve to execute project control on the apportioned tasks. In this chapter, emphasis is put on explaining decision analysis and interface management. Decision analysis is the process of making decisions based on research and systematic modeling of trade-offs. The objective of a decision analysis is to discover the most advantageous alternative under the circumstances. Decision analysis may also require human judgement and is not necessarily completely machine driven. In detail is demonstrated how to conduct the trade-off study. While interfaces are connection points between parties or elements, interface management provide a systematic methodology to handle with multiple parties or technical elements. Implementing an interface management process on a project identifies critical interfaces, streamlines communication, and monitors ongoing work progress while mitigating risks. With interface management, interface definition, identification and interface management tools are provided.

In Chap. 7, Peter Hehenberger, David Bradley, Abbas Dehghani and Patrick Traxler highlight the specific requirements of mechatronic and cyber-physical Systems within the domain of the Internet of Things. There has been a shift in emphasis within systems from hardware-oriented to more software-oriented topics integrated in an overlaying communication framework (e.g., cloud-based services). This chapter presents current research in the field of the interaction between mechatronic and cyber-physical systems. It presents design methods that are illustrated by some real-world applications. Four case studies (Smart Home, Bio-mechatronic Systems, Cyber-Physical Production System, Data-driven analysis) are discussed and provide illustration of applications involving different functional distributions of activity between the 4 key elements of people, data, mechatronics and cyber-physical system. The concepts present designers with the challenge of implementing structures within information rich environments where information and communications are increasingly the drivers of the design process. This in turn requires designers to have access to new and novel means of simulation capable of representing such situations.

In Chap. 8, Margherita Peruzzini and Stefan Wiesner introduce Product-Service Systems. Product-Service Systems (PSSs) are an emergent way to innovate traditional products and to extend the company portfolio, by reducing time and cost while offering high quality and meeting the expectations of both customers and stakeholders, which have to be considered during the design and development process. A further challenge is to close loops between Product Lifecycle Management (PLM) and Service Lifecycle Management (SLM) by providing feedback from service delivery to the beginning-of-life phase of products, or defining a structured procedure to coordinate product and service development activities. The objective of this chapter is to provide a common understanding on PSSs, to deepen the servitization process and its main features, and to understand how PLM and SLM can be integrated to define future organization of PSS-oriented companies. The final aim is to present PSS as a new business model, which companies can adopt to innovate their products and to enlarge their offer to the market, according to a consumer-oriented approach.

Part 3: Applications

In Chap. 9, Fabien Pfaender, Egon Ostrosi, Alain-Jérôme Fougères and Bin He draw a meta-model for intelligent engineering design of complex city. A city is a complex system, requiring the input of multiple disciplines for its (re)design. It shares some properties of two kinds of objects: empirical objects as well as theoretical objects. As city emerges as a complex object for multi-disciplinary studies, it is of the highest importance to adopt a systemic and global approach to bring new knowledge to this field. To master the growing complexity of cities and to consider at the same time heterogeneous ways of thinking of city, intellectual tools and models are needed. The goal of this chapter is to propose a model for describing engineering modelling knowledge with relationships and transformations between four domains: (1) citizen, (2) functional, (3) physical and (4) process. The proposed model is structured on four levels of modelling: (1) conceptual, (2) mathematical, (3) computational, and (4) experimental. These networks of models should be necessarily intelligent for managing the engineering design of a smart city. For overall city design, the paradigm should change from planner-centric to citizen-centric. However, while these models are potentially relevant, data that may feed these models is lacking most of the time. Moreover, filling and detailing each of the models, requires additional input from different experts and theories. In this chapter smart city engineering design is focused on three interrelated approaches: (a) data that should be gathered, (b) models that can be used by means of these data, and (c) interpretation methods and tools to elaborate knowledge and decision from the results these models can produce. The paper presents some findings from an application of the proposed meta-model.

In Chap. 10, Margherita Peruzzini and Stefan Wiesner conclude their treatment of Product-Service Systems (PSSs) with a description of its systematic development. The main problems occurring in PSSs, are due to an inadequate requirements analysis and lack of a strong PSS conceptual design. Problems vary from exceeding budgets, to missing functionalities, unsuccessful market launch, or even project abortion. Furthermore, the special characteristics of a PSS have to be considered already at an early stage of the development process. Requirements Engineering (RE) and design methodology as well as supporting Information and Communication Technologies (ICT) need to establish a common perception of the targeted PSS. At the same time, the inner complexity of PSS leaves requirements analysis, design activities and development tasks fragmented among many disciplines and sometimes conflicting, unstable, unknowable or not fully defined. In this context, a concurrent, transdisciplinary and collaborative design of PSS is required to create feasible and successful solutions. The objective of this chapter is to present a structured approach to face the specific challenges of PSS development in detail, to elaborate a general framework that features a systematic approach for PSS development, and to consider the effects of changes in specific product and service design on a systematic PSS development process.

In Chap. 11, John P. T. Mo and Songlin Ding explain the machining process engineering. Machining is the traditional product shaping process by removing materials from a block of original materials. Practically, the machining process itself has not changed much in the last couple of centuries but the accessories around the process have improved significantly, like data logging features in modern computer numerically controlled machines. The machining process is a system, the components of which should be considered as independent units, which work harmoniously with other systems in the enterprise. In this chapter a systems approach is adopted to examine methods and techniques that can improve five key performance indicators of the machining system, i.e. sustainability, accuracy, efficiency, precision and reliability. In particular, High Speed Machining, tool breakage prevention, thin wall deflection, tool geometry and chatter monitoring are studied in relation to five performance indicators, respectively. Application of these techniques has produced good machining outcomes showing strategic development direction leading to better performance of the machining system.

In Chap. 12, Timo Wekerle, Luís Gonzaga Trabasso, and Luís E. V. Loures da Costa discuss the concept of technology nationalization in the space sector from the Brazilian perspective. Brazil as an emerging country needs to catch up with technology to extend its position on the international market, especially in the space sector. The Technology Nationalization Framework (TNF) is a strategy for nationalization and industrialization of high technology products. The TNF is meant to assure that strategic technologies, that are currently lacking, will be designed, produced, and operated in Brazil as long as needed, without the risk of export bans or unavailability of components. The framework is based on reengineering with subsequent transfer to the national industry. The strategy starts with the identification of strategic technologies in relation to technologies already present in Brazil. For the nationalization process of these technologies, a decision-making process is needed taking into account available resources and competencies. In this chapter the TNF will be introduced and explained, while also a pilot project is described in which the TNF strategy is applied.

In Chap. 13, Alain Biahmou presents the application of Systems Engineering for emerging field of sustainable mobility. Nowadays, sustainability has established itself in the automotive industry and has evolved to an indispensable part of it. In contrast with its initial understanding as ecological improvement during development and production of vehicles, it has emerged to an advanced concept that considers much more, for instance the interaction of vehicles with the superordinate system in which they are included. Therefore, not only the reduction of the pollution as well as of resource consumption, but also the impact on the societal, economic and environmental development is of great importance. Current product development in many companies is still characterized by the fact that different disciplines create several partial models of the same product and provide much information in documents only. Periodic synchronizations of common parameters and models are performed. Information related to sustainability even when it exists is not consistent and not represented in models, which can be used for synchronization points. Therefore, sustainability is often not really taken into account along the product life cycle.

In order to master the complexity of smart products, which arises not only from customer behavior and requirements, but also from legal requirements related to sustainability, a proposal is made for Systems Engineering to integrate sustainability to a larger extent. Based on the main research directions in sustainability, such as innovative design concepts including alternative propulsions for less pollution, the safety and driver assistance for resource efficiency and life protection, the mastery of networked vehicles to control the interaction of a car with its superordinate system, adapted and even new methods as well as processes are needed to link Systems Engineering with sustainability. This chapter presents proposals of product development processes that take Systems Engineering methods into account as well as sustainable mobility. The prerequisites to realize such a product development process are described, whereby the whole product development cycle from the product concept down to disposal is taken into account.

Part 4: Future challenges

Chapter 14 summarizes the research and practical challenges that have been presented in the book. Systems Engineering (SE) is a well-established field of research and practice. Nevertheless, the theory underlying SE is experiencing significant development, directly and in association with advancements in closely associated research domains. In this final chapter, a socio-technical perspective is applied to identify and describe major trends in SE, as well as identifying future challenges in theory and application of SE. In doing so, trends are identified for (1) strategic issues from a product and process lifecycle perspective; (2) stakeholder representation and involvement; (3) current and future technologies employed to enable SE; (4) knowledge and skills as contributed by people and teams; and (5) structures to enable transdisciplinary activities supporting a socio-technical system perspective in systems development. Challenges remain present regarding these dimensions; SE requires methods and tools that are suitable to support the dynamic and evolving nature of the systems that need to be developed including the development system itself. Besides, management of SE projects for solving complex societal problems requires people with vision and power to motivate and mobilize the necessary people and value their respective input in the overall task. Transdisciplinary Engineering is introduced as an approach in which Systems Thinking and System Approaches interoperate, taking into account the different levels of abstraction of the system of focus.

1.7 Contributors of the Book

The editors have selected and invited contributors based on their recent contribution to TE conferences. Additionally, industry experts have been invited to contribute. The editors are grateful to all contributors for their excellent work.

References

1. von Bertalanffy L (1951) General systems theory: a new approach to unity of science. Hum Biol 23:303–361
2. Boulding KE (1956) General systems theory—the skeleton of science. Manage Sci 2(3):197–208
3. Nelson RJ (1976) Structure of complex systems. In: PSA: Proceedings of the Biennal meeting of the philosophy of science association, vol 1976, Volume two: symposia and invited papers. University of Chicago Press, Chicago, pp 523–542
4. Hofkirchner W, Schafranek M (2011) General systems theory. In: Hooker C (ed) Philosophy of complex systems. North Holland Elsevier, Oxford, pp 177–194
5. Wognum N, Bil C, Elgh F, Peruzzini M, Stjepandić J, Verhagen WJC (2019) Transdisciplinary systems engineering: implications, challenges and research agenda. Int J Agile Syst Manag 12(1):58–89
6. Marchal JH (1975) On the concept of a system. Philos Sci 42(4):448–468
7. Biahmou A (2015) Sustainable mobility. In: Stjepandic J, Wognum M, Verhagen JC (eds) Concurrent engineering in the 21st century: foundations, developments and challenges. Springer International Publishing, Switzerland, pp 779–803
8. Checkland P, Holwell S (1998) Information, systems, and information systems: making sense of the field. Wiley, Hoboken
9. Wang Z (2003) Systems intuition: oriental systems thinking style. J Syst Sci Syst Eng 12(2):129–137
10. Borsato M, Wognum N, Peruzzini M, Stjepandić J, Verhagen WJC (2016) Transdisciplinary engineering: crossing boundaries. In: Proceedings of the 23rd ISTE international conference on transdisciplinary engineering, Advances in transdisciplinary engineering, vol 4, IOS Press, Amsterdam
11. Chen CH, Trappey AC, Peruzzini M, Stjepandić J, Wognum N (2017) Transdisciplinary engineering: a paradigm shift. In: Proceedings of the 24th ISTE international conference on transdisciplinary engineering, Advances in transdisciplinary engineering, vol 5. IOS Press, Amsterdam

12. Jauhari TK, Maimun A, Siow CL (2019) Review of systems engineering methods, techniques and tools for ship design as large and complex systems. In: International congress and conferences on computational design and engineering 2019 (I3CDE 2019). http://www.appcom.co.kr/i3cde2019/i3CDE_proceedings.pdf. Accessed 22 July 2019
13. Tien JM, Berg D (2003) A case for service systems engineering. J Syst Sci Syst Eng 12(1):13–38

Part II
Methods

Chapter 2
Fundamentals of Systems Engineering—A Practitioner's Approach

John C. Hsu

Abstract In today's globalized environment, companies compete fiercely for business. They need world class product quality, no cost overrun and schedule slippage. Customer satisfaction is number 1 and cannot afford system development failures. Practicing systems engineering is the answer. It is an old subject but has been revitalized since the mid 90s. Systems engineering is now a major theme in this century has led to reduction in time-to-market, improving quality and reducing costs. However, systems engineering has not been sufficiently understood by the majority of workers (technical and nontechnical, professional and non-professional, and financial, etc.), evidenced by many failures been reported, specifically in US Government Accountability Office (GAO) reports. Therefore, continuous systems engineering education is still needed, which is the major theme of this chapter. In this chapter a history of systems engineering is introduced; including why it is needed, its evolution and revitalization. the fundamentals are presented. The requirement management process needs to be followed to analyze, derive, allocate and trace the requirements. Functional analysis can assist requirement hierarchy developed, vice versa; requirements can also feed function architecture development. This chapter is concluded by overview of the functional allocation and the system synthesis.

Keywords Systems engineering · Revitalization · Requirements management · Functional analysis and allocation · Trade studies · Interface management

J. C. Hsu (✉)
AIAA, Reston, USA
e-mail: john.hsu@csulb.edu

INCOSE ESEP, San Diego, USA

The Boeing Company, Long Beach, USA

California State University Long Beach, Long Beach, USA

System Management and Engineering Consulting Services, Cypress, USA

UK Royal Academy of Engineering, London, UK

Queens University, Belfast, UK

J. Stjepandić et al. (eds.), *Systems Engineering in Research and Industrial Practice*,
https://doi.org/10.1007/978-3-030-33312-6_2

2.1 Introduction

In today's globalized environment, manufacturing and designing companies compete for business. To be successful, companies need to practice strategies that minimize the possibility of degradation of product quality, cost overrun, schedule slippage, customer dissatisfaction and system development failures. Systems engineering is the answer to the above statement. It is different from mechanical, aeronautical, electrical and other engineering disciplines and yet bridges these traditional engineering disciplines. Systems engineering is focused on the system as a whole. While the primary purpose of systems engineering is to guide, it does not mean that systems engineers do not themselves play a key role in system design. Systems engineering is needed now and in the future to meet the following challenges:

- Constantly changing requirements
  - Requirements for new system are frequently changing
  - Changes in mission thrusts
  - Continuous introduction of new technologies
- More emphasis on "systems"
  - Greater emphasis on total system versus the components of a system
  - Functions need to be performed in an effective and efficient manner
  - Look at the system throughout its entire life cycle
- Increasing system complexities
  - System becomes more complex, such as, system-of-systems for network-centric applications
  - System design changes should be incorporated quickly, efficiently, and without causing a significant impact of the overall configuration of the system-of-systems, system, or subsystem
- Increasing globalization
  - More trading and dependency on different countries
  - Introduction of rapid and improvement communications
- Greater international competition
- Increasing globalization can also trigger more international competition
- More outsourcing
  - More suppliers associated with any given program
  - Needs early definition and allocation of system level requirements

Systems engineering fundamentals in this chapter will cover the following subjects:

1. Introduction to Systems Engineering.
2. Requirements Management.
3. Functional Analysis and Architecture.

Section 2.2 Introduction to Systems Engineering will explain why we need systems engineering; evolution of systems engineering; the history of systems engineering revitalization; and the role of systems engineers. The systems engineering methods and processes presented in this chapter can be directly applied to the job. Learn how to analyze and develop requirements in Sect. 2.3; and how to validate, trace and allocate requirements. Four (4) elements of functional analysis and Allocation are discussed in Sect. 2.3. In Sect. 2.3 the Functional Flow Block Diagrams and Integrated Definition for Functional Modeling are presented, as well as the relationships between functional allocation and system synthesis, and decision Analysis as needed during the design, development and manufacturing life cycle.

2.2 Introduction to Systems Engineering

In this section, we discuss in detail the poor performance of engineering projects in current industries, especially with respect to cost overruns and schedule delays. Practicing systems engineering (SE) may be the answer to correct these problems. The value of SE is presented. SE is not a new subject; therefore, the evolution of SE is discussed in this section. Many people may not know what SE revitalization is. It will be presented and discussed here. Finally, the role of systems engineers is explained.

2.2.1 Why Systems Engineering

This is often a question in people's mind. Why do I need to know and understand systems engineering? What is good for me? One obvious reason that we can give is [1]:

> Systems Engineering provides theory and methods for the management of complexity. Without Systems Engineering, we can expect additional developmental failures, cost overruns, schedule slippages, customer dissatisfaction, and environmental disasters.

Figure 2.1 is extracted from Metrics and Case Studies for Evaluating Engineering Designs, Moody et al. [2]. It shows that the cost overrun decreases as the systems engineering (SE) effort as a percentage of total program cost increases for most of the space programs in the past. Figures 2.2 and 2.3 [3] show that the cost ratio of actual to plan and schedule ratio of actual to plan, respectively, decreases as SE effort increases. The SE effort is defined as the product of SE quality and the ratio of SE cost to actual cost [4].

United States (US) Government Accountability Office (GAO) [5] have found problems related to quality that have resulted in major impacts to the 11 Department of Defense (DoD) weapons systems, billions of dollars in cost overrun and years-long delays, and decreased capabilities for the warfighter. GAO's analysis of 11 DoD

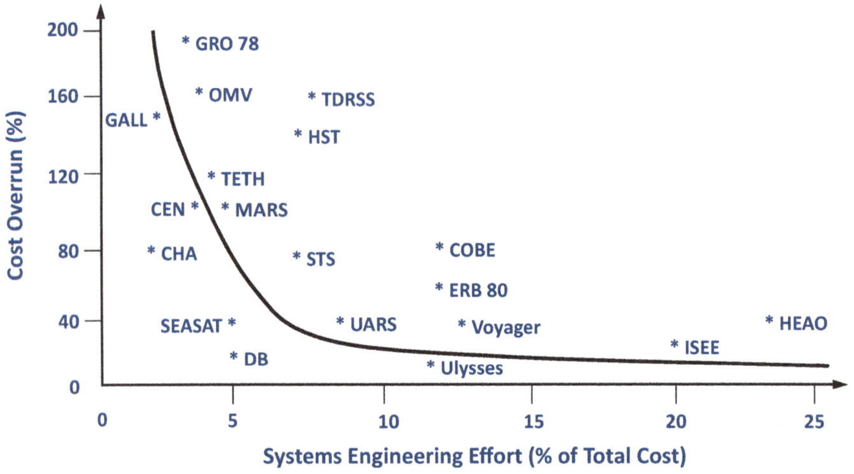

Fig. 2.1 Cost overrun versus systems engineering effort [2]

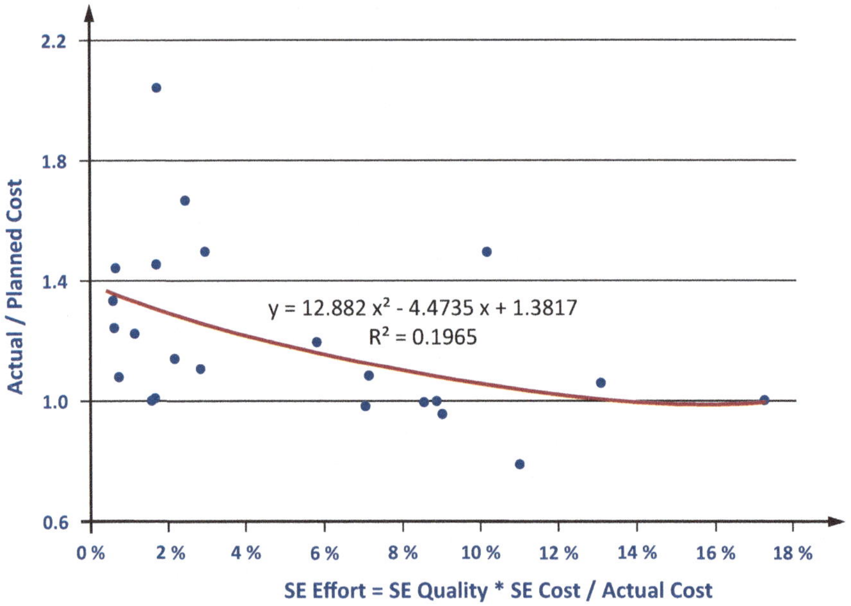

Fig. 2.2 Cost ratio of actual to plan [3]

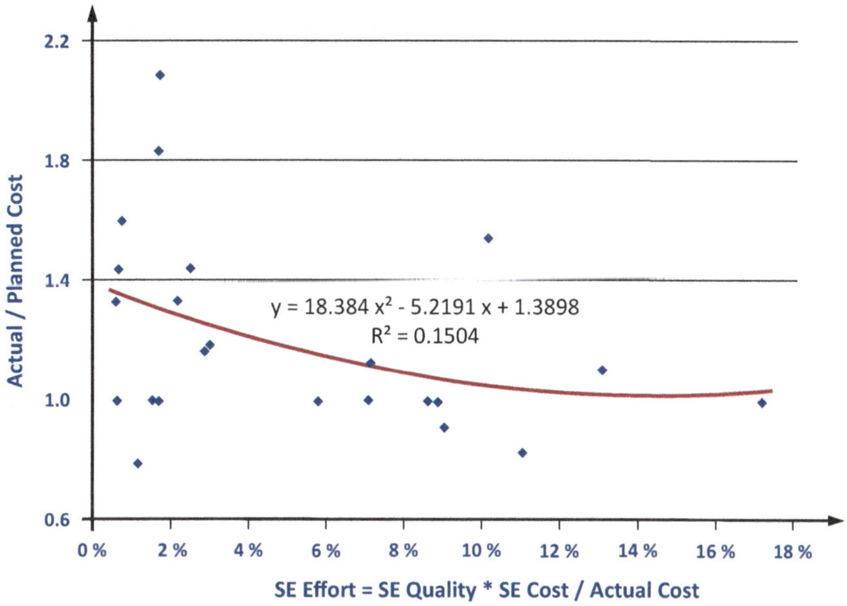

Fig. 2.3 Schedule ratio of actual to plan [3]

weapon systems illustrates those defense contractors' poor practices of systems engineering activities as well as manufacturing and supplier quality problems. A study [6] was undertaken why cost overrun and schedule delays have occurred and continue to occur in large-scale US federal defense and intelligence acquisition programs. One of the major reasons is inadequate systems engineering practices. Requirements redefinition and creep was discussed as a major problem. The acquisition strategy for programs must embrace the systems engineering processes and philosophies early in the program life cycle [7]. Systems engineering practices provide a program baseline where customer and stakeholder needs are satisfied, when diligently followed early and throughout the acquisition process.

It is emphasized by a customer, "…Imperative for all the programs is to focus more attention on the application of Systems Engineering principles and practices throughout the system life cycle" [8]. It is further directed, "Improve SE throughout the acquisition process, including workforce …education and training; tools … guidance; Provision for contractor's Board to consider contract performance when setting top executives' salaries/bonuses." One year later, "Application of rigorous systems engineering discipline is paramount to the department's ability to meet the challenge of developing and maintaining needed warfighting capability" [9].

The industries, especially defense and aerospace, are crying for the application of systems engineering practice in their companies, as described in the presentations of "Systems Engineering in Today's Competitive Environment" [10], and "Current and Future Trends in Systems Engineering" [11]. The four (4) key metrics for a company to achieve business targets are: Revenue Growth, Net Margins, Operating Cash Flow, and Return on Net Assets. The actions required to meet these four (4) metrics are all related to systems engineering practices. They are:

- Meet customer requirements
- Validate and verify requirements to final products
- Mitigate risks
- Make the best decision
- Continuously measure performance
- Shorten Cycle Times
- Execute to plan and schedule
- Best organization and budget control

With world-wide web communications nowadays, the newest technology will be most likely spread within two year or less; therefore, technology is a common commodity in most cases. The winner is not determined by the possession of the technology. It is the one who can first deliver the product to the market with the best quality and sell at the cheapest price. Then what are the challenges? They are cost and schedule.

There are three (3) basic arguments for the value of systems engineering: First, assurance that the system will accomplish its objectives; Secondly, the cost-time trade-off as shown in Fig. 2.4 [12] that the later problems are discovered, the more it costs to fix them; and thirdly, insurance against serious low-probability consequences. Well-practiced SE is for adequate upfront planning, adequate scope definition, and understanding customers intent, expectations and requirements definition at the early phase of a project. A well-executed team work can achieve the value of systems engineering.

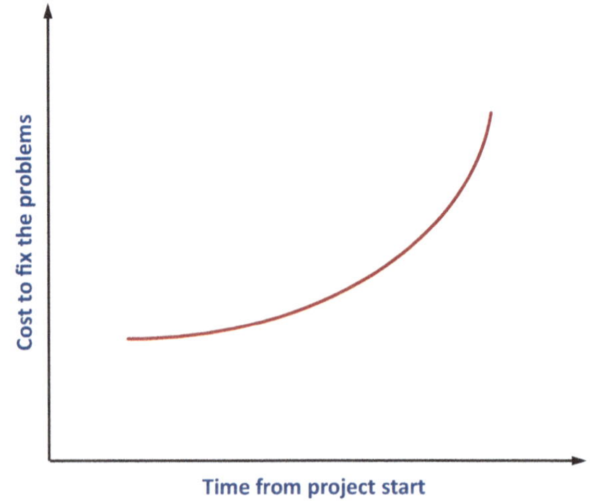

Fig. 2.4 Cost-time trade effect [12]

2.2.2 Evolution of Systems Engineering

SE is an old subject that started in ancient time; like Egyptian built pyramids, Chinese built the great wall, and Roman built the forum and bridge. During that time, it was not called systems engineering but they must used the system principle and method to build huge and durable structures, not to mention the tower of Pisa as a bad system example.

In 1957, the Soviet Union launched the human-made satellite orbiting around the earth and also had the intercontinental ballistic missile (ICBM), a guided ballistic missile with a minimum range of 5500 km (3400 mile) primarily designed for nuclear weapons delivery (delivering one or more thermonuclear warheads). The US suddenly fell behind. For national survival and space race, US government spent billions of dollars to build the satellite and ICBM. These are complicated systems that require a systematic approach with precision and sound management; therefore, SE was adopted and expanded rapidly. System performance for mission success was emphasized, as well as project management for technical performance, delivery schedule, and cost control. A driving force for high system reliability led to the development of parts traceability, materials and process control, change control, product accountability, formal interface control, and requirements traceability. These are all SE tasks and processes.

2.2.3 Systems Engineering Revitalization

After successful completion of the Apollo Program, US space race and ICBM program won over Soviet Union. The government significantly cut back funding for space and defense programs. The contractors suffered severely shortages of funding and forced major reductions in manpower. This was in the early 1970s. In the next ten (10) years, government slowly increased funding for space and defense programs. The contractors also gradually recovered and increased personnel hiring. In the late 1980s, more space and defense programs suffered cost overruns and schedule delays that raised government concerns. The newly hired employees were not aware of systems engineering principles, methods and processes. SE has faded away. In the mid 1990s, US Air Force started an initiative to revitalize systems engineering practices that were executed successfully in the 1950–1960 period to win the space race and ICBM competition. A year later DoD led the systems engineering revitalization until today. Michael W. Wynne, acting under the secretary of defense for acquisition, technology and logistics, and Mark D. Schaeffer, principal deputy, defense systems and director, systems engineering, Office of the USD (AT&L), called for the revitalization of systems engineering across the Department of Defense [13]. "Analyses of a sampling of major acquisition programs show a definite linkage between escalating costs and the ineffective application of systems engineering," Wynne and Schaeffer called for the "systemic, effective use of systems engineering as a key acquisition management planning and oversight tool" and said that, in addition, DoD would "promote systems engineering training and best practices among our acquisition professionals."

In its present form, the systems engineering process is broadened and combines elements of many disciplines:

- Operations research and analysis
- System modeling and simulation
- Decision analysis
- Project management and control
- Software engineering
- Specialty engineering
- Industrial engineering.

2.2.4 Role of Systems Engineers

Systems engineering (SE) is not a rocket science but it may be harder than rocket science. It deals with people, management and engineering. SE is with a project or a program from womb to tomb, and from cradle to grave. SE principles and methods can also be applied to individual's daily life.

Systems engineering differs from traditional disciplines in the following ways: It is focused on the system as a whole; it is concerned with customer needs and operational environment; systems engineering leads system conceptual design; and bridges traditional engineering disciplines and gaps between specialties.

The systems engineers can be:

- Deputy to Project Manager
- Customer Interface
- Requirements Owner
- System Architect
- System Analyst
- IMP/IMS Generator and Keeper
- Risk Management Administrator
- Trade Study Facilitator
- Interface Manager
- Verification Plan Owner and Administrator
- Process Owner
- Coordinator.

A Successful Systems Engineer should be a good problem solver, and welcome challenges, well-grounded technically with broad interests, analytical and systematic, but also creative, and a superior communicator with leadership skills.

Recommended systems engineers' role in a program is shown in Fig. 2.5 [14].

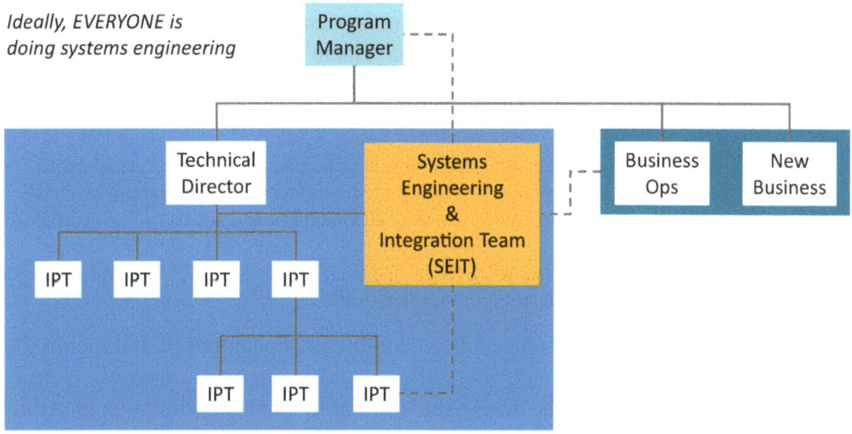

Fig. 2.5 Systems engineers' roles on the program [12]

2.3 Requirements Management

A general systems engineering process is shown in Fig. 2.6 [14]. On the left of the figure are customer requirements. As you can see that the customer requirements come in different forms, from highly sophisticated customers, like US government with fully developed user system specifications indicating performance, supportability, measures of effectiveness; and the constraints as affordability, interoperability, system evolution, and component reuse, etc., to casual customer requirements walking on the street with a few words. Regardless what forms customer requirements

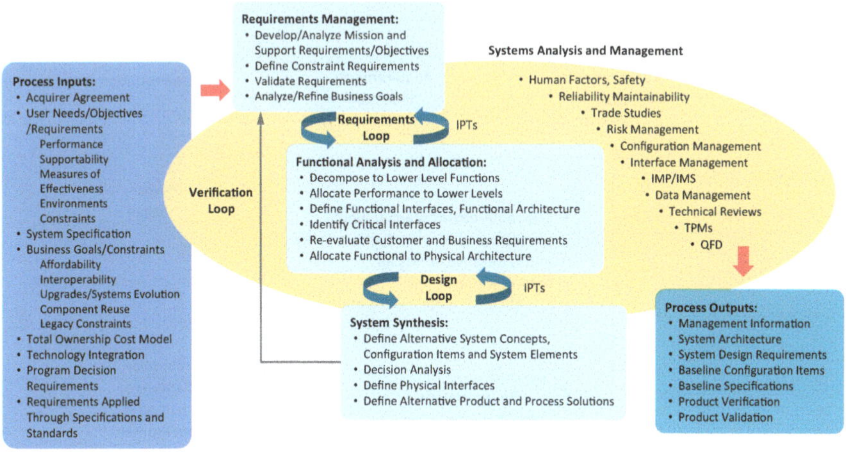

Fig. 2.6 The systems engineering process [14]

have, they have to be validated before deriving system level requirements. It leads to the middle top block "Requirements Management" which is the subject of this section.

The definition of Requirements Management is

> The identification, derivation, allocation, and control in a consistent, traceable, correlatable, verifiable manner of all the system functions, attributes, interfaces, and verification methods that a system must meet including customer, derived (internal), and specialty engineering needs.

The objectives of Requirements Management are:

- Ensure specified requirements have been completely decomposed and met in the design
- Ensure that any impacts to the design due to Requirements modification are completely understood
- Ensure no extraneous requirements (or components) have been introduced
- Ensure requirements have been completely verified.

Refer to the Requirements Management Process as shown in Fig. 2.7 [14], in the first-round of the requirement loop should work with customers to derive the top-level system requirements. For commercial products, derive the top-level system requirements based on validated customer requirements. Each system requirement needs to be allocated to function, organization or individual. This is called requirements allocation to ensure the ownership for each requirement. Under Implementation Process, there is no system design and supplier hardware/software development for the first-round. Each system requirement needs to have one or more verification requirements. For the second-round the systems engineers should also work with customers to derive the second-level system requirements. In sum, each requirement needs to be allocated and has one or more verification requirements. The second-round implies

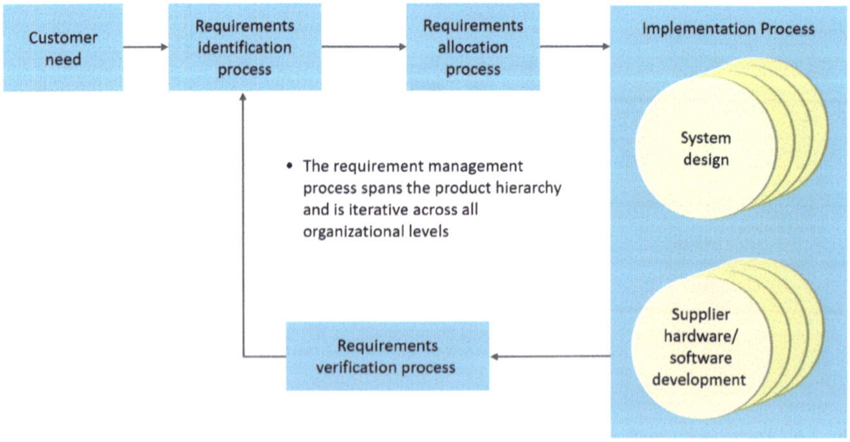

Fig. 2.7 Requirements management process [14]

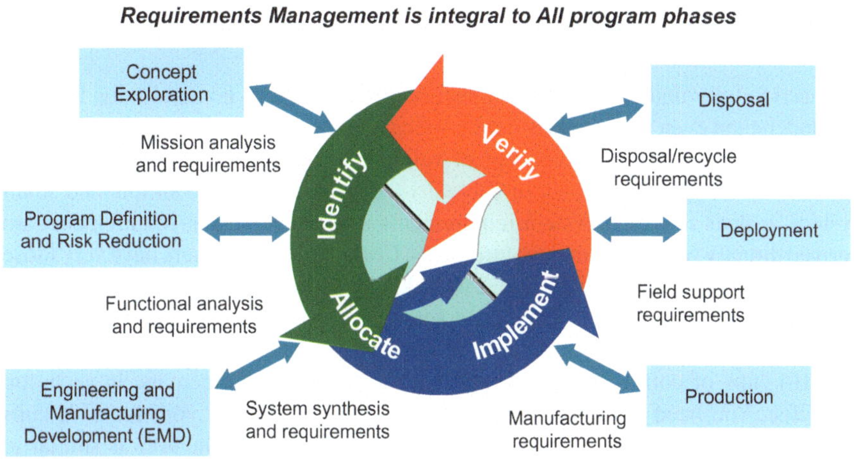

Fig. 2.8 Requirements management role in a system life cycle

conceptual design but not yet the hardware/software development under Implementation Process. With each more round, more lower-level requirements are derived for lower-level designs and hardware/software developments. Then the requirements hierarchy is formed and the requirement management process spans the product hierarchy and is iterative across all organizational levels. Requirements are central to all facets of the product. Requirements can be found everywhere and will depend upon each other; therefore, the continuous involvement and coordination with all interested parties must be accomplished. All impacted areas must be tied together to ensure an optimal solution.

The Requirements Management role in a system life cycle is illustrated in Fig. 2.8. The outside circle is a system life cycle from Concept Exploration to Engineering and Manufacturing Development (EMD) to Disposal. The middle cycle is the requirements for different phases in a system life cycle. The inner circle is the steps for each requirement from identification to allocation for the requirement, implementing the requirement by designers and verification requirement for insuring the product satisfying the requirement. At the beginning of developing requirements for Concept Exploration, the project team is formed on day 1, a concurrent engineering practice. All the required team members are gathered including representatives from each life cycle phase. The requirements will be developed for all phases of the life cycle at the beginning. Use X to represent the consideration and inclusion during the design of a product, such as, reliability, maintainability, survivability, safety, test, fabrication, assembly, operation and product support, etc. The goal of applying SE is to fulfil "Design for X" to cover the development and usage of a product in its entire life cycle.

2.3.1 Validate Requirements

We need to fully understand the customer requirements with no ambiguity. If we do not understand even with slightly doubt, we cannot derive and develop requirements. A better way to convey the concept of validating requirements is to use a cup of coffee as an example as shown in Table 2.1. A cup of coffee is a simple and small system. It does not need system requirements and derive design requirements are derived directly from customer requirements. Commercial customers have very loose requirements. If customers are asked what kind of coffee they would like to drink, they will say "taste good". You would scratch your head what does "taste good" means? You need to validate the requirement of "taste good". One way is to make ten (10) cups of coffee and you know exactly the flavour percentage in each cup of coffee. You need to get 90% of general population to taste your ten (10) cups of coffee. The number and kinds of people selected to satisfy the 90% is similar to polling process. If majority of the population like the taste of No. 5 cup of coffee. Then you have validated the "taste good" requirement that is the flavour percentage of No. 5 cup. The next customer requirement is "be hot". You can make five (5) cups of coffee with 70, 75, 80, 85, and 90 °C. Again, gather 90 percentile of general population to taste these five (5) cups of coffee. If the majority population think No. 4 cup is "be hot". Then you know 85 °C is the validate requirement for "be hot". Customers would like the coffee "wake me up" especially in the morning. You know "wake me up" relates to "quantity of caffeine"; however, the "quantity of caffeine" competes with "level of acidity", which corresponds to customers' "not upset my stomach". You make ten (10) cups of coffee. Each has different levels of caffeine and acidity. You know that the higher level of caffeine, the higher level of acidity. With 90 percentile of population to taste these ten (10) cups of coffee, you find out that the No. 2 cup of coffee the majority of customers feel enough "quantity of caffeine" to "wake me up" and yet "not upset my stomach". Then you have validated these two customer requirements with exact "quantity of caffeine" and "level of acidity". The last customer requirement "be cheap" competes with all the above customer requirements since better taste, higher temperature, and more quantity of caffeine will increase the base cost of a cup of coffee. You can validate this requirement by comparing the price of your cup of coffee with the price of other equivalent brand(s). When all these customer requirements are validated, you need to verify that each cup of your coffee will be made to No. 5 cup of flavour percentage, No. 4 cup 85 °C,

Table 2.1 Customer requirements driving the design requirements

Customer requirements	Design requirements
Taste good	Flavor components
Be hot	Serving temperature
Wake me up	Quantity of caffeine
Not upset my stomach	Level of acidity
Be cheap	Cost per cup

No. 2 cup of coffee "quantity of caffeine" and "level of acidity", and the price of "be cheap". You may have an idea now how to validate customer requirements and verify design requirements.

2.3.2 *Requirements Analysis*

The requirement and different kinds of requirements are defined below:

- Requirement—A statement of required performance or design constraint to which a product must conform.
- Customer Requirements—Statements of fact and assumptions that define the expectations of the system by customer.
- System Requirements—The necessary task, action or activity that must be accomplished. System Requirements identified in requirements analysis will be used as the top-level requirements.
- Design Requirements—The "build to", "code to", and "buy to" requirements for products and "how to execute" requirements for processes expressed in technical data packages and manuals.
- Derived Requirements—Requirements that are implied or transformed from higher-level requirement.

A requirement must be related to the function on which is to be performed, verifiable, which means testable, analyzable, demonstrable and inspectable (not applicable to customer requirements), precisely worded, and be unique.

First, we need to analyze customer requirements (CRs). After validation of customer requirements, we need to define customer functional and performance requirements. Also need to identify, understand, and define customer constraints. The specialty engineering requirements, such as, reliability, maintainability, availability, safety, survivability, and human factor. of the system are important to know at the onset of requirements analysis.

The next step is to analyze and derive the top-level system requirements. The sophisticated customers, like DoD (Department of Defense), NASA (National Aeronautics and Space Administration), and other government agencies, will provide systems specifications to contractors. If this is the case, we need to break down paragraphs in the system specification down to each sentence as a requirement. If the requirements are mandatory, remember to use the word of "shall" in front of the verb; otherwise, the designer or owner of this requirement does not have to follow the requirement. If we received compound and complex Statements from customers, we need to decompose these statements into a set of single requirements.

The design requirement drives the design as a one-way street. The design should be developed from these requirements. If a requirement is derived from the existing design, it is called reverse engineering that defeats the purpose of deriving requirements from CRs. Most likely, the design will not meet the CRs. Each requirement shall be a complete, simple, specific, and straightforward sentence with concise quantifiable meaning. Without quantity, how can the requirement be used for design or verification? It is often to read "TBD" (to be determined) included in the requirement writing. This is not allowed since quantity value have to be indicated clearly in the requirement statement. For compliance with military and commercial standards, specifications, or any documents, the section number of the document shall be cited in the requirement. The requirement sentence cannot contain any adjectives which will be interpreted differently by different people, for example, beautiful, comfortable, etc. Margin in a requirement will provide design flexibility and tolerance for ease in manufacturing. As a result, margin in requirements can reduce cost for design and manufacturing. How can you determine the margin in a requirement? The answer is to use trade-off studies if necessary to determine the optimum margin.

The requirement should be written in positive tense, not passively. Requirements engineers who write requirements can be categorized as "words engineer" since the words used in the requirement dominate the design and verification. Requirement documents are legal documents as part of the contract.

2.3.3 Requirements Allocation

Requirements should be allocated. Allocation means to set apart, assign, or allot for a particular purpose. Requirements allocation is the assignment of requirements to a responsible party. There are several ways of allocation. Partition a value assigned to a parent requirement into parts that are assigned to child requirements. Requirements can also be allocated to functions, organizations, or experts. Sometimes, a requirement is for a specific time period, for example, most of the people witnessed the countdown of rocket launch the last 10 s. In reality, the countdown starts at years or months ago. There are requirements associated with different year or time period.

Requirements allocation can be tabulated in a table, with requirement numbers, requirement sentences, associated verification requirements, verification methods, and allocation. This table is called Requirements Allocation Sheets (RAS).

2.3.4 Requirements Traceability

Each requirement should be traceable to the higher level requirement(s) all the way to customer requirements (CRs). If not traceable to a CR, it could be a new CR or this requirement is not needed. The top-level of requirements traceability hierarchy is CRs, followed by top-level systems requirements, then the next level systems requirements and continue to the lowest level requirements, usually component requirements. This total traceability is necessary to insure all the CRs are incorporated and complied. Requirements traceability for the top two levels of systems requirements should be developed jointly by customer and contractor. Requirements traceability is two-way protection between customers and contractors. The customer will know which CR or CRs are ignored; on the other hand, if the upward trace finds new CR or CRs, it will be out-of-scope and the customer has to increase the scope in cost and schedule; otherwise, the customers should delete the extra CR or CRs. Once the traceability hierarchy tree is established, any impact of top-level CR or CRs on the lower-level requirements can be easily found. Conversely, the impact of any middle-level requirement changes or removals can be easily found on higher level as well lower level requirements. Parent requirements flow down the hierarchy to the immediate lower level child requirements. Traceability means having clear knowledge of the ancestry of every requirement in terms of the parent requirements that make it necessary. CRs have child requirement. The lower level requirements have both parents and children requirements. Orphan requirement(s) that has no parent requirement is not allowed. Through a traceability tree one can spot those orphan requirements. The orphan requirements need parent requirement(s) or have to be deleted. Software is usually the source of problem for hardware systems as well as software-intensive system; therefore, it is important to have total traceability from CRs down to lines of coding.

There are three (3) ways to develop a requirements traceability tree. One way is to use block diagrams to show the traceability from top-level to lower-level requirements. A second way is a dedicated traceability matrix. The third way is to use computer tools. The most commonly one used in the systems engineering communities is DOORS [15].

2.3.5 An Example

After discussions of requirements analysis, allocation, and traceability in the above, an example may be necessary to reinforce the understanding. The example is to develop CRs, system requirements, design requirements, RAS and a traceability tree for a Mouse Trap [16].

The Mouse Trap CRs are shown in Table 2.2. As discussed above, CRs have no rules and restrictions. It could come in different formats and styles, sometimes, not even a sentence, just words or phrases. But shown here in Table 2.2, CRs are nicely written, at least understandable; therefore, the validation of CRs is not necessary. The

Table 2.2 Mouse trap customer requirements

No.	Requirement	Description
CR1	Simple device	There is a need to kill mice with a simple device
CR2	Simple to operate	This device should be simple to operate and affordable by the general population
CR3	Harmed instantaneously	The mouse will be harmed instantaneously
CR4	Price cheap	The price for the Mouse Trap device should be cheap

Table 2.3 Mouse trap system requirements

No.	Requirement	Description
SR1	Cost ≤ \$0.50	The cost of the simple device shall not be more than \$0.50
SR2	Mechanical design	This device shall be mechanically designed without any sensors and software
SR3	Attracted to enter	A mouse shall be attracted by cheese to enter the device
SR4	Harm when bait disturbed	The device shall harm the mouse in less than 1/2 s when the bait is disturbed
SR5	Operated by 90 percentile	The device shall be simple to be operated at least by 90 percentile of the general population

system requirements (SRs) are shown in Table 2.3. As can be seen from Table 2.3, the complete sentences of SRs are simple, specific and straightforward, using the word “shall”. SR1 is derived from CR4 and CR1 by comparing with the market prices of mouse traps. SR1 contains quantity of \$0.50. SR2 is derived from CR1 and CR2. Mechanical design is corresponding to simple device and simple to operate. SR3 is derived from CR3, the need to kill mice. SR3 specifically refers to cheese (requirement engineer’s choice). SR4 is derived from CR2 and CR3 to harm the mice in less than one-half second (this is a quantity) when the bait is disturbed and simple to operate. Less than one-half second is determined by trade-off studies or by building a prototype. We should strive to get the slowest possible action time since the slower the action time the less the cost. It is because a lower spring strength will give a higher action time with lower spring costs. This satisfies the margin for a requirement discussed above. SR5 is derived from CR2 simple to operate by meeting 90 percentile (a quantity) of general population. The design requirements (DRs) are shown in Table 2.4. DR1 is derived from SR1, SR2 and SR5 to hold the cheese firmly and in the meantime to set up the Lock Wire in its place. DR2 is derived from SR1, SR2, and SR4 to release the Kill Mechanism in one-half second or less. DR3 is derived from SR1, SR2, and SR4 to hit the mouse with 20 lb. force. We should strive to get the smallest possible impact force since the smaller the impact force the

Table 2.4 Mouse trap design requirements

No.	Requirement	Description
DR1	Hold cheese	The Bait System shall be designed to firmly hold the cheese and lock wire in its place in one step by meeting 90 percentile of population abilities
DR2	Release ≤ ½ s	The Trigger Systems shall be designed to release the Kill Mechanism in less than one-half second
DR3	20 lb force	The Power System shall be designed to hit the mouse with 20 lb force
DR4	Platform support	A platform shall be designed to house all the subsystems of this device

lowest the costs. It is because the lower spring strength will give a smaller impact force with lower spring costs. This is, again, to satisfy the margin for a requirement as discussed above. DR4 is derived from SR1, SR2, SR3, and SR4 to support the subsystems. Different individuals or teams may develop different set of SRs and DRs. For example, peanut butter may be used for bait rather than cheese; action time could be different from one-half second; and the impact force could be different from 20 lb. force, as may be known later in functional analysis. Different functional and system architectures are possible. We can apply synthesis analysis, basically through trade-off studies, to choose the best conceptual design from these different architectures. In a systems engineering process one can select the best conceptual design objectively and directly from CRs.

The Requirements Allocation Sheet (RAS) for the mouse trap SRs is shown in Table 2.5. RAS applies to any levels of requirements. For the example shown here the RAS applies to SRs. DRs can also have its RAS. Every SR has its verification requirement (VR). A VR follows the same requirement rules with a complete sentence. One may notice that a VR repeats words as used in SR, especially concerning quantities and compliance standards. This is necessary since a VR will verify these quantities and compliances with standards. Later on, the VRs and SRs will be dissected. SRs are given to a designer and VRs will be distributed to verification specialists. The verification methods are only for recommendation. Final methods will be determined

Table 2.5 Requirements allocation sheet (RAS)

No.	Requirement	Verification requirement	Method	Department
SR1	The cost of the simple device shall not be more than \$0.50	VR1: it shall be verified when the mouse trap is assembled that the cost for manufacturing shall be less than \$0.50	Demonstration	Manufacturing and design
SR2	This device shall be mechanically designed without any sensors and software	VR2: every component of the mouse trap shall be mechanically designed	Inspection	Quality
SR3	A mouse shall be attracted by cheese to enter the device	VR3: a cheese shall be placed on a holder with a clear visibility without any obstruction and easy to be reached by 90 percentile of mouse population	Demonstration	Human factor

(continued)

Table 2.5 (continued)

No.	Requirement	Verification requirement	Method	Department
SR4	The device shall harm the mouse in less than 1/2 s when the bait is disturbed	VR4: it shall be verified by test that a mouse shall be harmed in less than one-half second	Test	Design
SR5	The device shall be simple to be operated at least by 90 percentile of the general population	VR5: it shall be verified by demonstration that placing bait on Bait Holder and locking wire to the bait holder can be handled by 90 percentile of population	Demonstration	Human factor

by verification specialists. In this example the SRs are allocated to departments for ownership. One should be aware that requirements may also be allocated to functions or individuals, etc.

Both block diagrams and traceability matrices are applied to the mouse trap. Block diagrams are shown in Table 2.6. The traceability matrix is shown in Table 2.7. For a simple device like a mouse trap, the block diagrams for only three levels already appear complicated. For a larger device or system, it will be overwhelmingly complicated. A traceability matrix can contain more levels, i.e., more columns. If there are more requirements in each level, just add more rows can be added. A traceability matrix is more convenient and adaptable for large systems with more requirements and hierarchy levels.

Table 2.6 Mouse trap requirements traceability using block diagram

Customer requirements	System requirements	Design requirements
CR1: simple device	SR1: price ≤ \$1.00	DR1: hold cheese
CR2: simple to operate	SR2: mechanical design	DR2: release ≤ ½ second
CR3: harmed instantaneously	SR3: attract to enter	DR3: 20 lb force
CR4: price cheap	SR4: harm when bait disturbed	DR4: platform support
	SR5: operated by 90 percentile	

Table 2.7 Mouse trap requirements traceability matrix

Customer requirement	System requirement	Design requirements
CR1	SR1	DR1, DR2, DR3
CR1	SR2	DR1, DR2
CR2	SR2	DR1, DR2
CR2	SR4	DR3, DR4
CR2	SR5	DR1
CR3	SR3	DR1, DR4
CR3	SR4	DR3, DR4
CR4	SR1	DR1, DR2, DR3

2.4 Functional Analysis and Allocation

When referring to Fig. 2.6, Requirements Analysis is followed by Functional Analysis in an iterative way. If one recalls the RAS in which each system requirement can be allocated to a function, more than one system requirement can be allocated to the same function. After the two top-level system requirements have been developed, the top-level functional analysis can be performed using the functions allocated from the top-level system requirements. If there are inconsistencies with top-level system requirements, requirements analysis and functional analysis will be iterated to correct the inconsistencies. functional analysis will be continued to the next level using the functions allocated from the next level system requirements. Again, if there are inconsistencies between the next level functional analysis and the next level system requirements, they should be corrected iteratively. Then the top and next level functional architecture is formed. The lower level functional analysis and architecture can be continuously developed without waiting for the corresponding lower level requirements developed. The lower level functional analysis and architecture can assist the requirements analysis to develop lower level requirements. Some of the organizations, such as the Commercial Satellite Division of The Boeing Company, develop the functional analysis and architecture first. Then the developed functional architectures were used to develop requirements for all levels. The iteration between requirement analysis and functional analysis is called requirements loop as shown in Fig. 2.6.

When the top two levels of functional architectures are developed, through functional allocation, the two top levels system (product) architectures are developed. As discussed in Sect. 2.3.5, several system architectures may be developed. system synthesis can be performed to select the best system architecture to develop the best conceptual design. The developed system architectures can be checked against the functional architectures for consistencies since functional architectures are derived from customer requirements. These consistency checks will be iterative throughout the system architectural hierarchical levels between system synthesis and functional analysis and allocation. It is called design loop iteration as shown in Fig. 2.6.

2.4.1 Functional Analysis

Functional analysis is an important first step in determining system performance. It includes functions necessary for the product or service to operate properly. It is a structured approach for describing how a system might be used. The functional blocks will be defined through a series of functional analysis in all levels. The contractually specified usage modes are also included in functional blocks which are usually in the top or higher levels. These functional blocks in all levels hierarchically form a functional architecture for which system products and services can be designed. The operational sequence in time steps, for example, the last ten (10) seconds to launch a rocket, can be arranged as time sequence in functional blocks, that can be used to analyze time-critical requirements. The functional blocks, like the requirements, are arranged in a traceable and logical sequence.

Functions describe how users use a product or service. A functional statement begins with a verb and follows with a direct object, for example, fly airplane, surf internet, or enter password. As one moves away from user-interface level and into lower levels of details, functional descriptions become statements about what the system does, for example, compute coordinates, sense hydraulic pressure, or track target. Function name should identify the action or transformation accomplished by the function. Avoid the pitfalls of "provide" and "accept" functions since these two words cannot send out clear message. What does the function mean with either of these two words? For example, "provide diagnostics", what kind of diagnostics? A better way to write this function is "Perform BIT (Built in Test)". "Provide aircraft position", a more clear message is "Compute aircraft position."

2.4.1.1 Functional Flow Block Diagram

One of the often used functional analysis methods is Functional Flow Block Diagram (FFBD). Refer to Fig. 2.9 [17], the top-level function blocks are transformed from top level system requirements. There are two ways of transforming system requirements to functional blocks. One way to convert system requirements to functional blocks is through interpretation and judgement. If the system requirements have already been allocated to functions, it will be simply using the allocated functions as top-level functional blocks. The function blocks are connected in certain sequences, as shown in Fig. 2.9, Functions A to B to C to D, and Function A also to E to C to F. You would not want to be overwhelmed by too many blocks to perform complicated sequential relationships, one to one, one to many, and many to one, etc. It is recommended between five (5) to nine (9) blocks up to your preference. Each functional block is numbered, Function A as 1.0, Function B as 2.0, Function C as 3.0, Function D as 4.0, Function E as 5.0, and Function F as 6.0. Each block will have the next level functions. Let us choose Function E, 5.0. Continue to Fig. 2.10 for the next level FFBD. Since the Functions are all under Function E, aka 5.0, they will be numbered as 5.x. The functional sequences will be Function 5.1 to Function 5.3 to Function

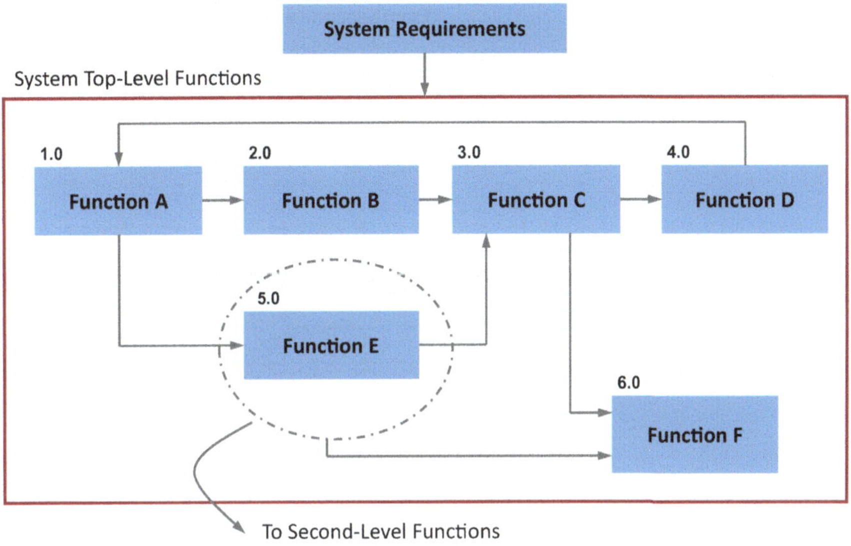

Fig. 2.9 Top-level functional flow block diagram [17]

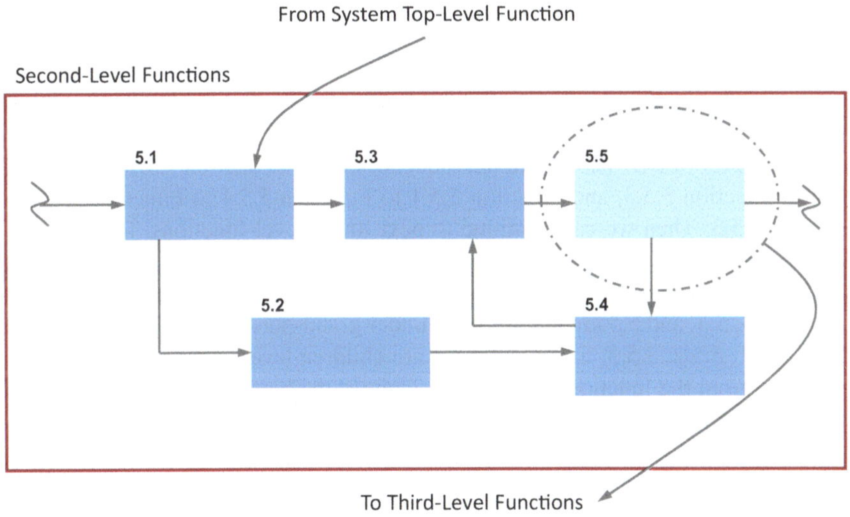

Fig. 2.10 Second-level functional flow block diagram [17]

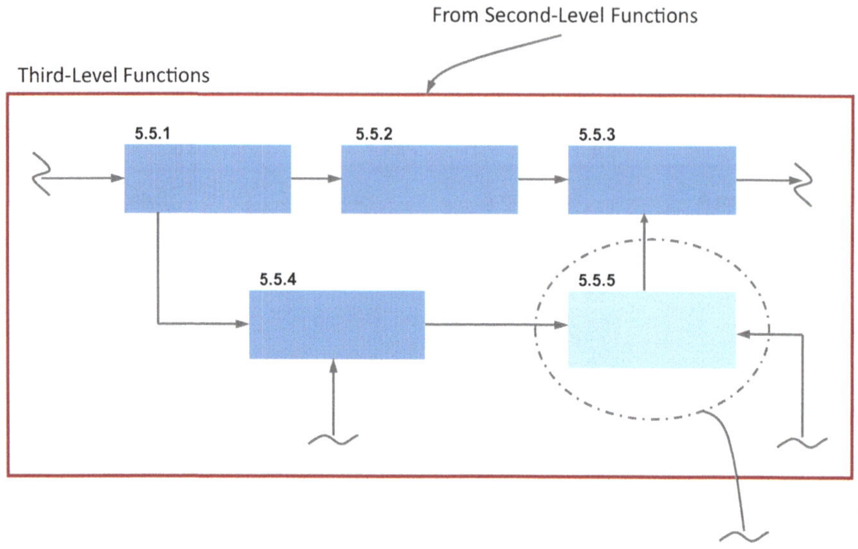

Fig. 2.11 Third-level functional flow block diagram [17]

5.5 to Function 5.4, Function 5.1 to Function 5.2 to Function 5.4, and Function 5.4 to Function 5.3. Now you may begin to appreciate why the functions are numbered. Each of 5.x Functions can have lower level functions. Let us choose Function 5.5, as shown in Fig. 2.10, to develop lower level functions. Continue to Fig. 2.11 to view the next lower level FFBD. The functional sequences will be Function 5.5.1 to Function 5.5.2 to Function 5.5.3, and Function 5.5.1 to Function 5.5.4 to Function 5.5.5 to Function 5.5.3. Then we can continue to next lower level functions, for example, from Function 5.5.5, as shown in Fig. 2.11. All the functional blocks 1.0, 2.0, 3.0, 4.0, 5.0, and 6.0 at the top-level are grand-parents level; all the functional blocks 5.1, 5.2, 5.3, 5.4, and 5.5 are parents level under grand-parent 5.0; all the functional blocks 5.5.1, 5.5.2, 5.5.3, 5.5.4, and 5.5.5 are children level under parent level 5.5. Therefore, from the function numbers can identify the hierarchy level under which function and above which functions. The top-level has only one FFBD. The second-level can have six (6) FFBDs under each grand-parents functional block. The numbers of FFBD in third-level will depend on how many functions under each functional blocks of second-level. For example, under second-level Function 5.5 will have five (5) FFBDs. It could be as many as thirty (30) FFBDs if each second-level function has five (5) FFBDs. All these functional blocks in each level piled in hierarchical layers, it forms functional architecture as shown partially in Fig. 2.12 focused on the branch of Function 5.0 to Function 5.5. As you can see that the functional architecture is fanned out from top-level to lower levels.

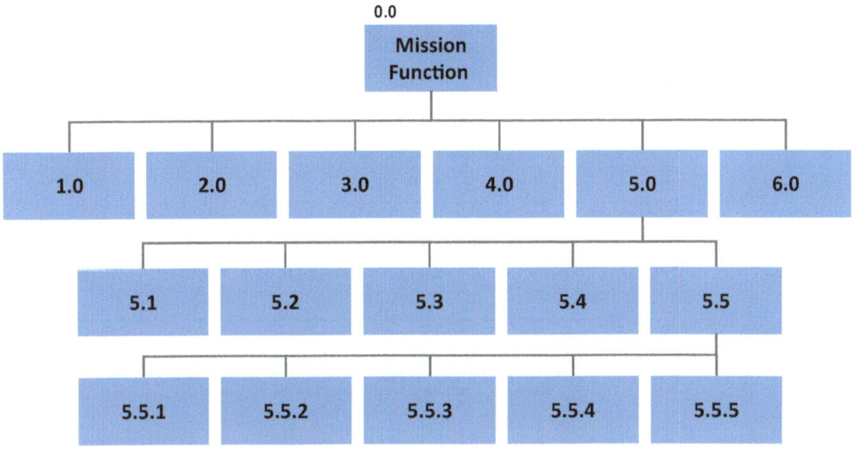

Fig. 2.12 Partial functional architecture [17]

Example 1 Use "Drive a Car", as shown in Table 2.8, as an example for FFBD. In the table lists the functions of driving a car. Each individual will drive the car in different ways. Shown in Fig. 2.13 is one way of driving a car. There are many functional sequences for driving a car pending on who is driving.

Example 2 Use "Dishwasher", as shown in Figs. 2.14, 2.15, 2.16, 2.17, 2.18 and 2.19 [18], as another example for FFBD. It is shown in Fig. 2.14 that through interpretation and judgement, top-level system requirements for dishwasher has been transformed to top-level FFBD for which Function 1.0 to Function 2.0 to Function 3.0 to Function 4.0. From the top-level four (4) functions generate four (4) second-level FFBDs. The

Table 2.8 Functions for driving a car

Drive a car
Accelerate car
Decelerate car
Turn car
Start car
Stop car

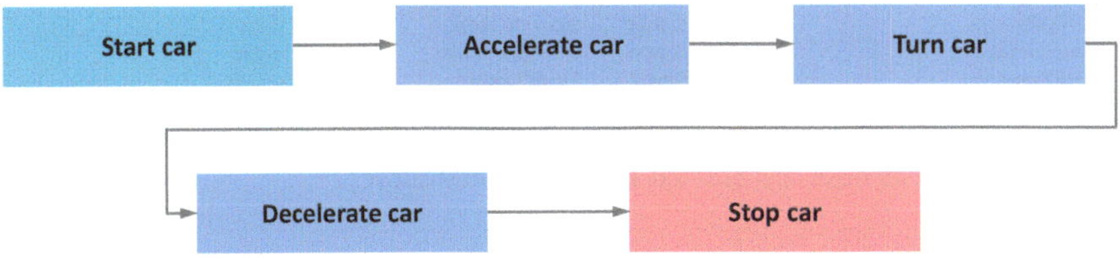

Fig. 2.13 Functional flow block diagrams—car

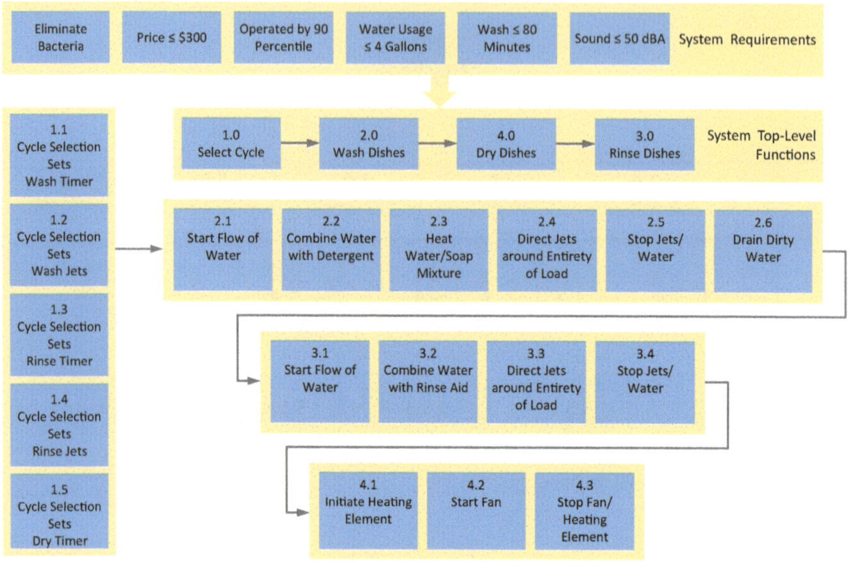

Fig. 2.14 Top-level and second-level functional flow block diagram—dishwasher [18]

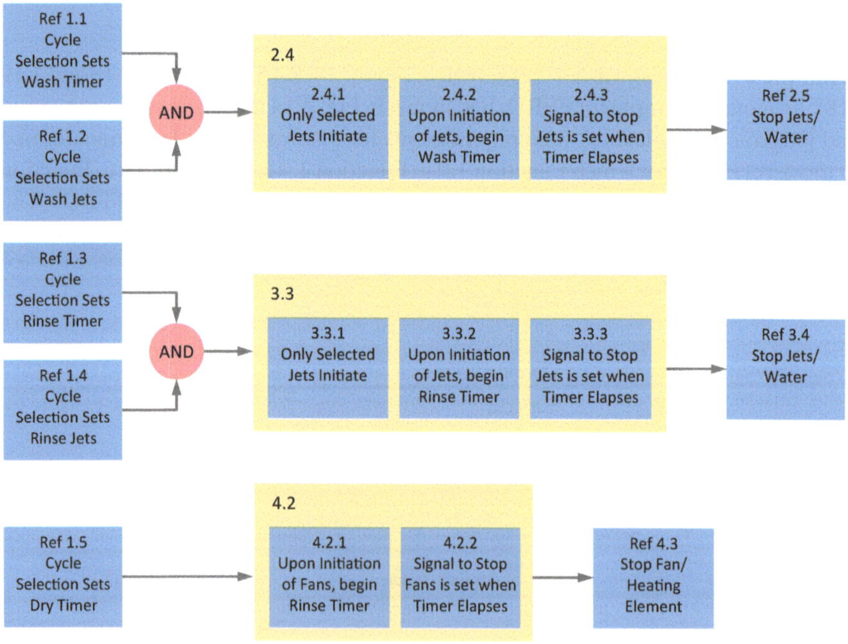

Fig. 2.15 Third-level functional flow block diagram—dishwasher [18]

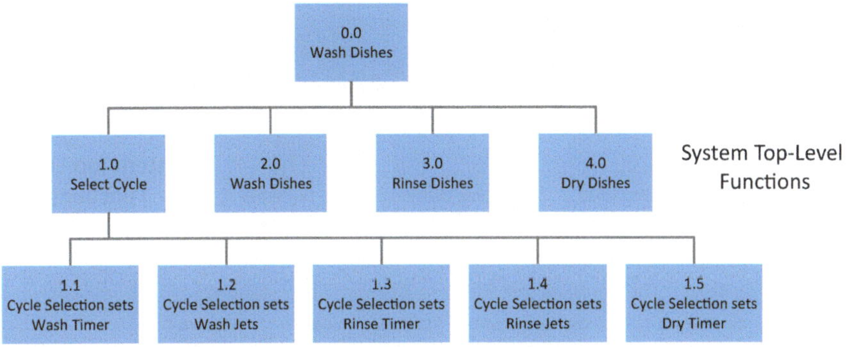

Fig. 2.16 Function 1.0 functional architecture—dishwasher [18]

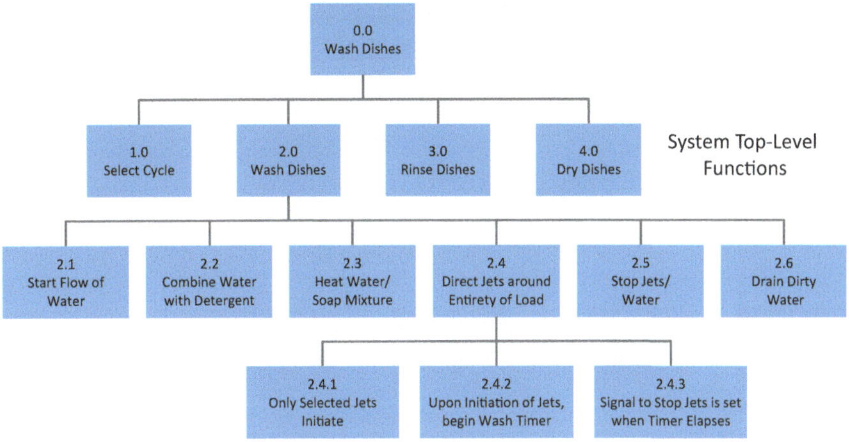

Fig. 2.17 Function 2.0 functional architecture—dishwasher [18]

functional sequences are shown in each of the second-level FFBD. The third-level FFBDs for the second-level Function 2.4, Function 3.3, and Function 4.2 are shown in Fig. 2.15. The external functions connected to each third-level FFBD are also shown in Fig. 2.15. It can be seen that the external interfaced functions can be in different levels. The word of "Ref" does not have to be included. When the FFBDs are completed, the generated functional blocks can be piled hierarchically in different levels to establish functional architecture. Function 1.0 branch, Function 2.0 branch, Function 3.0 branch, and Function 4.0 branch functional architectures are shown in Figs. 2.16, 2.17, 2.18, and 2.19, respectively.

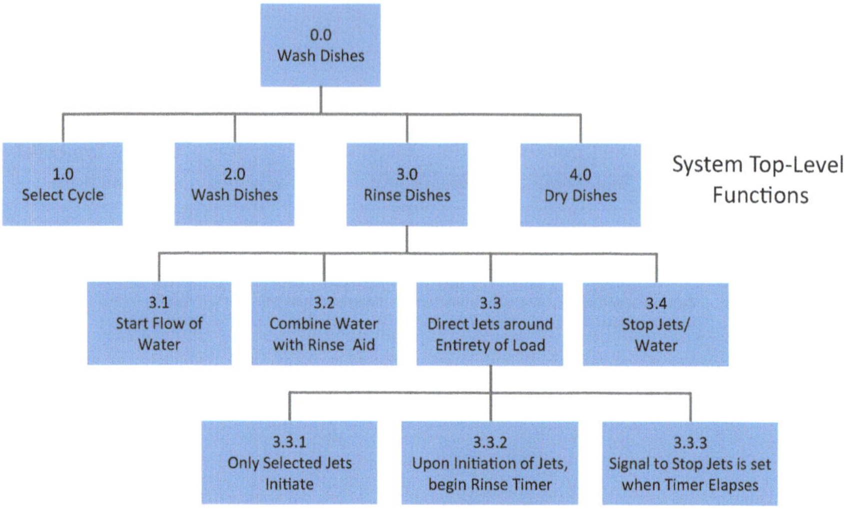

Fig. 2.18 Function 3.0 functional architecture—dishwasher [18]

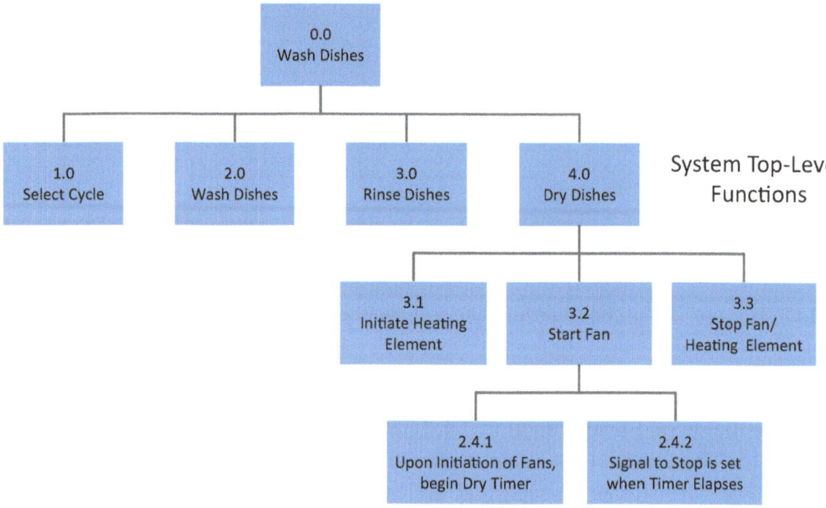

Fig. 2.19 Function 4.0 functional architecture—dishwasher [18]

2.4.1.2 Integrated Definition for Functional Modeling

There are as many as fourteen (14) Integrated Definition for Functional Modeling diagrams, i.e., IDEF0, IDEF1, 2, 3, 4, and 5, up to 14. IDEF0 is another commonly used functional analysis method. It is a functional model or process model of a system; a method designed to model the decisions, actions, and activities of an organization or system. It is useful in establishing the scope of an analysis, especially for a functional analysis. As a communication tool, IDEF0 enhances domain expert involvement and consensus decision-making through simplified graphical devices. As an analysis tool, IDEF0 assists the modeller in identifying what functions are performed, what is needed to perform those functions, what the current system does right, and what the current system does wrong.

The IDEF0 modeling diagram is shown in Fig. 2.20 [19]. There are two additional inputs as compared with FFBD. The Control enters the top of the box. The Mechanism points up to the bottom of the box to show the supporting means for performing the function. Use "Perform detail design" function as an example, shown in Fig. 2.21, to show how to form the IDEF0 diagram. The input data to the left-hand side of the box is Preliminary Design Data; the output data from the right-hand side of the box is Recommended Detailed Design; the control data from the top of the box is Design Requirements; and the mechanism data enters the bottom of the box is Design Engineer. The control input can also include the standards (industrial, commercial, and military), constraints, and processes, etc. The mechanism data can also include resources (facilities, computer, etc.), people, and tools, etc. IDEF0 can have functions at different levels same as FFBD, as shown in Fig. 2.22. The output data does not

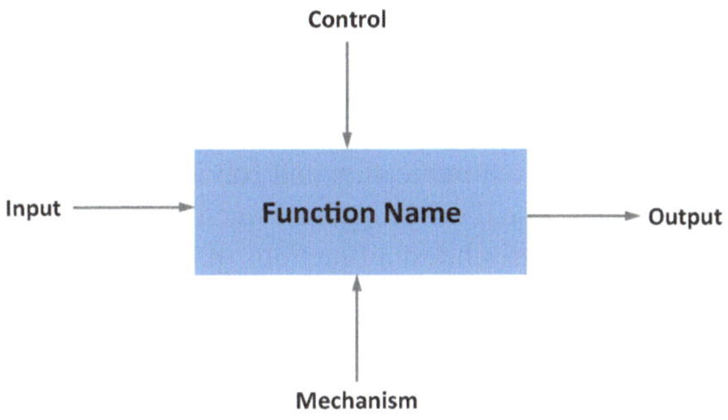

Fig. 2.20 IDEF0 modeling diagram [19]

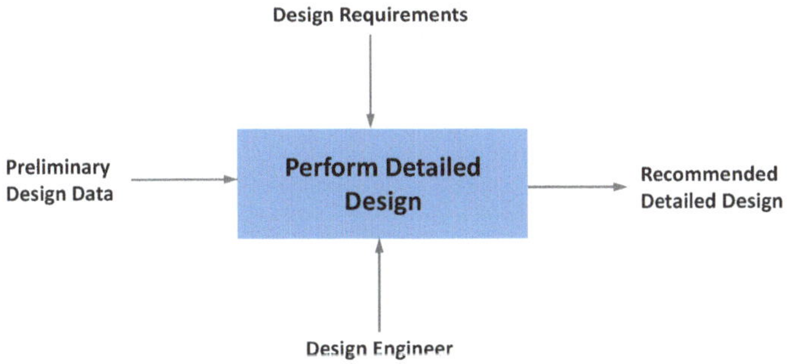

Fig. 2.21 IDEF0 modeling diagram—example

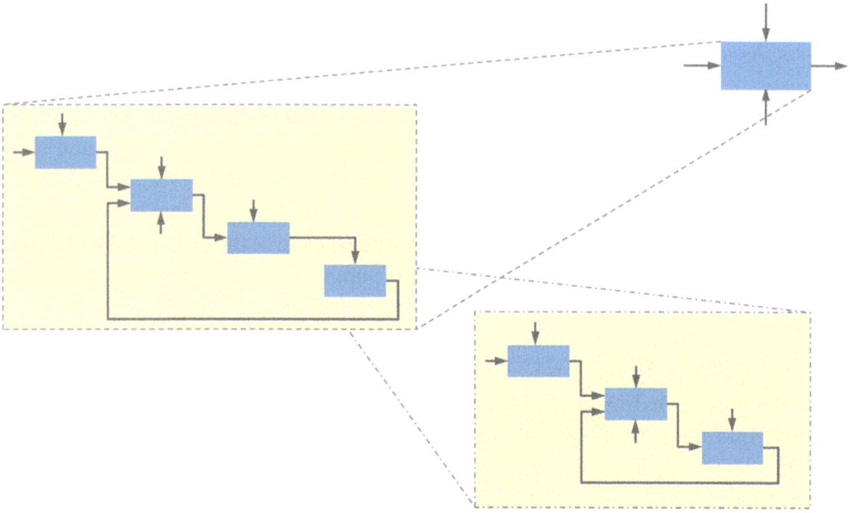

Fig. 2.22 Three levels of IDEF0 diagrams [19]

have to be the input data to the next connected diagram; instead, it can input as control data or mechanism data, as shown in Fig. 2.22. An example of IDEF0 diagram is shown in Fig. 2.23 [19]. IDEF0 can also use numbering system for each function same as that in FFBD, as shown in Fig. 2.24 [20], another IDEF0 diagram example included in Architecting the Communication and Navigation Networks for NASA's Space Exploration Systems. The IDEF0 diagram Function A1.3.1 to Function A1.3.2 to Function A1.3.3 sequence are children functions of A1.3.

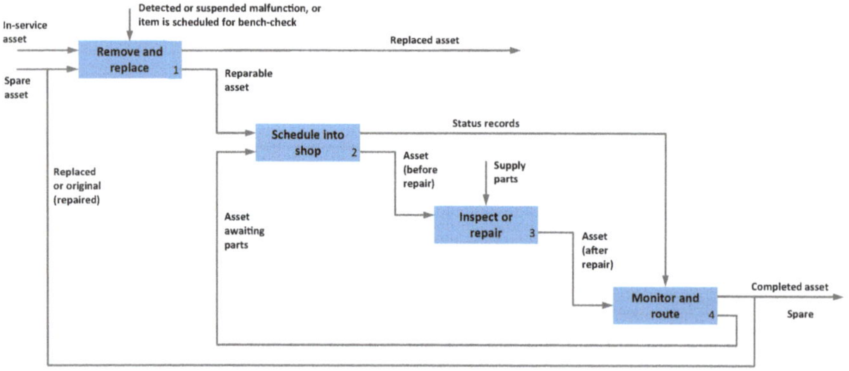

Fig. 2.23 IDEF0 diagram example—maintain reparable spares [19]

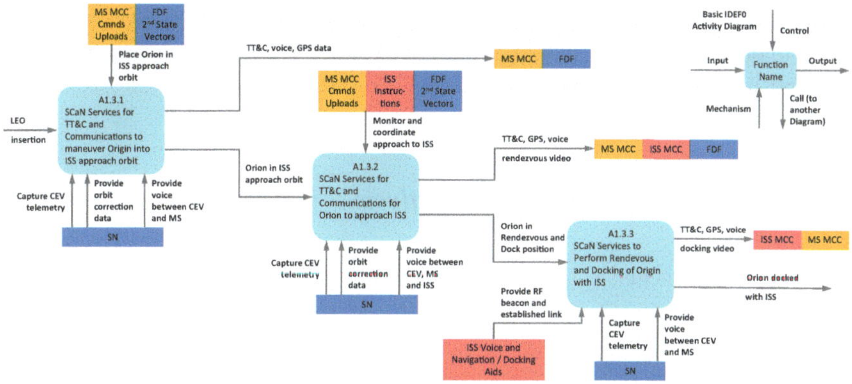

Fig. 2.24 IDEF0 diagram for operational activities flow diagram in support of Rendezvous and Dock [20]

2.4.1.3 Functional Decomposition

As discussed above, functions from top-level down to the lowest levels are developed by using FFBD or IDEF0 functional analysis method. Functions developed through this process are time consuming, especially for a large and complex system. A simpler way is to decompose top-level functions to the lowest levels in its most basic form a simple hierarchical decomposition of the functions. Functional decomposition is the breaking down of a high level function into smaller pieces of function that can be more easily managed and understood. Decomposing the top-level functions into sub-functions (i.e. Level 1 and 2, etc.) can also form functional architecture.

The primary steps for functional decomposition are:

1. Brainstorm functions performed.
2. Pick out the five to ten truly top level functions and arrange in sequence (if appropriate).
3. Place the other functions below the top-level functions.

A practical approach is to use a roll of white paper with 22 inches or similar width. Roll out the roll of white paper. Draw the top-level, second-level, third-level, and fourth-level regions, etc. on the roll of white paper. Team members write the names of functions on post-it notes. Remember to use the verb-noun naming function format. Team members can lay the post-it notes on different regions which already drawn on the roll of white paper. If there is a contention about where a function belongs, make a duplicate post-it note and put in both regions. When you get uncomfortable about further decomposition, it is usually the end of decomposition. Then the team members will align the functions in different regions. The top-level region can have only 5–9 functions. In the second-level region, group of 5–9 functions should be aligned to each function in the top-level region. In the third region, group of 5–9 functions should be aligned to each function in the second-level region. In the fourth region, group of 5–9 functions should be aligned to each function in the third-level region. In the same manner, work to further lower regions if there are more. The groups under each function should be between 5 and 9. The outcome from functional decomposition is a functional architecture.

2.4.2 Elements of Functional Analysis

There are four elements:

- Functional Decomposition
- Functional Sequencing
- Information/Data Flow
- Interface Definition

Function decomposition was discussed above in Sect. 2.4.1.3. Referring to Figs. 2.9, 2.10, 2.11, 2.13, 2.14 and 2.15, it can be seen that the functional blocks are connected in certain directions and sequences. This is the element of functional sequencing. When the two functional blocks are connected, information/data is transferred from one block to another block. It is directionally oriented. For example, Block A connects in the direction to Block B, asking "have you had dinner yet?" Block B connects in the direction back to Block A, answering "yes, I had dinner." This is the element of information/data flow. The information/data between the two blocks can be expanded to include all the necessary interface information/data. This is the element of interface definition where the interface requirement is defined and developed. The functional definition (requirement) shall be fulfilled by physical interface definition (requirement) that will be discussed under the systems engineering subject of Interface Management.

2.4.3 Functional Allocation and System Synthesis

The functions will ultimately be performed or accomplished through the use of equipment, personnel, facilities, software, or a combination. The functional architecture will need to be transformed into a physical, or software architecture by defining physical or software components needed to perform the functions. Functional partitioning is the process of grouping functions that logically fit with the components likely to be used, and to minimize functional interfaces. Functions at system, sub-system, segments and components levels should be allocated to the corresponding levels of physical or software system, sub-system, segments, and components. Functional analysis and allocation is repeated to define successively lower level functions and allocations, as shown in Fig. 2.25. When the allocations are performed to the lowest level, a system architecture, physical or software, is formed. The functional/physical matrix, shown in Fig. 2.26 [21], can be used to assist the functional allocation.

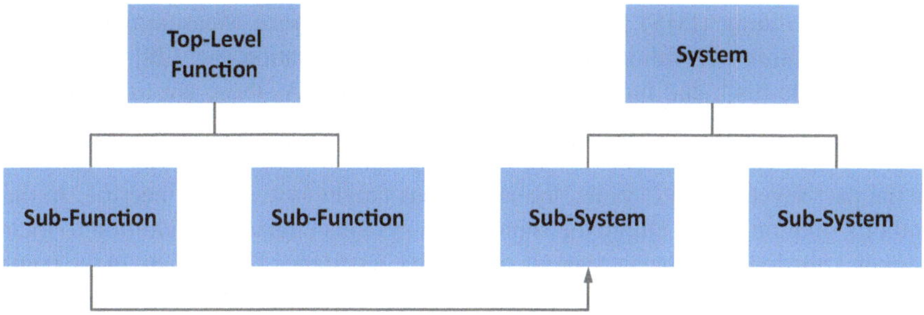

Fig. 2.25 Functional allocation

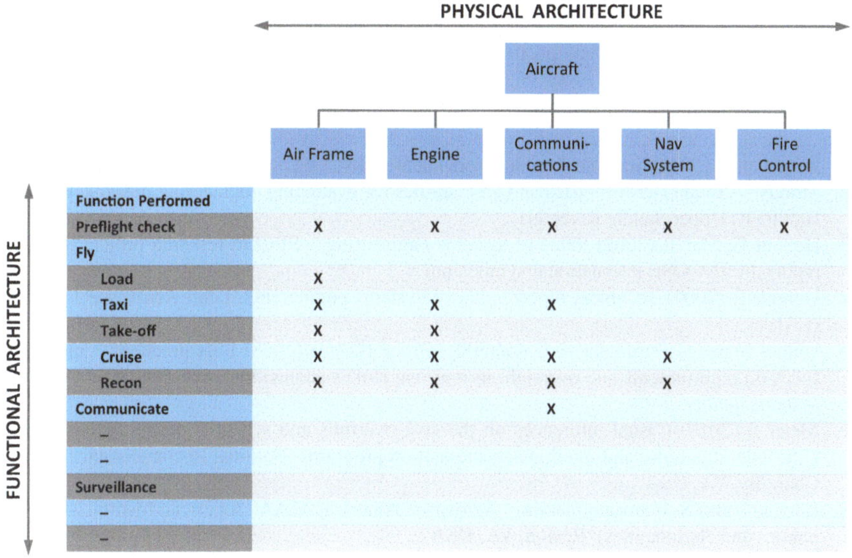

Function Performed	Air Frame	Engine	Communi-cations	Nav System	Fire Control
Preflight check	X	X	X	X	X
Fly					
Load	X				
Taxi	X	X	X		
Take-off	X	X			
Cruise	X	X	X	X	
Recon	X	X	X	X	
Communicate			X		
–					
–					
Surveillance					
–					
–					

Fig. 2.26 Functional/physical matrix

Since there may be several physical or software architectures developed from different functional architectures or from the same functional architecture through different ways of allocations, design synthesis sets the stage for trade studies to select the best among the candidate architectures.

2.5 Summary and Prospective

Referring to Fig. 2.6, there are more systems analysis and management subjects than what are presented here in this chapter, for example, risk management [22], Work Breakdown System (WBS) [23], Integrated Master Planning (IMP)/Integrated

Master Scheduling (IMS) [24, 25], Technical Performance Measurement (TPM) [26], Verification and Validation [27], and System Integration, etc. [28]. One could write a whole book and more on systems engineering. As these are fundamentals and important subjects they may be presented in another book.

It is worth discussing here how to educate people to become systems engineers. In the past seventeen (17) years, the author has taught systems engineering. In the author's opinion, every employee should be a systems engineer in addition to their domain knowledge. In Japan, there is no Quality Assurance Department in the company organization since every employee, whether engineer, manufacturing worker, software programmer, or subcontractor employee, etc., is equipped with quality knowledge, methods, and conscientiousness. The same strategy can be applied to systems engineering.

References

1. Friedman G (1994) Statement made as president of INCOSE
2. Moody JA et al (1997) Metrics and case studies for evaluating engineering designs. Prentice Hall PTR, Upper Saddle River, NJ
3. Honour E, Mar B (2002) Value of systems engineering – SECOE research project progress report. In: INCOSE international symposium
4. Honour E (2004) Technical report value of systems engineering. Lean Aerospace Initiative (LAI)
5. United States Government Accountability Office (GAO) (2008) Best practices – increased focus on requirements and oversight to improve DoD's acquisition environment and weapon systems quality
6. Meier S (2010) Casual inferences on the cost overruns and schedule delays of large-scale U.S. federal defense and intelligence acquisition programs. National Reconnaissance Office, Chantilly, VA, USA
7. Eiler J (2008) Systems engineering. Aerospace America, AIAA (American Institute of Aeronautics and Astronautics), Reston, VA, USA
8. Assistant Secretary of the United States Air Force (2003) Released letter to defense contractors, USA
9. Under Secretary of the United States Department of Defense (2004) Released letter to defense contractors, USA
10. Vice-President of The Boeing Company (2013) Systems engineering in today's competitive environment. Presentation for aerospace engineering seminar at California State University Long Beach
11. Senior Manager of Systems Engineering of The Boeing Company (2014) Current and future trends in systems engineering. Presentation for aerospace engineering seminar at California State University Long Beach
12. Jackson S, Hsu J (2002) Systems engineering lecture notes. University of Southern California, California State University Long Beach and University of California, Irvine
13. Defense AT&L (Acquisition, Technology and Logistics) (2005) Revitalizing systems engineering - how six components are meeting the acting USD (AT&L) imperatives
14. Hsu J (2016) Systems engineering lecture notes. California State University Long Beach, University of California Irvine, AIAA continuing education, USA
15. IBM Rational Dynamic Object Oriented Requirements System (DOORS) (formerly Telelogic DOORS) is a requirement management tool

16. Hsu J (2006) Applying systems modeling language to a simple hardware systems. In: INCOSE international symposium, Orlando, FL, USA
17. Jackson S (2003) Systems engineering lecture notes. University of Southern California, Los Angeles, USA
18. Lessmueller B, Scribner V, Vilchez H, Rodriguez Y (2006) Systems engineering class projects. California State University Long Beach
19. Defense Acquisition University (2001) Systems engineering fundamentals. Supplement 5-B
20. NASA (2007) Architecting the communication and navigation networks for NASA's space exploration systems. In: IEEE international conference on system of systems engineering
21. Defense Acquisition University (2001) Systems engineering fundamentals, chapter 6
22. Aven T (2016) Risk assessment and risk management: review of recent advances on their foundation. Eur J Oper Res 253:1–13
23. Sharon A, Dori D (2015) A project-product model-based approach to planning work breakdown structures of complex system projects. IEEE Syst J 9(2), 6748857:366–376
24. Sofian S, Li X, Kusumawardhani P, Widiyani W (2015) Sustainable systems integration model-metrics in design process. Procedia Soc Behav Sci 184:297–309
25. Hsu J, Raghunathan S (2007) Systems engineering for CDIO - conceive, design, implement and operate. In: 45th AIAA aerospace meeting and exhibit, 8–11 Jan 2007, Reno, Nevada, AIAA 2007-591. https://doi.org/10.2514/6.2007-591
26. Sun J-F, Yuan J-H, Zhao Y, Pu H-B (2013) Technical performance measurement method for technical management of system development projects. Binggong Xuebao/Acta Armamentarii 34(suppl. 2):44–47
27. Dabney JB, Arthur JD (2019) Applying standard independent verification and validation techniques within an agile framework: identifying and reconciling incompatibilities. Syst Eng 22(4):348–360
28. Wognum N, Bil C, Elgh F, Peruzzini M, Stjepandić J, Verhagen WJC (2019) Transdisciplinary systems engineering: implications, challenges and research agenda. Int J Agile Syst Manage 12(1):58–89
29. Kepner CH, Tregoe BB (1997) The new rational manager. Kepner-Tregoe Inc, Princeton, NJ

Chapter 3
New Challenges for Ideation in the Context of Systems Engineering

Wojciech Skarka, Katarzyna Jezierska-Krupa and Ryszard Skoberla

Abstract The chapter presents the usage of Ideation Methods in the Systems Engineering context. The first part presents the classification of ideation methods. This classification is based on different criteria to allow designers take a look from different points of view. There is also a discussion about presented division of ideation methods and how to choose the right method for a given case. The second part deals with background and review of the most popular methods with special attention paid to the analysis of Ideation Methods in complex and multidisciplinary projects over the last few years. The definitions of discussed methods are given in this part. This description concerning with the classification can be a quick way to get to know the issue of ideation methods. This approach allowed us to elaborate guidelines concerning the usefulness of the analyzed Ideation Methods in Systems Engineering. Special attention has been put to methods that have been commonly used in the recent case studies. The presented two case studies describe the usage of selected methods in given examples from the field of designing and development of medical devices as well as automotive industry. The main theme of this chapter is classification of ideation methods and their characteristics. It should help to choose the right ideation method to realize the ideation phase in context of Systems Engineering. The reading of this chapter should result in the evaluation of the value of ideation methods depending on the application. The conclusions sum up guidelines concerning the usage of Ideation Methods and indicate the direction of changes and improvements in these methods.

Keywords Ideation · Design thinking · Concept generation · Ideation methods classification · Technology readiness level

W. Skarka (✉) · K. Jezierska-Krupa · R. Skoberla
Silesian University of Technology, Konarskiego 18A, 44-100 Gliwice, Poland
e-mail: wojciech.skarka@polsl.pl

K. Jezierska-Krupa
e-mail: katarzyna.jezierska-krupa@polsl.pl

R. Skoberla
e-mail: ryszard.skoberla@polsl.pl

J. Stjepandić et al. (eds.), *Systems Engineering in Research and Industrial Practice*,
https://doi.org/10.1007/978-3-030-33312-6_3

3.1 Introduction

Nowadays it is becoming unavoidable to search for new methods that can enable solving more and more challenging engineering tasks and adapting to dynamically changing design approach. This is the implication of the constantly growing complexity of currently faced design challenges and their multidisciplinary nature. Systems Engineering (SE) defined as a methodology of complex technological problems solving responds to these needs. SE is based on procedural systematic model, focused on specific goal of problem solving.

'Systems engineering is a methodical, disciplined approach for the design, realization, technical management, operations, and retirement of a system. [...] It's a way of looking at the "big picture" when making technical decisions. It's a way of achieving stakeholder functional, physical, and operational performance requirements in the intended use environment over the planned life of the systems. In other words, systems engineering is a logical way of thinking' [1].

This chapter is about ideation methods and their classification, that can be useful in systematic approach to many tasks in the context of Systems Engineering. Both newly designed systems and already existing ones, that are just being modified, require the application of advanced ideation methods. These systems consist of multiple subsystems, that are characterized by various levels of complexity [1] and mutual relations. Such methods should ensure the necessary level of innovation and a defined structure enabling to integrate multidisciplinary issues. Systems Engineering enables to apply the procedures to organize the ideation process, decoupling it from influence of external factors. Nevertheless Systems Engineering methodology should include the ideation methods selection, being one of the systematic approach elements in terms of system design. Therefore, also these methods should be systematic methods, defined and followed formally.

Ideation is the creative process of generating, developing, and communicating new ideas, where an idea is understood as a basic element of thought that can be either visual, concrete, or abstract [2]. Ideation comprises all stages of a thought cycle, from innovation, to development, to actualization [3]. As such, it is an essential part of the design process, both in education and practice [4].

3.2 Classification of Ideation Methods

Ideation methods are techniques that are used during the design process and focus on idea generation. Organizing knowledge about available ideation methods in the context of determining the suitability of each technique in order to solve a specific group of problems is a very difficult task. It seems that it is possible to define only general recommendations, which allow finding the proper ideation method in a quick way. First of all, it is due to a lack of possibility to classify the problems precisely. Identification of existing methods is not a severe problem.

However, in the case of characterizing an issue, which is the object of ideation methods, the field of possibilities is basically unlimited. Especially in the case of complex systems, features may differ significantly from each other. These differences result from a multitude of aspects that must be considered, the existing interdependencies, and frequent contradictions. This fact causes the need to consider these problems in an individual way as well as large difficulties with developing an algorithm which allows designation of the most suitable method to solve them. The second significant obstacle is difficulty in formulating a complete classification of ideation methods. Each division is made on the basis of specified criteria. The way of formulating these criteria may not consider aspects that are important for a specific group of problems or may be ambiguous.

Therefore, it is important to make a decision about the significance level of criteria based on which classification will be made. In other words, it should be specified what kind of division will facilitate selection of the ideation method in a particular case. As a result the classification can be made in different ways and it is necessary to have several kinds of divisions to make at least a general view of available methods and domains in which they can be used. The problem of classification of ideation methods is also associated with the difficulty of verification of several aspects. An example would be the division of idea generating techniques into individual methods and group methods. This division, in a procedural sense, does not cause more significant troubles. In most cases a procedure of a particular method determines the assignment to one of two categories. The situation will be more complicated when the classification will not be connected to formal aspects but instead it is connected with the quality of ideas and mechanisms of their generation, depending on the number of people involved in the ideation process of a particular method. No data on this subject, as well as difficulties of their gathering, make it impossible to divide idea generation techniques in a way that would be significant from the point of view of choice of method, elaboration of organizational assumptions, and organization of the design team.

Ideation is the process of searching for a solution to particular design issue. In narrower terms, it concerns finding as many solutions as possible to the particular issue. This is not the same as indicating the target solution which usually requires use of additional methods of evaluation and optimization. The evaluation and selection stage is most frequently an integral part of ideation, but it can be considered as independent process based on the results of ideation. Then ideation is considered only a procedure of generating a set of solutions which are said with high probability to be the right ones. In many cases, the decision on the selection of the concept does not occur directly but is done in an indirect way during the implementation of the algorithm of a specific ideation method [5].

Figure 3.1 is a graphic representation of two approaches to the ideation process:

- The way, where the result is a more or less formalized final concept,
- The way, where the result is a set of possible variants of the solution.

The first way is based on the generation of one specific concept, which is potentially a solution to the initial problem. This result may be achieved through the use

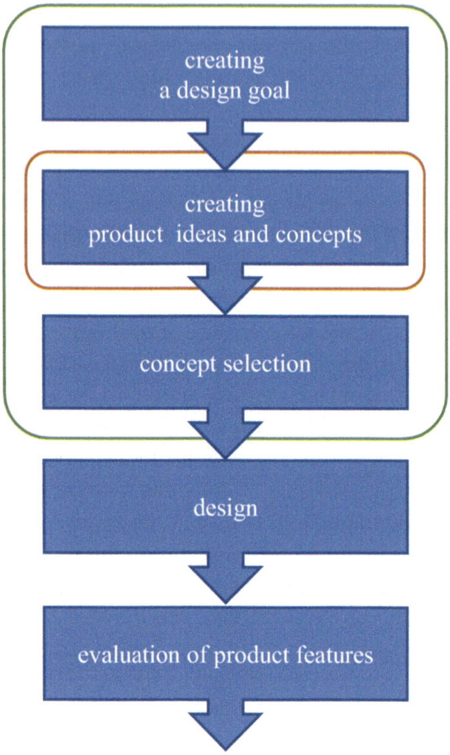

Fig. 3.1 Steps of design method

of different ways of evaluation or through the features of the method itself, which leads to the direct solution or to a narrow set of solutions. The solution is verified, and next, depending on the result of verification, another idea is generated or the ideation process ends. The specific concept appears frequently when the solution is elaborated upon and that improves an existing object or system, especially in the situation that the set of possible solutions is limited.

The result of the second approach to the ideation stage in design process is a certain set of potential solutions. These solutions have to be analyzed to enable decision making about the choice of suitable concept, frequently using criteria.

The presented division may seem a little artificial, but it answers the important question, namely, at which point the ideation process ends and when the optimization process of the concept begins. Does verification belong to the ideation process or it is a separate stage? The answer to those questions may seem not that ambiguous. For example, you can imagine the development of the solution using brainstorming or a C-Sketch method. In these cases, with probability nearing certainty it can be assumed, that a significant number of solutions will be generated. In the case that the TRIZ method is used, it may turn out that one solution will be found. Additionally, while using this method, there is a kind of self-acting verification of subsequent versions which is the result of the implementation of the principles and recommendations from the TRIZ method.

Generally, the classification and characteristics of ideation methods should help to systemize the knowledge of potential possibilities associated with the initial stage of the design process. This phase is the ideation process. Classification should also help with the choice of ideation method depending on the characteristics of the considered issue. Therefore it is important to look at the set of available ideation methods from different points of view and classify them due to their selected aspects.

The choice of the suitable ideation method corresponding to the particular issue may save a lot of undesirable problems during further stages, for example: the omission of certain aspects (i.e. technological) and results from the wrong relationship between method and problems specification, or a too lengthy conceptual process, which can interfere with further stages.

The ideation process is focused on creative thinking and generating ideas. Very often it is thought that the success of this process is determined by the ability of developers only, but in fact it is possible to learn this process and its progress depends on many factors, among others:

- designer's talent
- designer's knowledge
- designer's commitment workload
- workshop
- working method.

It is worth noting that in the above list only *designer's talent* is an independent factor. Therefore it can be assumed that the deficiency of talent can be compensated in the simplest way by selecting the appropriate method. Moreover, the use of an ideation method is not necessarily connected with a lack of talent or skills but it should be the way to use either aspect better.

The basis of the choice of the ideation method, beyond its knowledge, is the appropriate formulation of the design task. The identification of design requirements is the stage before the ideation process and is connected with this process of a two-way relationship. They may be those assumptions, which are the input data to develop the concept, later they will be modified at the ideation stage. Therefore, the identification of the requirements is particularly important, but at the same time very difficult. There are methods supporting the identification and definition of these requirements.

The input data for the ideation process should be:

- the defined purpose or function, which has to satisfy the objective
- design requirements (constraints)
- criteria with its hierarchy
- knowledge necessary for the creative process
- the ideation process constraints.

Usually the aim is to define the design assumptions in as much details as possible. Ideation must be goal-oriented and this goal is the central element of the ideation process. This approach minimizes the risk of development of projects discordant with requirements and enables verification of the emerging concepts. It may be that too many constraints with a high level of detail can block freedom to generate ideas at an early stage of the ideation process. It seems that it is particularly important in the context of complex systems and objects. Let us assume that we design an object and the constraints connected with this object were identified in a lot of details. As a result, new ideas would not be generated during the ideation process. If the constraints were formulated in a more general way, more different ideas would probably appear. These ideas would not satisfy more detailed conditions, but would be adjusted to them through small modifications that do not change their main intention. This hypothetical consideration shows a large importance of the design requirements and inclines us to think about the reduction of risk connected with them.

One way may be the connection between the design requirements and the criteria at the beginning of the ideation process, before choosing a method. Thanks to this, the most important requirements appear at the input of the process and the insignificant issues are temporarily omitted. The final verification and modifications will be conducted with a whole list of requirements. Nowadays the development directions are focused on creative thinking methods, which are contradictory to too detailed or defined requirements or constraints. Creativity in this sense is understood as the ability to generate a number of alternative solutions, which should be novel and functional [3]. This issue is associated with the choice of characteristics of the ideation process.

There are two available models. The first one involves development, modification, and improvement of existing solutions. It is an iterative procedure operating on a specific object, therefore it has all the limitations associated with this process. The latter one is based on looking for an ideal solution, and next, on its adaptation to constraints and requirements. It can be made through modifications causing the smallest possible differences in relation to the ideal solution. The results of the first way are called adaptation or improvement. The effects of ideation according to the second model are called inventions. The choice of the methodology depends primarily on the specification of the design task and on the description of the problem. Currently, the ideation process is conducted most frequently as opened and limited only to the extent sufficient to achieve the expected results. As a result, it becomes more creative and more directed at innovation.

The efficiency of achieving this goal frequently depends on the proper choice of a method. In the next section of this chapter, the authors try to classify known ideation methods with their characteristics and areas of their application.

The general division of methods may be represented as follow:

Presented in Fig. 3.2 is a very general and wide classification that involves all the known ideation methods. Each method may be assigned to at least one category. This division and knowledge of the application area, in which a given group of methods gives the best results, can be the basis of the appropriate choice of method depending on the considered problem.

One of the more comprehensive classifications has been defined by Shah [6]. He divided ideation methods into two main groups: intuitive and logical. Intuitive

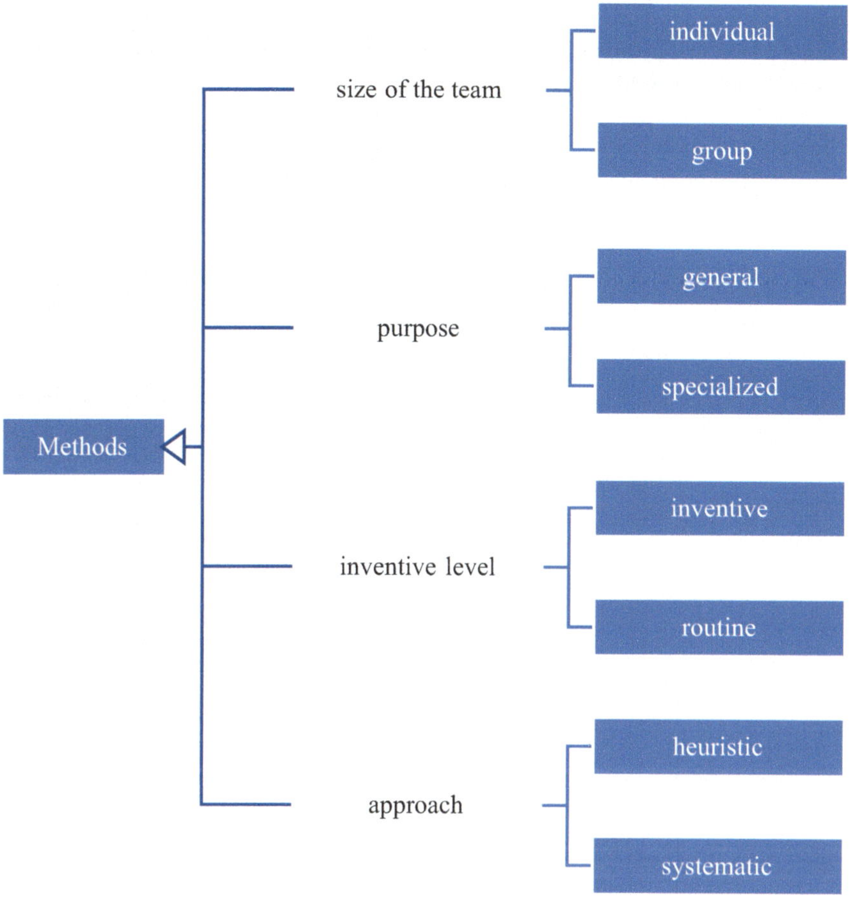

Fig. 3.2 General classification of the ideation methods

methods are based on thinking processes only, which are not supported in any way from the outside. They are based on the designer's creativity or on the creativity of other people taking part in the ideation process. Therefore, it is important to create the conditions for the implementation of these methods. First of all, the number of obstacles that inhibit creative thinking should be minimized. Stimulation is also necessary. The way to stimulate may be by building a connection network, identifying elements for broadening the narrow field of view and, sometimes, identifying new directions of thinking.

Logical methods are based on specific algorithms and include a systematic approach associated with procedures and often very well-defined operating modes. The path to solve the problem is realized step by step using analysis, decomposition, and knowledge from sources such as databases, catalogs, experience and existing solutions. Logical ideation methods are formalized in terms of the mechanism of conduct so that they are relatively easy to implement and ensure stability of the ideation process. It is important to pay attention to the fact that based on the external knowledge sources used, the effectiveness of a logical method depends on the quality of data.

Taking into account the classification presented in Fig. 3.3, it should be said that this classification mainly includes methods for general use. There are no methods in

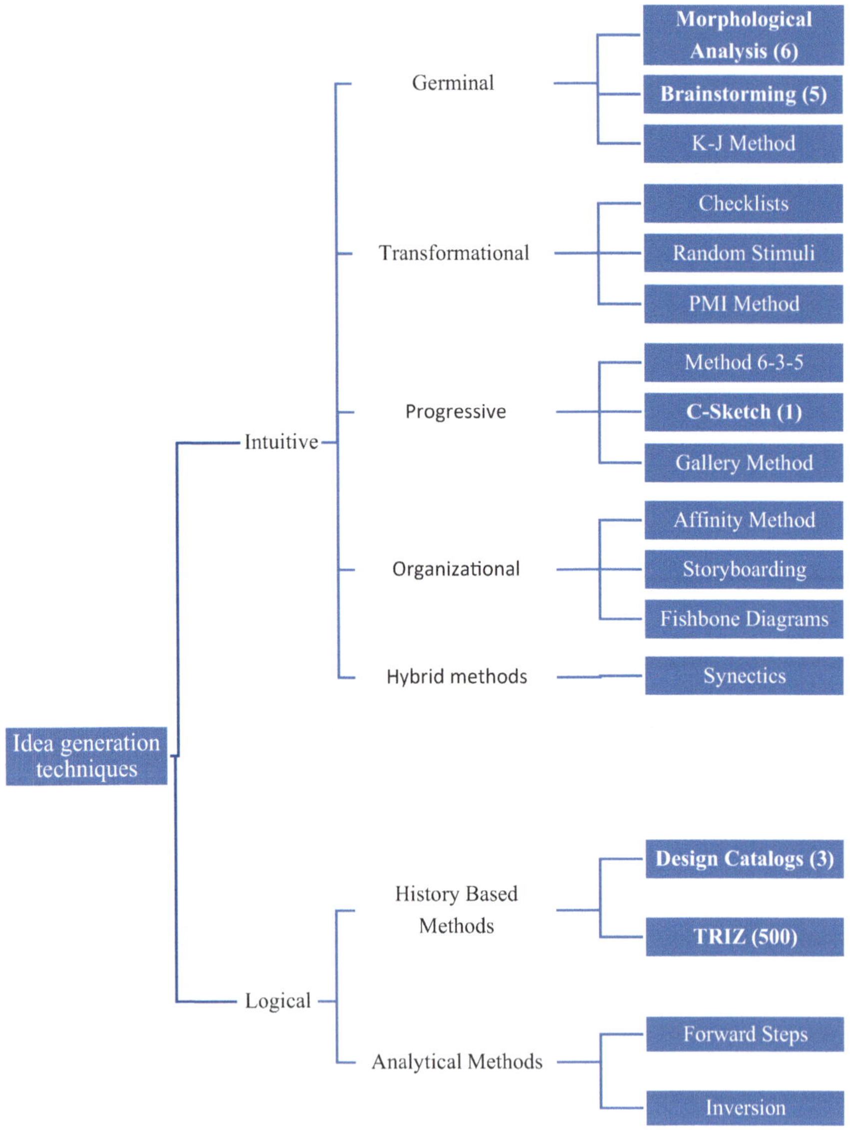

Fig. 3.3 Classification of structured idea generation methods [15]

this classification that are specialized or that are used in directly specified problems. There are, however, methods which achieve the best results for directly specified issues. For example, Inversion Method is dedicated especially to problems associated with kinematics.

Another very important division is classification of the number of team members necessary to conduct the ideation process using a given method. The division presented below (Fig. 3.4) looks at the majority of ideation methods included at Fig. 3.3 from another point of view. It allows us to modify and narrow down the search area depending on human resources and to predict the necessity of modifying the design team.

Clear division to individual and group methods is possible only in the case of intuitive methods. These methods use open and creative thinking and in general, are group methods. This approach is justified by the conclusion from observation that the work of an individual in the group is more effective than individual work. It is associated with mutual motivation of team members and with their mutual impact stimulating to develop new concepts. Logical methods cannot be divided into group methods or individual methods due to their specific characteristics. Formal structure allows the use of them both, either one or many people working on the idea. In contrast to intuitive methods, the number of people does not determine the efficiency and quality of solutions obtained by logical methods. The most important in this case is the ability to think analytically and skillful using procedures and logical methods.

The most difficult part seems to be assigning the particular method to a group of inventive methods and a group of routine methods. The first problem is how to define the terms "inventive" and "routine". The term "inventive" refers to originality and novelty of the solution to the technical issue. The result is the technical object including elements, that were previously unknown and do not come directly from the known state of art. The term "routine" may be related to the design processes which are restricted to the modification of the existing solutions and their combinations. Therefore they are associated with the methods based on known elements and using them to create new technical means. The second problem results from the fact that assignment to routine methods or inventive methods frequently will be depend on the achieved result. This applies particularly to inventive methods, which, in a large number of cases, give solutions unacceptable as an invention. The example of the strictly routine method is Morphological Charting. In this method, solutions are generated as combinations of fragmentary concepts of each element or system of the considered problem. Inventiveness in this method is limited mainly to the need to define subsystems of the considered system. This does not mean that the results obtained using Morphological Charting cannot be solutions having the features of invention. Another example, classified as an inventive method, is the TRIZ method. It is based on ways of overcoming the technical contradictions appearing during elaboration of the concept of solution which is called the ideal solution to the problem. Using only the set of recommendations and procedures and the preliminary generalization of the considered issues significantly increases the probability of finding an innovative solution.

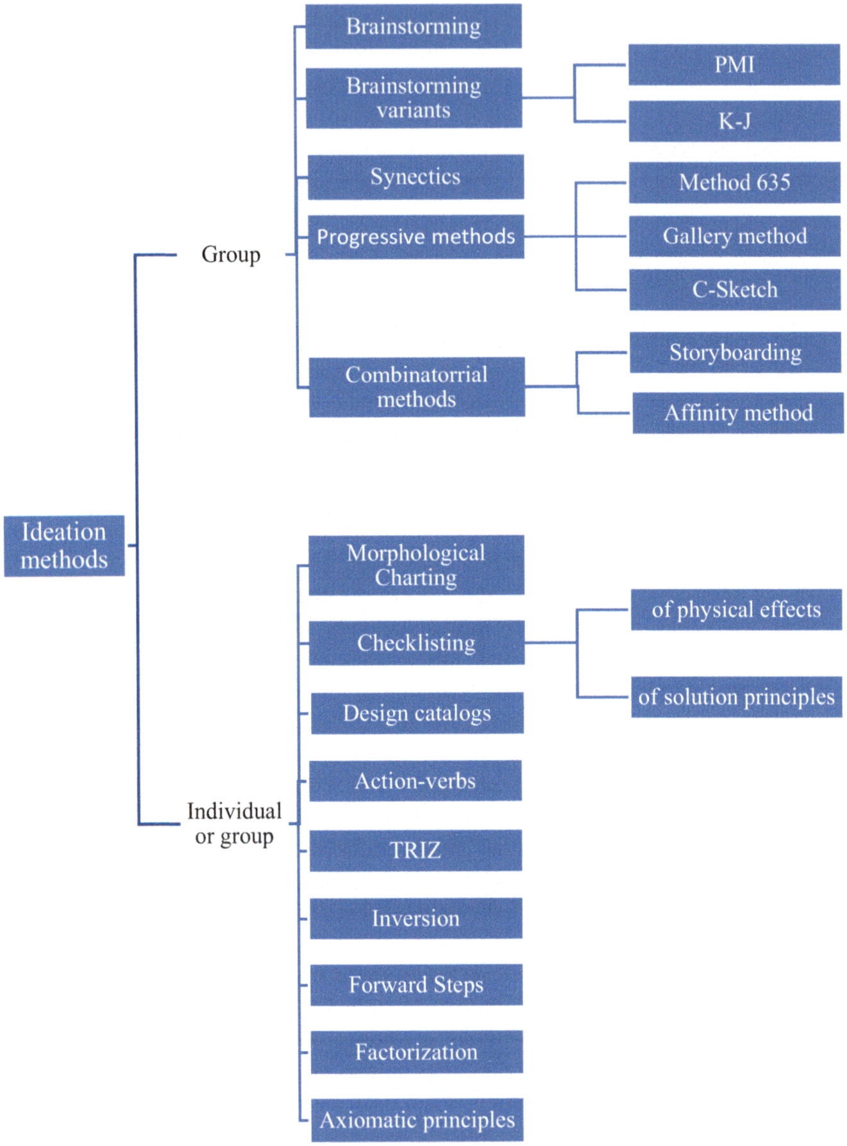

Fig. 3.4 Ideation methods according to the number of people in the design team [6]

It is apparent from the presented literature review and authors' observations that the TRIZ method is especially popular among ideation methods. This fact can be justified by many cases in which this method leads to the solution which is correct due to the realization of the defined purpose. It is based on thorough knowledge and experience gathered for many years, what is its unquestionable advantage. On the other hand, the TRIZ method can be considered as not that creative. In opposition to it, we have brainstorming and C-Sketch. In these methods, creativity leads most frequently to achieving good results. The last method that is useful for systems or objects which are imaginable and possible to represent graphically, as well as in cases, is a method in which design is very important. With regard to Systems Engineering, it seems reasonable to consider the possibility of dividing the system into smaller parts and elaborating on them in the ideation process, taking into account their connections with the system.

3.3 Background/Review of Ideation Methods

Among many ideation methods, there are several that are most commonly used in the design process. The popularity of these methods results from their versatility and quality of effects achieved with their help. One way to determine the popularity of the method and domains in which it is most commonly used is literature research. The chart (Fig. 3.5) shows the number of publications concerning different ideation methods. This chart has been created on the basis of data from Web of Science base from 2000 to 2016. In the case of the TRIZ method, the chart includes data only from 2016, because the total number of publications associated with this method in the mentioned range of time is over 500.

TRIZ is "a problem-solving, analysis and forecasting tool derived from the study of patterns of invention in the global patent literature" [7]. In English the name is typically rendered as "the theory of inventive problem solving" [8, 9], and occasionally goes by the English acronym TIPS. TRIZ presents a systematic approach for understanding and defining challenging problems: difficult problems require an inventive solution, and TRIZ provides a range of strategies and tools for finding these inventive solutions. One of the most popular tools which evolved as an extension of the 40 principles was a contradiction matrix in which the contradictory elements of a problem were categorized according to a list of 39 factors which could impact on each other. The combination of each pairing of these 39 elements is set out in a matrix. Each of the 39 elements is represented down the rows and across the columns (as the negatively affected element) and based upon the research and analysis of patents: wherever precedent solutions have been found that resolve a conflict between two of the elements, the relevant cells in the matrix typically contain a sub-set of three or four principles that have been applied most frequently in inventive solutions which resolve contradictions between those two elements [10].

The **Axiomatic Design** [11] provides a systematic search process through the design space to minimize the random search process and determine the best design

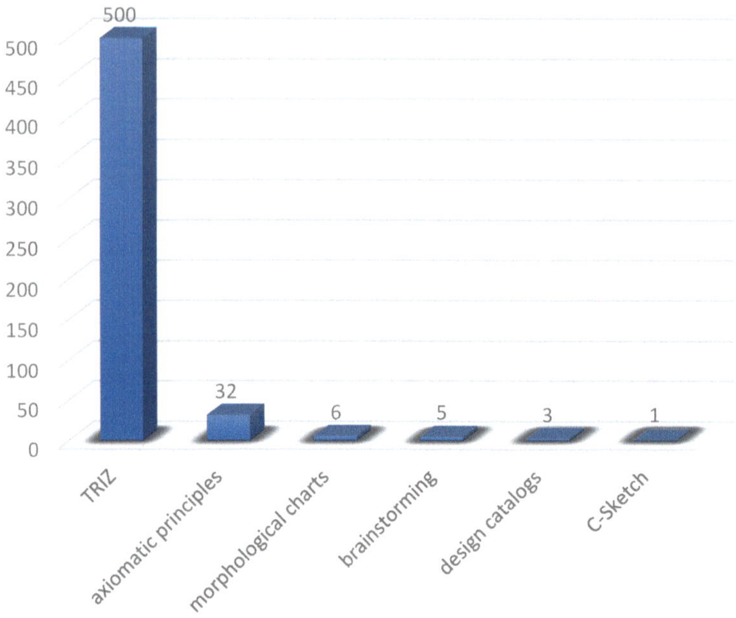

Fig. 3.5 Ideation methods according to the amount of appearances in literature

solution among many alternatives. The most important concept in axiomatic design is the existence of the design axioms. The first design axiom is known as the Independence Axiom and the second axiom is known as the Information Axiom:

- Axiom 1. The Independence Axiom. Maintain the independence of functional requirements.
- Axiom 2. The Information Axiom. Minimize the information content.

Mathematical relationship between the FRs and DPs:

$$\{FR\} = |A|\{DP\}$$

where:

{FR} is the functional requirement vector
{DP} is the design parameter vector, and
|A| is the design matrix that characterizes the design.

A **morphological chart** is a visual way to capture the necessary product functionality and explore alternative means and combinations of achieving that functionality. For each element of product function, there may be a number of possible solutions. The chart enables these solutions to be expressed and provides a structure for considering alternative combinations. This can enable the early consideration of the product 'architecture' through the generation and consideration of different combinations of 'sub-solutions' that have not previously been identified. Used appropriately, it can help to encourage a user driven approach to the generation of a potential solution [12].

Brainstorming is a group creativity technique by which efforts are made to find a conclusion for a specific problem by gathering a list of ideas spontaneously contributed by its members.

In one version it consists of two stages. In the first one participants are encouraged to submit ideas and freely exchange their views. The requirement is that an this stage no ideas are criticized and all of them are recorded. In the second stage an expert or a group of specialists who did not participate in the previous phase evaluate and select valuable ideas [13].

Design catalog is a method using catalog, which is a set of known and tried solution of specified design tasks. The proper solution is selected on the basis of representation of different properties, for example: characteristic dimensions, characteristics, number of elements [14].

In the **C-sketch method** (Collaborative sketching) designers work on graphical representation of considered problem. This method is suitable for well-defined problems. Designers work independently, developing sketches of their proposed solutions to the problem. At the end of cycle, the sketch is passed to the next person. Designers working on this issue may add, remove or modify elements of the design solution. The limitation is that the entire idea (sketch) cannot be deleted and the communication between team members is possible only through the sketches. In this manner, the sketches are passed sequentially through the design team. The end of this process takes place when the number of sketches is equal to the number of team members [15].

The aforementioned complexity of the developed systems leads to the necessity of using advanced methods on each stage of making a new object. This dependence also concerns ideation methods. An inappropriate choice of idea generation technique may have significant consequences. There is a risk that the necessity for the repetition of the ideation process or that the quality of the developed solutions will be lower in comparison with solutions that are developed when using the appropriate method. It is very important to analyze the specifics of the considered problem and the goal which should be achieved.

K-J Method lets to specify whether the designer working on the most important parts of the problem.

Checklist is a method used for developing new ideas by combining and altering existing elements. The creation of this idea is based on actions such as: substitute, combine, adapt, magnify, put to other use, eliminate and rearrange.

A random stimuli is a class of creativity techniques that explores randomization. In a random creativity technique, the user is presented with a random stimulus and explores associations that could trigger novel ideas. The power of random stimulus is that it can lead to explore useful associations that would not emerge intentionally [16].

The PMI technique is a thinking technique to find the Plus Points, Minus Points, and Interesting Points about the issue before forming an opinion.

Method 6-3-5 is a group-structured brainstorming technique. It consists of 6 participants supervised by a moderator who are required to write down 3 ideas on a specific worksheet within 5 min, this is also the etymology of the methodology name. The outcome after 6 rounds, during which participants swap their worksheets passing them on to the team member sitting at their right, is 108 ideas generated in 30 min [15].

In the **Gallery Method** the participants move past the ideas (as in an art gallery) rather than the ideas moving past the participants. Several flip chart sheets are posted around the room and participants circulate and record their ideas. As participants move around the room and read the ideas of others, they often get other, related ideas that they add to the list. The distinctive feature of the Gallery Method is that group members are permitted to move about during the break period (incubation period) [17].

The fishbone diagram identifies many possible causes for an effect or problem. It can be used to structure a brainstorming session. It immediately sorts ideas into useful categories.

Synectics is a problem solving methodology that stimulates thought processes of which the subject may be unaware. Synectics is a way to approach creativity and problem-solving in a rational way [12].

3.4 Ideation and Design Creativity

No matter how we understand the idea of conceptualization or generation of a set of new solutions, it is always assumed that the solutions will represent the level of superiority equal or outrating the current state of technology. In development works of complex systems, a certain innovation level of development works is usually assumed. Typical division of designing process locates the process of concept development after the elaboration of engineering specification and before product development (Fig. 3.6).

However, development works are not carried in a single process but in a repetitive way for different degree of technical development of a newly created system, especially if we deal with a complex, interdisciplinary technical system.

The formal division into industrial research and experimental development which has been introduced for product development allows to state what kind of innovation we can expect at a given phase of development (Fig. 3.7). Technology Readiness Level, which has been introduced, forms an additional way of formalization of this level. It can be generally stated that for earlier level of commitment and lower level of technology readiness and at the initial phase of industrial research it is more important to create a more innovative concept. Nevertheless, at the consecutive phases, where key solutions have been verified, more detailed solutions are valued which rely more in improvement and choosing the best out of the known solutions. In other words, engineering skills are more important than innovation or creativity.

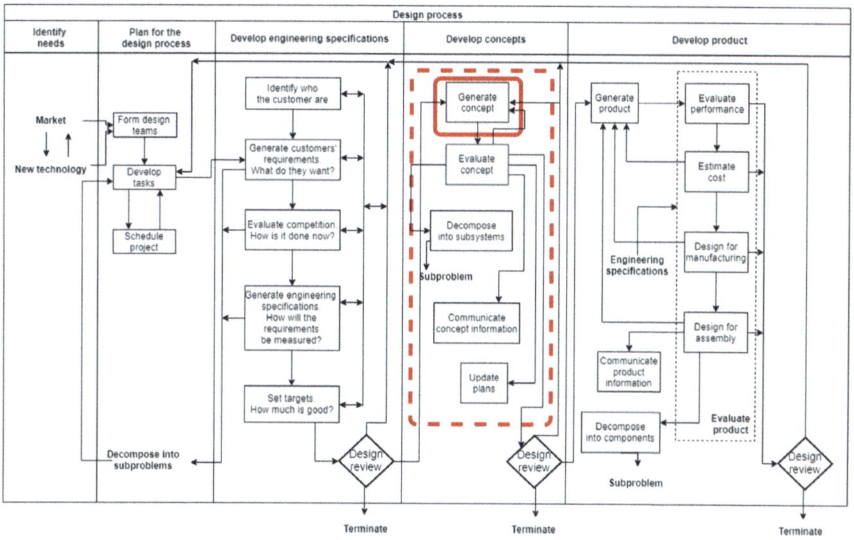

Fig. 3.6 General structure of design process [18]

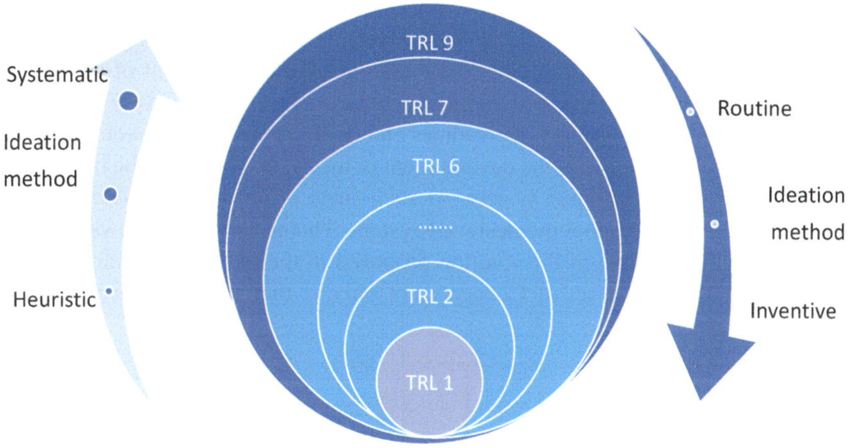

Fig. 3.7 The dependence of ideation process on technology readiness level

Where a topic description refers to a TRL, the following definitions apply [19]:

Fundamental research

- TRL 1—basic principles observed.

Industrial research

- TRL 2—technology concept formulated
- TRL 3—experimental proof of concept

- TRL 4—technology validated in lab
- TRL 5—technology validated in relevant environment (industrially relevant environment in the case of key enabling technologies)
- TRL 6—technology demonstrated in relevant environment (industrially relevant environment in the case of key enabling technologies).

Experimental development

- TRL 7—system prototype demonstration in operational environment
- TRL 8—system complete and qualified
- TRL 9—actual system proven in operational environment (competitive manufacturing in the case of key enabling technologies; or in space).

The expected design creativity at different phases of system maturity changes its meaning depending on the level of product development. At the early stages of development it is of innovative character but gradually with its growth it evolves to improvement or adaptation. The methods and techniques of ideation should be chosen accordingly to this.

Regardless of the generation of the concept for the correct evaluation of this concept in the case of a simple system sophisticated methods are not necessary to apply. On the other hand, in the case of a multidisciplinary complex system, the influence of a fragment of this system on the operation of the whole is difficult to determine. Typically, computer methods are used in this case, and Model-Based Design (Fig. 3.8) is a common solution [20, 21]. The MBD assumes the development of numerical simulation models of the designed multidisciplinary system from the very beginning of design concept. Additionally, it presumes elaboration and current development of this model in accordance with the development of the system and with the increasing level of detail of the designed system. At each stage of design, the model reflects the state of knowledge about the designed system. The use of MBD allows to make a quantitative assessment of the solution proposed in the Ideation phase or system evaluation of the whole set of Concepts not only at the stage of conception but

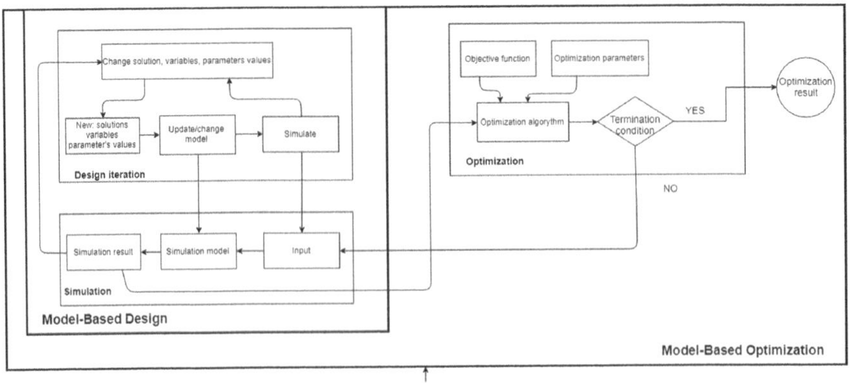

Fig. 3.8 Model -based design and model-based optimization diagram

also at any stage of iteration in the product development process regardless of the TRL degree. This is especially important for complex and multidisciplinary systems whose operation is difficult to predict. The simulation shows if the model (developed system) works correctly according to the assumptions (Fig. 3.8).

What's more, the same method can be used to optimize the features of the proposed solution in the context of a complex multidisciplinary system by integrating into the simulation the model the system used in MBD optimization methods. Then, simulation numerical model together with optimization methods are used directly for calculating optimal design features and the method is called Model-Based Optimization (Fig. 3.8) [21–23]. Optimization criteria are described in objective function and they are implemented by optimization algorithm for the sake of evaluation of simulation results and to introduce proper changes in optimization variables. As a result, optimal set of design features is obtained. This method is used for subassemblies optimization however, with complex objects due to the difficulty of defining objectives function and complexity of calculations it is usually not used.

3.5 Assessment of Methods in Context of Systems Engineering

For the assessment of Ideation methods it is important to clarify the notion of Idea Generation model. Researches indicate [5] that the inspiration for new ideas can appear at any stage of designing. IR3 Idea Generation Model is mentioned where additional three stages are differentiated i.e. researching, representing and refining. Each of the existing methods supports these stages in different degree.

In order to apply the various creativity techniques effectively, a creative process needs to be obeyed. A very simple model of the creative process has been known for a long time [24]: preparation, incubation, illumination, and verification. In the preparation phase the problem is defined. During the incubation phase, the problem is left and attention is focused on other more demanding aspects. In the illumination phase a start is (suddenly) found, from which an approach is developed to deal with the problem immediately. During the verification phase the idea is tested and evaluated. Tassoul [25] has modelled the creative problem-solving process in a more elaborated model, called the CPS model popular. This model consists of three phases:

1. problem statement,
2. idea generation,
3. concept development.

In these stages divergence of problem, clustering ideas and convergence into concepts take place.

Cognitive scientists have identified a set of generative and exploratory processes which occur in various combinations during the creative process.

Generative processes [26] include: memory retrieval, association, mental synthesis, mental transformations, analogical transfer and categorical reduction.

Exploratory processes are as follows: attribute finding, conceptual interpretation, functional inference, contextual shifting, hypothesis testing, and searching for limitations. These processes are supported by mental representations called cognitive structures, such as: visual patterns, object forms and category exemplars. These cognitive structures often have the conflicting properties of ambiguity and direct meaningfulness.

It is not possible to determine out of many Ideation methods, which of the given methods are suitable for designing and developing complex technical systems will be used especially in interdisciplinary field of engineering. The cause of that can be found in the following features of the process, namely:

- Complex systems are usually developed by multi actions which result in a system in the consecutive phases of maturity
- Subsets and subsystems are usually developed individually and independently of the whole system
- Development actions are carried in teams which are technically, organizationally and even culturally different and which realize their tasks based on their experience, tools and tradition and imposing other methods on them can be in many cases pointless
- At various stages of development different ideation methods are necessary, depending on the number of members in the team, the results, used assets or expected results.

Taking these into consideration different Ideation methods can be used in the process of Concept development, and in particular in concept generation.

Having the table (Table 3.1) and basing on the assets and expected features of the method and its results, it is possible to choose the most suitable method. It is also possible to rely on the usage frequency of the used method. However, the table includes only the most important and the most commonly used methods, since the number of Ideation methods is too big to analyze them all.

Takahashi [27] indicated that more than 300 idea generation methods have been invented, but only a few are applied regularly. Smith [28] identified 172 idea generating methods additionally he put these methods into a smaller set of active ingredients that represent the core functionality behind each method. Though many methods have been identified, there is a limited number of research that have dealt with the frequency of use of idea generation techniques and their applicability during constrained situations [29].

Knowledge of the importance of these techniques is the key for choosing Ideation method for specific application and creativity support tools design, because only a few of them are often used in practical design situations. Therefore, in the process of Concept development and in particular concept generation, different Ideation methods can be used and first of all the ones which are used and developed in companies with good practices. Depending on expectations and the stage of development, the general table below which shows particular methods can be used to choose the appropriate one. It is also advisable to take into account the frequency of usage of a

Table 1 Ideation methods features

Features	Size of the team		Purpose		Inventive level		Approach		Promoters
Ideation method	Group	Indywidual	General	Specialized	Inventive	Routine	Heuristic	Systematic	SJ/DA/CFR/MUC/PAS/FR/EQ
TRIZ/TIPS	√	√		√	√			√	SJ/DA/CFR/MUC/PAS/-/EQ
Axiomatic Principles	√	√	√			√		√	SJ/-/-/-/-/-/EQ
Morphological analysis	√	√	√			√	√		SJ/-/-/MUC/-/-/EQ
Brainstorming	√		√		√		√		SJ/DA/CFR/MUC/PAS/-/EQ
Design Catalogs	√	√		√		√	√		-/-/-/MUC/PAS/-/-
C-Sketch	√			√	√				SJ/-/-/MUC/PAS/FR/-
K-J Method	√		√		√		√		SJ/DA/CFR/MUC/PAS/-/EQ
Checklist	√	√	√		√	√	√		-/DA/CFR/MUC/PAS/-/-
Random Stimuli			√		√		√		-/-/-/MUC/PAS/-/EQ
PMI Method	√		√		√		√		-/DA/CFR/MUC/PAS/-/EQ
6-3-5 Method	√		√		√		√		SJ/DA/CFR/MUC/PAS/FR/-
Gallery Method	√		√			√	√		-/-/-/MUC/PAS/FR/-
Storyboarding	√		√		√		√		SJ/DA/CFR/MUC/PAS/-/EQ

(continued)

Table 1 (continued)

Features	Size of the team		Purpose		Inventive level		Approach		Promoters
Ideation method	Group	Indywidual	General	Specialized	Inventive	Routine	Heuristic	Systematic	SJ/DA/CFR/MUC/ PAS/FR/EQ
Fishbone Diagrams	√	√	√			√		√	-/DA/CFR/MUC/PAS/-/EQ
Synectics	√	√	√		√		√		SJ/DA/CFR/MUC/PAS/FR/-
SIT (Systematic Inventive Thinking)	√	√	√		√				SJ/DA/CFR/MUC/PAS/-/EQ
Forward step	√	√	√			√		√	-/-/-/-/-/FR/-
Inversion	√	√	√		√	√		√	SJ/DA/CFR/MUC/PAS/-/-
Factorization, systematic combination	√	√	√			√		√	SJ/-/-/MUC/PAS/-/EQ

(continued)

Table 1 (continued)

Features	Blocks	Idea generation phase support	Representation form	Additional assets	Additional features	Specialized personel	Referencing
Ideation method	DF/R/PJ/GO/IFC	I/RE/RP/RF	TX/MX/SK/DI/TR/CH/MM	(S/DB) Necessary (N), Available (A) Desirable (D)	G/T/F/O/H/HB/A	(Designers/ Multidomain Experts) D/ME	WoS/ER
TRIZ/TIPS	-/R/-/-/-	I/RE/-/RF	TX/MX/SK/-/-/-/-	DB, N, A	HB	D	500/H
Axiomatic Principles	-/R/-/-/-	I/RE/-/-	TX/MX/-/-/-/-/-	S, A	T	D	32/L
Morphological analysis	-/R/-/-/-	I/RE/-/-	TX/MX/-/-/-/-/-	S, A	G	D	6/M
Brainstorming	-/R/-/-/-	I/RE/RP/RF	TX/-/-/-/-/-/-	-	G	ME	5/H
Design Catalogs	DF/-/PJ/GO/IFC	I/RE/-/RF	TX/MX/SK/-/TR/-/-	DB, N, A	HB	D	3/H
C-Sketch	DF/-/-/GO/-	I/RE/RP/RF	-/-/SK/-/-/-/-	-	P	ME	1/H
K-J Method	-/R/-/-/-	I/RE/RP/RF	TX/-/-/DI/-/-/-	-	G	ME	H
Checklist	DF/-/PJ/GO/IFC	I/-/RP/-	TX/MX/-/-/-/-/-	S, A	T	D	H
Random Stimuli	-/R/-/-/-	I/RE/-/-	TX/-/-/-/TR/-/-	-	T	ME	L
PMI Method	-/-/PJ/-/-	I/RE/RP/RF	TX/-/-/-/-/-/-	-	T	ME	H
6-3-5 Method	DF/R/-/-/-	I/RE/RP/RF	TX/-/SK/-/-/-/-	-	P	ME	H

(continued)

Table 1 (continued)

Features	Blocks	Idea generation phase support	Representation form	Additional assets	Additional features	Specialized personel	Referencing
Ideation method	DF/R/PJ/GO/IFC	I/RE/RP/RF	TX/MX/SK/DI/TR/CH/MM	(S/DB) Necessary (N), Available (A) Desirable (D)	G/T/P/O/H/HB/A	(Designers/ Multidomain Experts) D/ME	WoS/ER
Gallery Method	DF/-/PJ/GO/-	I/RE/RP/RF	TX/-/SK/-/-/-/-	-	P	ME	H
Storyboarding	-/R/-/GO/-	I/RE/RP/RF	TX/-/SK/DI/TR/-/-	-	O	ME	H
Fishbone Diagrams	-/R/-/GO/IFC	I/RE/RP/RF	TX/-/-/-/TR/-/-	-	O	ME	M
Synectics	_-/-/-/-/-	I/RE/-/-	TX/-/SK/-/-/-/-	-	H	D	M
SIT (Systematic Inventive Thinking)	-/R/-/-/-	I/RE/RP/RF	TX/MX/SK/-/-/-/-	S, A	G	ME	H
Forward step	DF/-/PJ/GO/IFC	I/RE/RP/RF	TX/MX/SK/DI/TR/CH/MM	-	A	D	L

(continued)

Table 1 (continued)

Features	Blocks	Idea generation phase support	Representation form	Additional assets	Additional features	Specialized personel	Referencing
Ideation method	DF/R/PJ/GO/IFC	I/RE/RP/RF	TX/MX/SK/DI/TR/CH/MM	(S/DB) Necessary (N), Available (A) Desirable (D)	G/T/P/O/H/HB/A	(Designers/ Multidomain Experts) D/ME	WoS/ER
Inversion	DF/R/PJ/GO/-	I/RE/RP/RF	TX/-/-/-/-/-/-	-	A	D	L
Factorization, systematic combination	DF/R/-/-/IFC	I/RE/-/-	TX/MX/-/-/TR/-/-	S, A	T	D	M

Promoters: (suspended judgment , detachment/ abstractions , change of the frame of reference, making unexpected connections, provocative action or stimuli, friendly representation, emphasis on quantity) SJ/DA/CFR/MUC/PAS/FR/EQ

Blocks: (Design fixation, Representation (textual, mathematical), Premature judgement, Goal orientation, Imposing fictitious constraints) DF/R/PJ/GO/IFC

Representation form: (text/matrix/sketch/diagrams/trees/charts/mindmaps) TX/MX/SK/DI/TR/CH/MM/

Additional assets: (S/DB Software/Database) Necessary (N), Available (A) Desirable (D)

Idea generation phase support: (inspire/research/represent/refine/)I/RE/RP/RF

Additional features: (Germinal/Transformational/Progressive/Organizational/Hybrid/History Based/Analytical) G/T/P/O/H/HB/A

Specialized personel: (Designers/Multidomain Experts) D/ME

Referencing: (Web of Science, Experts referencing) (WoS/ER) ER 100% >=H>=60%>M>=30%>L>=0%

given method and their applicability in different contexts. It is very noticeable that the TRIZ/TIPS method significantly differs in this matter (Fig. 3.5).

In order to evaluate Ideation methods, the types of features mentioned below will be helpful as they can assist the choice.

For preliminary evaluation it is good to put the method in a suitable class. According to the descriptions presented in the introduction the following classes and subclasses of methods have been distinguished:

Size of the team (Group, Individual), Purpose (General, Specialized), Inventive level (Inventive, Routine), Approach (Heuristic, Systematic)

Promoters (suspended judgment, detachment/abstractions, change of the frame of reference, making unexpected connections, provocative action or stimuli, friendly representation, emphasis on quantity)

Short names: SJ/DA/CFR/MUC/PAS/FR/EQ

Blocks (Design fixation, Representation (textual, mathematical), Premature judgement, Goal orientation, Imposing fictitious constraints)

Short names: DF/R/PJ/GO/IFC

The various idea generation methods have some constituents that are common in other methods. These parts are embedded in these methods since they are thought to assist the idea generation process. These ingredients are Promoters and Blocks. Promoter are defined as components designed to aid in the idea generation process. Blocks are defined as conditions that work against the idea generation process.

Following groups of promoters have been identified:

Suspended judgment, detachment/abstractions, change of the frame of reference, making unexpected connections, provocative action or stimuli in [6] and friendly representation, emphasis on quantity in [15].

Whereas the following groups of Blocks have been identified [15]: Design fixation, Representation (textual, mathematical), Premature judgement, Goal orientation, Imposing fictitious constraints).

Representation form

(text/matrix/sketch/diagrams/trees/charts/mindmaps)

Short names: TX/MX/SK/DI/TR/CH/MM/

In particular methods different forms of record are used, which is important for evaluation of the method as well as easiness or difficulty of application and ability to generate associations. It has been proved that higher efficiency of Ideation methods where visual form of record is used, especially drafts [15], have greater stimulation efficiency and also play an important role in designing [18]. Other graphical forms of record are also beneficial such as diagrams or maps, while charts and matrixes offer great ease at processing. However, difficulties can be expected with text form of record and the use of mathematical formulas. They are called Blockers of these processes and they significantly reduce the number of associations and achieved concepts.

Additional assets Software/Database (S/DB) Necessary (N), Available (A) Desirable (D)

It may happen that for a given Ideation method no additional assets are required. There are cases where the key to the method is the access to additional assets for example the data base of solutions. Naturally, the quality of the assets influences the result of the method. Having the access to aiding software e.g. software, which aids elaboration of a proper representation such as matrix, mindmap etc. can be an additional facilitation. In some cases such an access is a must for task realization.

Idea generation phase support (inspire/research/represent/refine/)

Short names: I/RE/RP/RF

Studies show that [28] an inspiration for new ideas generation can come at any stage of designing. What is more there are three phases in Idea generation Model which are: research, represent and refine and each of the existing methods supports them.

Specialized personnel (Designers/Multidomain Experts)

Short names: D/ME

Apart from the division into individual and group methods it is also important which requirements must the staff meet while realizing a given Ideation Method. With some methods it is necessary to have many various specialists from different branches which are far-off from the designing domain. It has great influence on the quality of the obtained results and hinders realization of a given method.

Referencing

The number of references and actual frequency of use constitute a very important factor in choosing a suitable method. The assessment presented in Table 3.1 covers the results of our own research carried on the basis of scientific publications in Web of Science from 2000 to 2016. The striking majority of TRIZ method is the most commonly referred to as Ideation method in these types of scientific papers. Additionally, the second discriminant based on the research was given [13, 15, 28, 29]. The quality evaluation gives the frequency of references of specialists surveyed on the use of methods. High frequency (High—H) of references covers the range of 100–60%, medium (Medium—M) covers the range of 60–30% while low (Low—L) covers the range of 30–0%. The evaluation does not take into account the results of applicability due to big differences in methodology and the scope of research as well as relatively small number of results which makes it impossible to compare.

Independently to official scientific classifications, interesting results can be achieved in researching popularity of Ideation Methods. If a question refers to a used method the terminology differs from official methods in [5]. 19 methods of Idea Generation were given with the most popular referred to by over 50% of the surveyed specialists, namely Active Search (100%), Sketching (100%), Expert Opinion (90%), Critique (90%), Brainstorm (80%), Empathy/User Research (80%), Prototyping (70%), Collaborate (60%), Documenting (60%), Passive Search (60%) and Reflect (60%).

It should be noticed that the categories were set in the process of interview and they were called Ideation Techniques although they refer to methods.

In [29] the top 10 techniques selected are in the decreasing popularity in Taiwan: brainstorming (44%), checklist (27%) 1H5W (12%) 5Why (8%) mind mapping, Delphi, TRIZ, SCAMPER, KJ method and NGT. It should be noticed that according to the research each method is of particular suitability for a given context. In particular brainstorming is considered as a "very applicable" technique when the idea generation process is characterized by differences among participants, knowledge background of participants, opportunity of try and error, democratic process in meetings, lively intercommunication, or variety of ideas. It is also considered as a "passable" method when the process is characterized by availability of information, constructive dialogues, positive phrasing, thoroughness of ideas, or elaboration of ideas. However, TRIZ method which has been particularly popular in recent years, is identified as an "applicable" technique when the idea generation process is characterized by knowledge background of participants, opportunity of try and error, thoroughness of ideas, or elaboration of ideas. It is considered as a "passable" method when the process is characterized by time constraint, differences among participants, availability of information, democratic process in meeting, constructive dialogues, lively intercommunication, positive phrasing, or variety of ideas. Lower general evaluation of the use of the TRIZ method contradicts its great popularity in the recent years. It can be explained by different time and area of the research i.e. earlier time and area limited to Taiwan.

Various methods can be simultaneously used for Idea Generation at different stages of Ideation Process and at different stage of product development. It is also quite common to use hybrid methods which combine two or more methods. It allows exact usage of properties of assets and in particular facilitates unlimited search for ideas and systematic adjustment of valuable ideas. It can be seen from the evaluation in Table 3.1 that there is a tendency for favoring some kinds of methods such as Brainstorming, C-Sketch, Gallery method [13] but on the other hand methods from Logical group such as TRIZ/TIPS or Axiomatic principles also share great interests. All in all, based on these results it is difficult to evaluate decisively the efficiency of the methods since in majority of cases the quality was assessed based more on the level of its familiarity [28], number of generated Ideas or its expert evaluation at the concept level. Nevertheless, evaluation of these methods based on commercial result or ability of patent protection seems to be more adequate. The last issue results in noticeable popularity of TRIZ/TIPS method in publications in Web of Science.

TRIZ/TIPS method relies on some noticeable principles which form a base of patent solutions. Altshuler [30] by means of his research on patent solutions noticed that 40 rules identified by him form the base of a set of analyzed inventions. It seems to be enough to follow these rules and, in analogy to already existing inventions, make a discovery which would constitute a base for a researched problem. This method is mainly used for bridging designing contradictions which unable elaboration of satisfactory solutions. This course of actions becomes an inspiration for novel elaboration of solutions with great potential since it was created based on methods used in previous inventions. As we know from practice it guarantees in many cases elaboration of groundbreaking solutions. Moreover, a method which is constantly developed gains big support in a form of software tools and data bases and in their latest versions it is supported by more inspirational graphical representations of a concept.

Popularity of Brainstorming results from both its great potential of results and vast familiarity with it. In layman's terms any meeting of a few people trying to solve a problem is called brainstorming but the correct Brainstorming procedure requires obeying some important limits which significantly increase possibilities of the method as compared to a casual meeting.

In the evaluation process of usefulness of particular Ideation methods, decision and selection process of methods from the given elaborated Ideation set of methods, has not been taken into account since this part of the process is not so important. Depending on the chosen method, the process itself can vary since the result of Ideation can be a set of many concepts or just a few. The selection process will also be of different course of actions. There are many methods which aid the decision making on the choice of correct concept with [31]. Weighted Objectives Method, C-Box, Itemised Response and PMI, vALUe, Harris Profile and Datum Method being at the top.

3.6 Use Cases

In order to bring various aspects of the selection of the method of Ideation closer, two use cases were presented. In the first one, the case of selection of the method of Ideation for a specific complex device for medical therapy for which the context of many disciplines, also those far away from each other, is particularly visible. For the assumed Design Thinking methodology, the selection of alternative methods for Ideation of an innovative device design has been analyzed. In the next use case, a design task regarding the automotive application was implemented. Two different methodologies were used and different methods were compared in the same application, paying particular attention to the use of the Artificial Specialist Team (AST) and its suitability for Ideation in group work. The conclusions include a discussion taking into account additional aspects which can decide upon the suitability of a given method, which can not be easily taken into account in tabular form (Table 3.1). The presented discussion together with previous information systematizing the approach to Ideation significantly simplifies the task of choosing the proper Ideation method.

3.6.1 Place of Ideation in Physiotherapeutic Devices Design Process

In the following section a draft of methodology aimed at supporting the design process of physiotherapeutic devices is presented. The goal of this short description is to present the universalism of ideation methods in terms of applying them at various design process stages. This approach was referred to design thinking matter.

Physiotherapeutic systems

Complexity of nowadays constructions make it crucial to analyze not only their usability and reliability, but also the way they interact with the environment. This includes: how they support operators in performing their tasks correctly or minimize the adverse effects of tasks performed incorrectly. It is particularly crucial for physiotherapeutic engineering domain, where device incorrect use and errors usually imply that patient's health is at risk. This is why, when searching for an optimum design solution, one needs to consider the whole system which the designed device is an element of and, therefore, make use of knowledge of the problem characteristics taken from various domains perspectives. Such an approach enables to consider not only the device construction itself but also a whole range of human and environmental factors that influence the course of design application which are a part of physiotherapeutic system. These factors shall not be ignored by device designer, as many of them can be influenced by the design and can lead to its improvement.

Ideation in systems design

Currently, one of the most popular strategies for searching for novel and creative solutions is *design thinking*. It is assumed, that the strategy distinguishes five stages, repeated iteratively: empathizing, defining, ideation, prototyping and testing. It can be concluded, that ideation considered here is an act of generating ideas understood as specific solutions e.g. constructional. There is also another meaning of term *design thinking* though, introduced by R. Buchanan is *Wicked Problems in Design Thinking* [32]. Here, the discussed term expresses a broader view of design thinking as addressing intractable human concerns through design. In the following considerations we want to return to the original meaning of *design thinking*, which includes ideation as an approach to searching for solutions to problems which are difficult to define through design. This is the reason why we deviate here from five-stages *design thinking* scheme, placing ideation right at the earlier stages of strategy—stages of defining problems. Such an approach is connected to specific kind of purpose of a designed device and to the nature of the design improvement process (in opposite to searching for a completely new solution). In case of physiotherapy, device design can have crucial influence on patient health, both positive and negative. Early identification of potential hazards of using physiotherapeutic device is crucial for preventing an occurrence of treatment adverse results.

Searching for such hazards cannot yet base only on a designer elicitation of implicit knowledge on treatment. It must also enable performing hypothetical considerations—considerations of events, that cannot be anticipated just yet, and gathering tacit knowledge about system operation. In other words, we aim to elaborate ideas and assumptions of possible adverse results of applying physiotherapeutic device. This means, that the subject of ideation in this case should be the searching for adverse results itself, not the solutions for problems they entail. Therefore, ideation must be conducted as early as empathy stage of currently adapted *design thinking* model.

Object

The structure, that is discussed here, is a triaxial spine traction table (Fig. 3.9). The research goal was to identify potential adverse results of using the table, thus defining the problems that need to be solved at later design process stages.

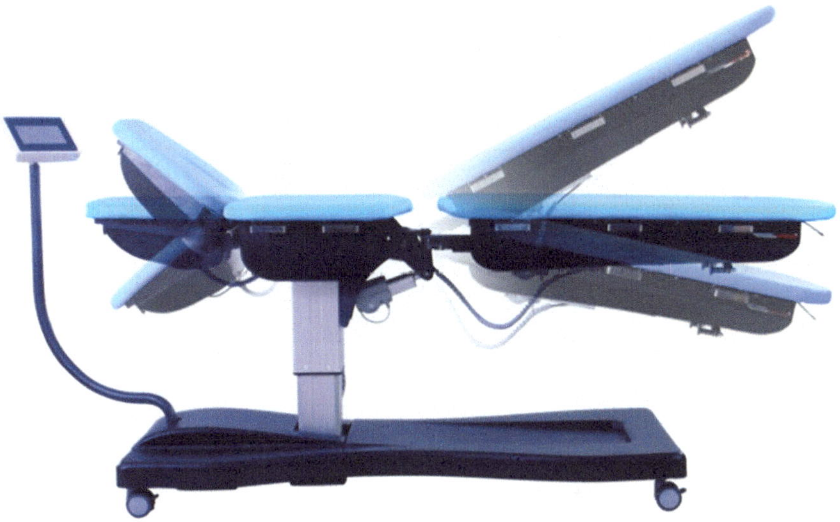

Fig. 3.9 Spine traction table

Ideation session participants selection

As the knowledge required for designing physiotherapeutic device extends across many various disciplines, naturally it is hardly possible for an engineer himself/herself to be able to perform a complete analysis. In [33] Hall mentions an *Ideal Systems Engineer* that has specialist knowledge of multiple domains, as a mean to find a perfect solution. Buchanan, on the other hand, highlights the need for multidisciplinary communication in early design stages [32]. This approach we found much more feasible. We assume, that reaching for collaboration methods and using the knowledge and ideas of specialists in various domains, enables to create a complete understanding of physiotherapeutic system.

Following this assumption, we defined the first step to build a collaboration model, namely identification of knowledge and ideas sources. These are the kinds of specialists, and from their contribution the design could benefit (Fig. 3.10).

It can be seen, that we find essential to go beyond the physiotherapeutic system and reach for knowledge of each participant of the design process. Important source of knowledge on potential adverse results are specialists and experts, including contrarians. Knowledge sources that could share their insights on potential hazards originating from technical characteristics of the device are people responsible for construction, manufacturing and service of the device.

Ideation methods choice

Having defined knowledge sources one can consider ideation methods possible to be applied. Methods selection can be based on classification presented in previous sections of this chapter and particularly in Table 3.1 which integrates and compares main features of the most popular methods (Figs. 3.1, 3.2, 3.3, 3.4). Referring to,

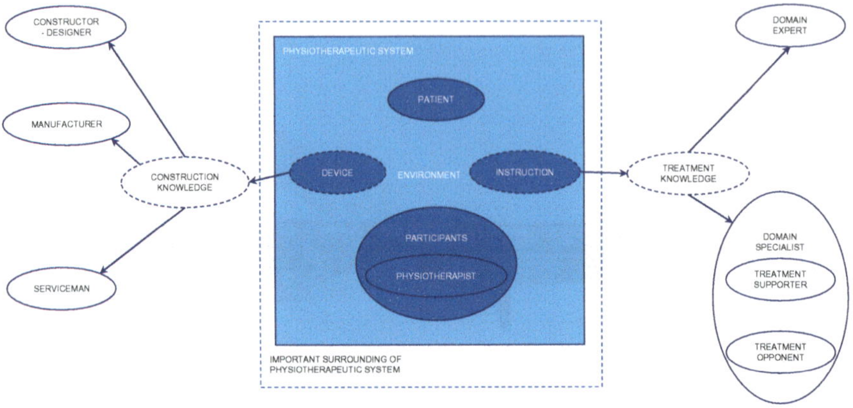

Fig. 3.10 Sources of important knowledge for potential adverse results identification process

it must be assumed, that to the set of considered methods following types would belong: both group and individual, specialized, inventive and routine ones. The lack of possibility to predict the outcome of ideation makes it sound to focus on heuristic methods.

Considering the goal of ideation process—identifications of potential adverse results and therefore design problems to be solved, it can be treated both as routine task and inventive. In case of routine-task approach, effective methods would be transformational methods. In other case, it is recommended to reach for organizational methods. An effective method would be e.g. storyboarding, as a method that enables gaining deep understanding of researched process (in this case—physiotherapeutic treatment process) and by that identification of moments of potential adverse results occurrence. Due to the abstract form of the ideation outcome (an assumption of an adverse result), it would be impossible to apply progressive methods, that involve graphical representation of a solution such as Gallery or C-Sketch.

Conclusions

When designing physiotherapeutic devices, one should think of them as parts of physiotherapeutic system. Going even further, we propose an analysis that goes beyond the system itself. One of the elements of such analysis should base on collaboration of specialists in various domains. The domains should be determined as a result of system and systems environment analysis. Goal of specialists collaboration would be to build a complete understanding of physiotherapeutic system and therefore to give a possibility of foreseeing potential hazards and adverse results early enough to include preventive measures early at the design process.

3.6.2 Approaches to Ideation in Context of Automotive Design

Automotive design concerns complex objects and systems. In order to acquire a full spectrum of their complexity and design possibilities, it is required to adopt a multidisciplinary approach at the very first stages of the design process, namely research and ideation [34, 35]. This will enable to use knowledge from different, seemingly not related domains. That is a big challenge though, as it requires gathering and integrating information, requirements and constraints from many scientific fields, which sometimes happen to be ruled by conflicting priorities. Important is also the aspect concerning situation, in which during preparation to the ideation process there is a lack of specialists of specific domains. These are cases that require a specific approach. In this subchapter two approaches to ideation are discussed.

Object and methodology

During the preparatory work of the teams for the development and modernization of racing electric vehicles [36–38] it was decided to investigate the usefulness of the Ideation Methods in multidisciplinary design issues. The study consisted on carrying out the design process in two separate groups in a strictly moderated way. Each group was divided into two design teams for increasing the set of data gathered during the research and differentiation of the kind of considered object. In the first group the method of profiling of the project team has been used. In the second group the design process was freely. At the earlier stages, before the ideation process, some elements of Pin Card Technique and Gallery Method have been used. The ideation process was based on the C-Sketch method [15].

The C-sketch method was selected due to the specificity of the problem (design problem), group type of work on the project, the need to combine knowledge from different fields, clear representation of the developed solution and the need for a high level of stimulation for innovative thinking and modifying emerging concepts. Taking into account the specification of the method (Table 3.1) as well as the possibilities and existing limitations, it was decided to choose the C-sketch method.

To ensure the correctness of the process, the selected object should have a significant complexity, allow to generate a wide field of possible solutions and require a multidisciplinary approach. Because of that the vehicle and the airplane were selected. The airplane is not directly associated with automotive matters, but the problems of airplane design can be considered as similar to problems of automotive design. It was decided, that the attention will be focused on the door and the door will be considered in context of the whole system (vehicle or airplane).

Specialized approach

The first approach to the design process which was adopted was to compare and at the same time to test the elaborated method is the manner called specialized approach. In many cases of design of the new solutions, specialization can give a good effect. Having an in-depth domain knowledge allows to develop advanced solutions that meet the complex criteria. In particular, the complex monodisciplinary issues require a specialist knowledge, without that finding a solution to the problem becomes extremely difficult. Due to these arguments, the authors decided to develop a method that allows the use of quasi specialist knowledge at the time, when the project team lacks specialists. The method was called AST (Artificial Specialists Team) and it consists in assigning defined specialization to each member of the project team. This manner gives a large flexibility and allows to match the team profile precisely to the considered issue. Each artificial specialist focuses on his own field and collects knowledge from a narrow range of limited specialization. The number of artificial specialists should be related to the number of the main areas of knowledge, that are the subjects of exploration.

In this case, the C-sketch method allows for great opportunities to innovate and modify, considering the specialist view of a specific project area. This method is very flexible in this respect, allowing smooth changes without the need to interfere with the basic layers of the project.

Dispersed approach

The second approach, used for research and comparison of the achieved results, is the way called dispersed approach. In this case, the project team is moderated to work together without a clear and intentional division of specialists. Team members are encouraged to approach the issues in a multidisciplinary way, and explore many fields of knowledge in a limited range. An important difference compared to the AST method is the pursuit of a broad perspective on the problem by each team member. In addition, dispersed approach aims to provoke overlap of gathered knowledge to confront generated ideas.

Regardless of the approach, the stage before the correct ideation phase was the search for useful information at the stage of generating ideas, collecting data to formulate design assumptions and analysis of the state of the art. Multidisciplinary nature and significant complexity of considered issue create a need to analyze many areas of knowledge. Their elements tend to be opposites of each other or are connected with the necessity of solving the existing contradictions. In principle, the process of gathering knowledge and generating initial ideas about the considered object or system is different for each of the mentioned approaches to the design process. The specialized approach focuses on a specific area, without more substantial analysis of the links between the different areas. This approach guarantees good organization of work, it allows to minimize the risk of missing important aspects during the step of generating design assumptions and ensures having in-depth knowledge in specific fields. In the dispersed approach, knowledge acquisition takes place in a less structured way. The designers are responsible for certain areas of the project and gather information from many fields at the same time, not focusing specifically on any of them. This results in the acquisition of knowledge, that is more superficial than in the case of specialized approach, but on the other hand allows to diagnose at the earlier stage the risks and to avoid them in further process steps. In this case, the C-sketch method enables the integration of the work of many designers with multidisciplinary knowledge. They have the freedom to supplement solutions created in the iterative process with their ideas and modify solutions with "joint forces". Thanks to this, the C-sketch method becomes a highly inventive tool.

Conclusions

According to the conducted research, one of the most important aspects of the design process concerning the multidisciplinary issues, is the management of the team, profiling tasks for individual team members and knowledge management at every stage and, in particular, at the stage of generating ideas and design assumptions. In the case of projects where the impulse to define the needs is a very general external factor, the impact of the organization of the project team is visible already at the stage of needs recognition. In the experiment a general need existed and the project team was created. It formalized and clarified the need and directed the stage of defining the assumption and directed them to the established group of recipient. or realized step of defining a set of assumptions on their audience. Therefore, the stage of defining needs and indirectly determining the direction of development of project assumptions depend on the findings of the project team. Most of the following steps are directly determined by the method of defining the project goal and objectives. Therefore, it the influence of team organization and selection of the profile of duties and responsibilities of the team members on the development of the whole design process may be noticed.

The applied approach to profiling the project team, in many cases, may be difficult to implement due to social engineering factors. It should be associated with the natural specialization of obligations of the participants of the project and evolve on the basis of natural personal characteristics. In practice, it is difficult to retain a complete specialization of team members. This is due to personality and mental factors as well as the multidisciplinary nature of the considered issues. Typically, the team members have multidirectional interests and overlapping areas of knowledge related to the analyzed problem are difficult to separate. This results in a limited possibility of division of knowledge into specific domains. There is no doubt that introduced specialization of team members resulted in a more detailed analysis of issues. This causes the concepts to be more refined and feasible. The responsibility for a particular area of specialization creates the need for expert opinion on the appearing ideas. It is a good solution at the stage of refining the concepts.

Nevertheless, a disadvantage of the specialized approach is a narrow way of thinking. The ideas generated by artificial specialists and modifications made by them were much less diverse than in the case of dispersed approach. It can be a weakness of the solution, particularly at the early stages of ideation when a large variety of ideas is desired.

Last aspect worth mentioning is the natural inclination and desire to quasi-specialization for dispersed approach. Specific areas of knowledge explored by each team member create the need to focus on specific domains, resulting from the necessary division of work.

Based on the above conclusions it seems desirable to include AST method in the second step of ideation after achieving a large variety of ideas. This gives an advantage over the design team using real specialists who naturally are burdened in a positive and negative sense of the word. They have knowledge that gives them an advantage over other designers but also limits the scope of the sought concept.

It can also be said that the C-sketch method works well both when used by a project team consisting of specialized professionals, as well as by people with broader, though not so deep knowledge. This is due to the specificity of the method, which can be called "interactive" and giving the possibility of close and natural cooperation over a given problem. This conclusion coincides with the conclusions drawn from the specification set out in Table 3.1.

3.7 Conclusions and Further Work

Currently Ideation Methods meet new challenges due to multidisciplinary character, complexity and magnitude of the realized projects. Systems Engineering imposes adaptation of Ideation Methods to work in complex technological problems solving environment which extends in time and different phases of design development as well as many complex interdependent relations from various domains. Results of traditional methods are not quite reliable or they fall behind SE methods. As a result we have the lack of diversity of final products and their mimicry. It is also caused by intensive use of formal optimization methods which with similar scope of space lead to similar solutions. Proper attention paid at the ideation phase improves diversity of solutions and can result in surprisingly positive results, which explains great interest in TRIZ/TIPS methods. Great potential of these methods also increases their development and adaptation to different design stages and also elaboration of simplified version of the method. Due to big differences in the methods themselves it is essential to match potential of the method with expectations and the character of the problem. The chapter presents the analysis and a list of most popular methods which can be useful in making proper choice of a method to solve a given problem. The huge number of methods and their variations (about 300) is not conducive to the ease of application of these methods, especially that there is no comprehensive comparison of the most important of these methods. Therefore, the presented comparison together with the descriptions of these methods and a summary table with the features of the Ideation Methods makes it very easy to choose methods for a given application. This is important because in multidisciplinary design tasks, the assessment of the concept is significantly impeded and their impact on the whole system is difficult to determine. The proposed method of assessing MBD and MBO facilitates quantitative assessment of the concept and selection of the best of the proposed set of concepts.

Next important tendency which can lead to improvement of Ideation effects is to conform well-chosen methods to current usage. However, it requires particular attention so that advantages of a given method are not diminished. As an example we can look at the use of AST (Artificial Specialist Team) which aids Ideation in multidisciplinary project which is described as one of use cases.

The choice of proper Ideation Method systematically carried improvement of the method and adaptation to current project conditions can enhance the result. Correct use of the method can lead to interesting and innovative results which can be the key to the market success of a product.

References

1. Moser HA (2014) Systems engineering, systems thinking, and learning. A case study in space industry. Springer, Berlin
2. Michalko M (2006) Thinkertoys: a handbook of creative-thinking techniques. Paperback
3. Jonson B (2005) Design ideation: the conceptual sketch in the digital age. Des Stud 26(6):613–624
4. Graham D, Bachmann T (2004) Ideation: the birth and death of ideas. Wiley, Hoboken
5. Herring SR, Jones BR, Bailey BP (2009) Idea generation techniques among creative professionals. In: Proceedings of the 42nd Hawaii international conference on sciences. IEEE
6. Shah JJ (1998) Experimental investigation of progressive idea generation techniques in engineering design. In: Proceedings of DETC98. Atlanta
7. Hua Z, Yang J, Coulibaly S, Zhang B (2006) Integration TRIZ with problem-solving tools: a literature review from 1995 to 2006. Int J Bus Innov Res 1(1–2):111–128
8. Barry K, Domb E, Slocum M (2010) Triz—what is Triz. Triz J [Online] [Cited: October 06, 2019]. https://triz-journal.com/archive-what-is-triz/
9. Sheng ILS, Kok-Soo T (2010) Eco-efficient product design using theory of inventive problem solving (TRIZ) principles. Am J Appl Sci 7(6):852
10. Gadd K (2011) TRIZ for engineers: enabling inventive problem solving. Wiley, Hoboken
11. Kulak O, Durmusoglu MB, Tufekci S (2005) A complete cellular manufacturing system design methodology. Comput Ind Eng 48(4):765–787
12. Cross N (2005) Engineering design methods. Strategies for product design. Wiley, Hoboken
13. White CK, Wood KL, Jensen D (2012) From brainstorming to C-Sketch to principles of historical innovators: ideation techniques to enhance student creativity. J STEM Educ 13(5):12
14. Pahl G, Beitz W, Feldhusen J, Grote K (2005) Pahl/Beitz Konstruktionslehre. Springer, Berlin
15. Shah JJ, Varggas-Hernandez N, Summers JD, Kulkarni S (2001) Collaborative sketching (C-Sketch)—an idea generation technique for engineering design. J Creative Behav 35(3):168–198
16. Goldberg D (2002) The design of innovation: lessons from and for competent genetic algorithms. Springer, Berlin
17. Petersson AM, Lundberg J, Rantatalo M (2016) Ideation methods applied in a cross-functional inter-organizational group: an exploratory case study from the railway sector. Res Eng Des. https://www.Springerlink.com
18. Ullman D (1997) The mechanical design process. McGraw Hill, New York
19. Horizon. Horizon 2020 work programme—Technology Readiness Level (TRL). [Online] [Cited: January 15, 2017]. https://ec.europa.eu/research/participants/data/ref/h2020/wp/2014_2015/annexes/h2020-wp1415-annex-g-trl_en.pdf
20. Skarka W (2018) Methodology for the optimization of an energy efficient electric vehicle. In: Proceedings IRF2018: 6th international conference integrity-reliability-failure, Lisbon/Portugal, pp 415–422
21. Skarka W (2018) Model-based design and optimization of electric vehicles. Transdisciplinary engineering methods for social innovation of industry 4.0, IOS Press, pp 566–575
22. Bil C (2015) Multidisciplinary design optimization: designed by computer. Concurrent engineering in the 21st century. Foundations, developments and challenges. Springer, Berlin
23. Targosz M, Skarka W, Przystałka P (2018) Model-based optimization of velocity strategy for lightweight electric racing cars. J Adv Transp. 2018: 20

24. Wallas G (1926) The art of thought. Penguin, Harmondsworth
25. Tassoul M (2006) Creative facilitation: a Delft approach. VSSD, Delft
26. Sternberg R (1989) The nature of creativity: contemporary psychological perspective. Cambridge University Press, Cambridge
27. Takahashi M (1993) Dictionary of creativity. Mo To Publishing, Tokyo
28. Smith GJ (1998) Idea-generation technique: a formulary of active ingredients. J Creative Behav 32(2):107–134
29. Lin C-L et al (2006) A study of the applicability of idea generation techniques. Think Skills Creat
30. Altshuller G (1999) The innovation algorithm: TRIZ, systematic innovation, and technical creativity. Technical Innovation Center, Worcester
31. Boeijen A, Daalhuizen J (2010) Delft design guide. Delft
32. Buchanan R (1992) Wicked problems in design thinking. Des Issues 8(2):5–21
33. Hall AD (1964) A methodology for systems engineering. Van Nostrand, New York
34. Skarka W (2015) Reducing the energy consumption of electric vehicles. Advances in transdisciplinary engineering, vol 2. IOS Press, pp 500–509
35. Cichoński K, Skarka W (2015) Innovative control system for high efficiency electric urban vehicle. Tools of transport telematics. Springer, 121–130
36. Sternal K, Cholewa A, Skarka W (2012) Electric vehicle for the students' shell eco-marathon competition. Design of the car and telemetry system. Telematics in the transport environment. Springer, 26–33
37. Jezierska-Krupa K, Skarka W (2016) Using simulation method for designing ADAS systems for electric vehicle. Advances in transdisciplinary engineering, vol 2. IOS Press, pp 595–604
38. Jezierska-Krupa K, Skarka W (2018) Design method of ADAS for urban electric vehicle based on virtual prototyping. J Adv Transp. 2018:19

Chapter 4
System of Systems Modelling

John P. T. Mo and Ronald C. Beckett

Abstract The design, manufacturing and through-life support of modern engineering systems such as an aircraft or a frigate are complex, multifaceted and may change over time. These engineering systems are working in an environment that has multiple individual users, complicated supply chain, many government and socially affected stakeholders. In essence, these systems are working as a system of interacting semi-autonomous systems each of which are governed by their individual set of rules and could operate with different enterprise structures. Engineers trying to apply the theory of systems engineering to "design" a system of systems find the outcome often unpredictable and uncontrollable, as the linked systems operate with high degree of independence. System operations are embedded in business networks that are evolving and changing all the time. Individuals and organisations participate voluntarily in the networks. They can come and go at any time without warning. This highly uncertain relationship requires a different approach. This chapter will address the modelling requirements to design, develop, implement and operate a complex system that interacts with many socio-technical systems. The methodology is illustrated by two case studies.

Keywords System of systems · Network · Socio-technical system · Network-centric approach · 3PE model

4.1 Introduction

The emergence of systems of systems (SoS) has been recognized by the United States (US) Department of Defense (DoD) at first to fulfil their user capability needs. The DoD Guide for Systems Engineering of Systems of Systems (SoS SEG) provides a definition of SoS as a "collection of systems, each capable of independent operation,

J. P. T. Mo (✉)
Manufacturing Engineering, RMIT University, Melbourne, Australia
e-mail: john.mo@rmit.edu.au

R. C. Beckett
Swinburne University of Technology, Melbourne, Australia

J. Stjepandić et al. (eds.), *Systems Engineering in Research and Industrial Practice*,
https://doi.org/10.1007/978-3-030-33312-6_4

that interoperate together to achieve additional desired capabilities" [1]. The "independence" characteristics of individual systems in the "collection" distinguishes from the traditional systems thinking that the sub-systems in a system will have certain dependency relationship with the overall system. While a sub-system has certain autonomy in its operation, its inputs and outputs need to be tightly integrated with sub-systems so that the totality of the system can achieve a predicted outcome. A SoS doesn't necessarily have this level of coupling requirement.

SoS doesn't comprise only technical aspects of integrating systems as such and, furthermore, considers also the economic, social, and political aspects. Subsequently, while many complex systems are working in concert there to satisfy user needs, the field of defense always was the favourite application domain of SoS.

Due to the fact that the constituent systems (that build up an SoS) are in service before the SoS is either built or recognized, SoS ideation is an evolutionary process running from the bottom up, by developing systems that can interoperate with other systems, or less preferred, by treatment of interoperability issues (e.g. the authority relationships between the constituents and the SoS can overlap considerably) after the systems are fielded. SoSs differ from traditional systems in ways that require tailoring of systems engineering (SE) processes. The distinctive characteristics of SoS have implications for the application of test and evaluation (T&E) to SoS [2].

SoSs tend to have distributed control and component systems tend to choose by themselves to participate or not in a SoS (i.e. they decide to consume resources to achieve the goal of the SoS). Thus, SoS architecting tends to be dynamic and focuses on interactions between component systems. Subsequently, SoS architecture is one of the main problems for developing SoS. This assertion comes from the classical system architecting that is really far from SoS architecting. In SoS, the emphasis on SoS concerns interface architecting to foster collaborative functions among independent systems and the concentration is on choosing the right collection of systems satisfying the requirements. So it can be noticed that contrary to classical systems, SoS architecting focuses on collaboration between component systems to get the right organization [3].

4.1.1 Complex System Support

Complex system support requirements can vary over the life-cycle of the system being supported, and the total cost of ownership can be many times the initial acquisition cost. Systems of systems life cycle is an evolution with time of an SoS. A recent trend around the world among the owners of complex engineering systems such as an aircraft or oil refinery is to include consideration for the sustainment of the system at the very early stages of system development. According to the Defence Materiel Organisation in Australia [4], the asset acquisition project is considered a continuum of four phases, which can be generalised as a capability systems lifecycle as shown in Fig. 4.1. The goal is to ultimately attain desired capability levels that can be measured as a performance outcome of systems in-service.

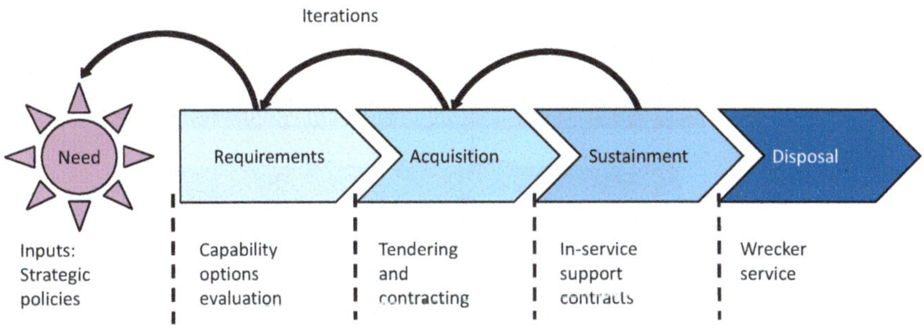

Fig. 4.1 Capability systems lifecycle

There are two different contracting regimes displayed in Fig. 4.1:

1. System acquisition agreements including functional and performance specification of the final system, with associated training arrangements—i.e. the tendering and contracting activities in the Acquisition phase.
2. Sustainment agreements specifying outcomes and performance requirements for in-service support, i.e. linked owner responsibilities and in-service support contracts in the Sustainment phase.

Type 1 contracts are normally handled by systems engineering lifecycle modelling methodology [5]. However, type 2 contracts involve not only the original equipment manufacturer (OEM). Suppliers of consumables and services to support operation of the system are essential partners to deliver this type of contracts.

4.1.2 System of Systems Engineering Lifecycle

Traditionally, management of sustainment services after commissioning of the system is the responsibility of the asset owner, after the product is commissioned. Most asset owners simply take the recommended schedule of the manufacturer, either by in house service department or by a maintenance services contractor [6]. However, the application context and operating environment may change over the long service life of the asset. Many service decisions on assets are therefore made on rules of thumbs rather than using analysed system performance data [7].

A representation of the essential characteristics of the System of Systems Engineering (SoSE) life-cycle process is depicted in Fig. 4.2 [8]. It consists of steps that are implemented in an iterative fashion, with each step providing feedback into the ongoing, evolutionary process. The key features of this life-cycle model reflect the nature of systems of systems and its impact on systems engineering [9]:

- **Multiple Overlapping Iterations of Evolution** reflect the fact that most systems of systems leverage developments of their constituent systems, and consequently, systems of systems are characterized by incremental development.

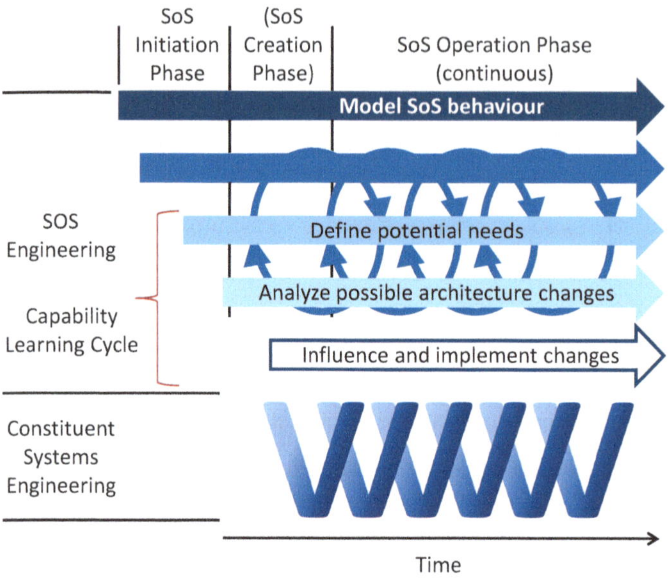

Fig. 4.2 Essential characteristics of the SoSE life-cycle process lifecycle

- **Ongoing Analysis** provides an analytic basis for each iteration of SoS evolution. Unlike traditional systems engineering in which upfront analysis drives development, engineering of systems of systems requires continuous analysis to address the dynamic nature of the SoS and its context.
- **Continuous Input from External Environment** is key for SoSE, since any manager or engineer of an SoS has control over only a small part of the environment that affects the SoS.
- **Architecture Evolution** is also important. While the architecture of an SoS ideally provides a persistent framework for the SoS evolution over time, the planned SoS architecture is typically implemented incrementally and may itself evolve.
- **Forward Movement with Feedback** drives the evolution of an SoS, which typically adopts a "battle rhythm" driven by elements in the SoS context (e.g., the development plans of a key constituent system or the unit fielding schedule) that are not under the control of the SoS. These external driving events effectively "pace" the SoS evolution. While there may be feedback within an evolution, many systems of systems adopt a "bus stop" approach, where they deliver those changes that can be implemented during an iteration and defer the rest to subsequent evolutions (or the next time the bus stops).

4.1.3 Lifecycle of Service Systems

Traditional systems engineering lifecycle management theory is an essential constituent in the design and development of component systems. However, operation and maintenance of these systems are amalgamated into a single phase that is in fact at least 10 times longer than all design and development phases combined. Changes in the regulatory rules, environment, technology and culture of the SoS all point to the need for the SoS to adapt to new and ever changing circumstances. An adaptive, interactive, agile approach to development and evolution is the proper choice in today even changing world.

The design, manufacturing and through-life support of these systems must be flexible and adaptable to suit the environment of individual user requirements, complicated supply chain and many other stakeholders influencing ongoing level of performance. In essence, each of the entities in the supply chain is designed as an autonomous system that can operate in specified conditions and procedures. However, these systems are also working within a larger socio-technical system that is governed by a different set of rules and operates with different types of structure. Engineers trying to apply the theory of systems engineering to "design" a system of systems find the outcome often unpredictable and uncontrollable, as the linked systems behave in their own way. Last but not least is that these systems are constrained by the business and sociological environment in which they are operating. The environment includes regulatory and institutional requirements, and their operation must make business sense. Some of these constraints are not visible until later in the system's lifecycle.

To understand the complexity of the interactions among the component systems, a new model of system of systems that orients towards adaptable network thinking is required. Furthermore, the processes and procedures are managed as soft systems that can be readily changed, modified and refined. Operation of a system of systems depends on the understanding of people and roles in the component systems as well as their interaction within the system of systems. This chapter will address the modelling requirements to design, develop, implement and operate a complex system of systems that encapsulates many engineering and socio-technical systems. The outcome of this new modelling concept for system of systems can enable engineers to design and integrate systems more suitably for both types of system acquisition contractual regimes. Taking the even higher importance of services into account, the remainder of this chapter will have a particular emphasis on service systems.

The outline of the chapter is as follows. Section 4.2 contains characteristics of service system thinking. Section 4.3 offers a value proposition in services and support of systems. Section 4.4 describes network-centric modelling. In Sect. 4.5 The 3PE Model is introduced. Section 4.6 discovers system of systems modelling using 3PE with 2 use cases from practice. We conclude and plan some future works in Sect. 4.7.

4.2 Characteristics of Service System Thinking

Users and owners of complex engineering systems are demanding more value out of the system. Availability, readiness, sustainment, cost savings and many other attributes are valued. Most system owners consider the maintenance function as simply following the recommended schedule of the manufacturer, either by in house service department or by a maintenance services contractor. Therefore, classical services and maintenance plans are designed on the principle that mean time between failure is a constant and hence the focus is to replace components before it is expected to fail. Typically, service activities including inspection, adjustment and replacement are scheduled in fixed intervals [10]. Due to multifaceted relationship between operating context and characteristics inherent in the complex system, these intervals may not be optimised [11]. In addition, many other factors are also influencing the operations of the system [12]. These changes quickly render the initial design of services and support system ineffective.

New industry practices such as performance based contracting force some fundamental changes required in the manufacturing and service operations of complex engineering systems in terms of tangibility, perceptions of performance and quality, the lag between production and consumption, capacity for storage, the nature of customer contact, and geographical proximity considerations [13].

Enduring systems may change over time in response to re-purposing or technology change, but these changes are likely to be slow compared with changes in the business networks their support systems are associated with. Political agendas, world events, competitive pressures and various kinds of enterprise re-organisation can change what makes business sense on both the demand and supply sides. At a more local level, the incorporation of new technology can enhance product capabilities, and people and procedures can change norms and priorities [14]. Likewise, Teece [15] described the concept of a business model as an articulation of the logic and provides data and other evidence that demonstrates how a business creates and delivers value to customers. It also outlines the architecture of revenues, costs, and profits associated with the business enterprise delivering that value.

One of the key questions emerging from this approach is how to adapt to the uniqueness of service system requirements. Every complex engineering product is different. The same product may operate in different environments. Hence it is fair to say that each service system is customised. Johansson and Olhager [16] examined the linkage between goods manufacturing and service operations and developed a framework for process choices that enable joint manufacturing and after-sale services operations. Other studies showed that moving into services oriented business could have significant financial implications to the company [17]. In a performance oriented service system, decisions for optimization can be quite different from maintenance oriented service concepts. For example, in order to reduce time to service to customers, Shen and Daskin [18] suggested that a relatively small incremental inventory cost would be necessary to achieve significant service improvements. Hence, to develop service systems that can handle this type of business requirements, companies should build common business functionalities as shared services so that they can be reused across lines of business as well as delivery channels [19].

The Product Service System (PSS) concept extends, on the basis of an existing complex product, the provision of support services when the product is in operation [20]. It is obvious that there are commercial benefits for companies to move into continuous services and support operations of complex products utilising both their knowledge and physical assets. A PSS comprises people and technologies that adaptively adjust a system's value of knowledge while the system changes in its lifecycle [21].

When compared to traditional support arrangements, the PSS concept changes contractor roles and responsibilities, requiring a stronger customer focus. Under service oriented arrangements, the service provider may carry a wide range of responsibilities, possibly including ownership, sustainment and operation of assets. Furthermore, contracting arrangements may include incentives and penalties against levels of support service or delivery, influencing both *what* has to be done and *how* it is done. The service provider will need to think differently and design the output solutions that deliver the desired performance as well as generating profit [22].

4.3 Value Proposition in Services and Support of Systems

Development and delivery of an engineering system to customer does not necessarily involve many partnering companies. Many large scale projects are done by international conglomerates with a few companies only. However, provision of after commissioning service (type 2 contract) is a different story. A service contract often requires active interaction of local service providers with the customer. A new service enterprise is required to be formed from the multi-international companies as well as several local specialist companies. The local companies act as the front line, on the ground partners maintaining or carrying out engineering change with the support the multi-international companies.

There are many risks in this strategy, for example, risks exist in collaboration, confidentiality, intellectual property, transfer of goods, conflicts, opportunity loss, product liability and others. To minimise the risks for the new service enterprise, enterprise engineering researchers have introduced the concept of an enterprise architecture framework as a common starting point. The study of enterprise architecture in the last couple of decades has been on how enterprises can be designed and operated in an environment when the missions and objectives of the enterprise are clear. In that case, the enterprise can follow well-established common engineering practice: from design, implementation, operation to decommission phases [23].

However, the operational and business requirements of the system may change over time. The enterprise architecture approach promotes planning, reduces risk,

Table 4.1 Categorised service domains

Service	Characteristics	Example
Professional service	High degree of direct contact or customisation of service and high labour intensity	Consultant
Service shop	High degree of direct contact or customisation of service and low labour intensity	Restaurant
Service factory	Low degree of direct contact or customisation of service and low labour intensity	Airline
Mass service	Low degree of direct contact or customisation of service and high labour intensity	Call centre

implements new standard operating procedures, controls and rationalizes manufacturing facilities [24], but the approach is too rigid to be used in the dynamic environment like through-life-support.

Several research attempts have been made to understand how enterprise architecture methodology can be adapted to engineering services. Bernus and Nemes [25] postulated a generic enterprise reference architecture framework after extensive research of the state-of-the-art enterprise architecture in the 1990s. Chattopadhyay and Mo [26] modelled a global engineering services company as a three column progression process that was centred on human engineering effort. Mo and Nemes [27] introduced the concept of enterprise DNA to enable a more flexible enterprise architecture that could adapt to changing modelling requirements in different stages of the system's development.

Although many companies, particularly the financial sector, like to summarise all activities and procedures into a package and call it a "product", services are not products. Services can't be stored for later use. Services have to be consumed together with the user in real time. Bessant and Tidd [28] categorised the nature of the service domain influences in operational sense as in Table 4.1.

Provision of services need partnering with many companies and depends heavily on the supply chain. A service contract will be delivered to the customer in a less well defined set of conditions, and these conditions are always changing due to different people and time, irrespective whether the same "product" (the service definition) is offered.

The unpredictable nature of service demands frequent innovations in service offerings. Miles [29] suggests four focus areas applied singly or in concert may characterise a service innovation as in Table 4.2.

Table 4.2 Innovation characteristics

Innovation	Example
A service new to its particular market—a new value proposition	New information services
Changes in the client interface	Self-service takeaway fast food restaurant
Changes in the service delivery system—changes in the ways service workers perform their jobs	On-line supermarket, order picking and home delivery
Application of a new technology to facilitate the way service is delivered and/or what is delivered	Home security system connected to police station

These service innovations have inherent impetus to change the system while the service is being delivered. Frequent system adaptation and changes in value proposition, plus other characteristics that are not aforementioned, impose the need to consider the system of systems approach to design such systems.

The concept of system of systems offers the prospect of improved performance, however they exist in a broader ecosystem where seemingly simple changes in one part of the system can have an unintended impact elsewhere. Peruzzini et al. [30] proposed a new methodology to support ideation and preliminary design of service and support systems. The methodology captured customer needs and allowed evolution of industry networks for sustainable operations. It has been suggested that the dynamic capabilities of an enterprise can influence its ability to adapt to change, and that this is a set of "specific and identifiable processes such as product development, strategic decision-making and alliancing" [31].

To be sustainable, a system of systems must represent a good value proposition, but there may be differing stakeholder views about what is most important. Figure 4.3 outlines the functional relationships between stakeholders (who is the target beneficiary), system architecture (how is value delivered) and rationale (what is the value proposition).

Systems of systems can be designed and evolved if they are viewed as configurable multi-level networks like the internet where operating rules and procedures can be recognised and used for controlling the interaction of functional elements. This leads to the notion of 'architecting' as a design process where key system architecture elements are identified, along with how they fit together. The value proposition of these configurations is a better promise.

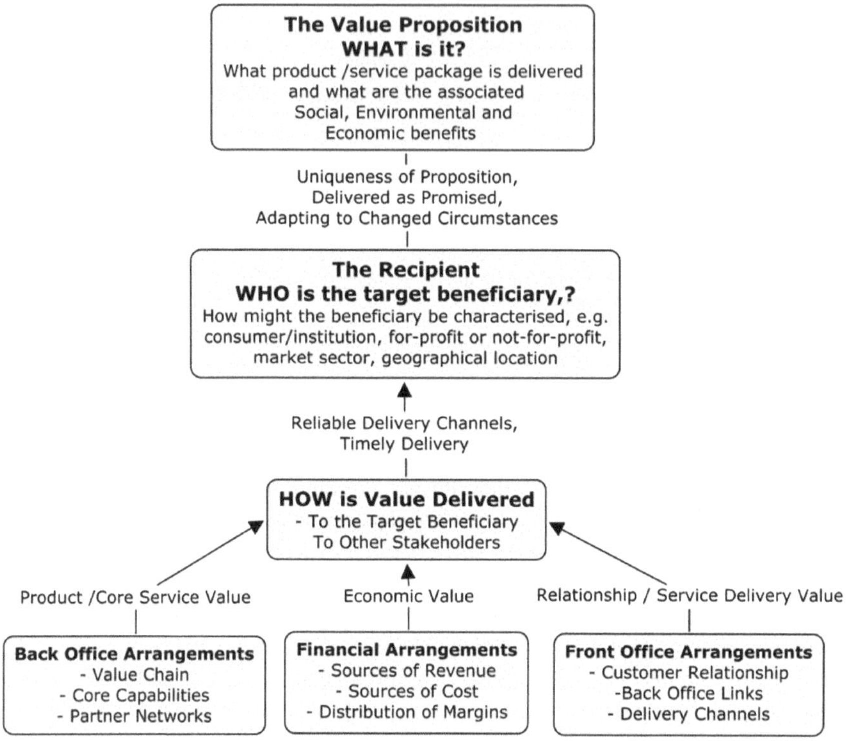

Fig. 4.3 Functional relationships leading to the value proposition of the system

4.4 Network-Centric Modelling

Systems of systems (SoS) are constantly evolving to address new user needs. This type of systems no longer has a single controlling authority. The component systems are developed independently and changed continuously over their operation lifecycle, typically extending a long period of time. As a result, system engineers cannot specify the system of systems at its formation phase by a top-down set of requirements. The methods for designing and operating the system of systems need to be modified from the methods for engineering traditional systems. Based on their observations, Lewis et al. [32] identified the characteristics of SoS and proposed a SoS lifecycle for analysis. An initial set of requirements for engineering the systems in an SoS environment was also suggested. Their observations promoted the development of a new modelling approach that does not rely on tightly coupled connections and fixed enterprise architectural constructs.

Gezgin et al. [33] designed large scale systems of systems which were collaborative and distributed safety critical systems. The systems participating in a system of systems follow both global as well as individual goals, which may be contradicting and change over time. There are parallels in the business world of marketing and procurement, where researchers have found it convenient to frame the business ecosystem as interlinked networks of actors, activities and resources [34]. In this world view, actors may be individuals, groups or enterprises (people); activities are commonly generic functional activities (processes); and resources may be any element of supporting infrastructure (product/platform). Bondar et al. [35] found that the supplier networks could be triggered to work concurrently. On the operational level, the main pitfall was the complexity of communication rules and system could be a barrier to system transitioning.

Researchers have observed that whilst the actor, activity and resource elements relating to a particular situation may not change very fast, the connections between them can. The international standard ISO/IEC 42010:2007 provides architecture description advice for software intensive systems. The international standard has in concert with a business model view as a tool for framing service system concepts that blend both tangible and intangible components. A representation of the standard is shown in Fig. 4.4. The standard calls for the integration of multiple viewpoints, and different models can help tease these out. In practice, we have used enterprise architecture models referred to earlier along with any existing representation of related systems to prompt consideration of different viewpoints (see lower part of Fig. 4.4).

Maier [36] postulated five key characteristics of Systems of Systems:

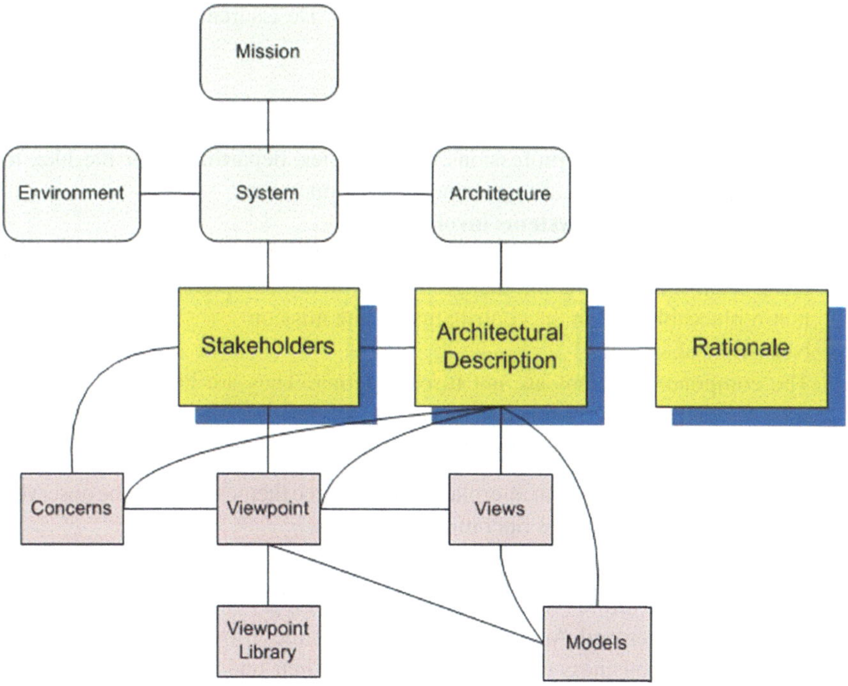

Fig. 4.4 A representation of ISO 42010

(a) **Operational independence of component systems**
If the system-of-systems is disassembled into its component systems the component systems must be able to operate independently otherwise they are not "systems" by themselves.

(b) **Managerial independence of component systems**
The component systems are managed separately by different authorities. The management of one component system does not rely on or come under control of any other component systems.

(c) **Geographical distribution**
The geographic extent of the component systems is large such that the requirement for communication is critical to the operation of the system of systems. The extent to which the component system are separate is large such that exchange of substantial quantities of mass or energy is difficult.

(d) **Emergent behavior**
The system of systems performs functions that do not normally reside in any component system such that these behaviors could not be characterized to any component system.

(e) **Evolutionary development processes**
The formation of system-of-systems is an eventual process in which functions and purposes are added, removed, and modified with experience.

Using a transportation system as the case study, De Laurentis [37] added three more characteristics:

(f) **Inter-disciplinarity**
The system of systems consists of component systems that are combining or involving two or more professions, technologies, departments, or the like, to cover a broad spectrum of requirements in its operation.

(g) **Heterogeneity of the systems involved**
Heterogeneity signifies diversity. The component systems in a system of systems come from lots of different backgrounds so every component system plays a non-replaceable role in the contribution of the mission.

(h) **Networks of systems**
The component systems are not fused together. They are linked in networks such that they are free to change their interface relationships or re-configure to form new networks for fulfilling new requirements.

These characteristics drive an interplay between three themes that may be observed in both military and in business operations:

- **The shift in focus from the platform to the network.**
In the business world, we see a change from linear supply chains to more complex supply nets. Möller and Arto [38] proposed that the effective management of different types of business net was dependent on their underlying value creation logic. Based on this notion, they proposed a value creation framework of three generic net types—'current business nets', 'business renewal nets', and 'emerging new business nets'.

- **The shift from viewing actors as independent to viewing them as part of a continuously adapting ecosystem.**
  In the business world, this may require cooperation between enterprises that normally operate at arm's length. Osborne discussed a number of frameworks for the evaluation of the management of business system of systems [39]. Bengtsson and Kock [40] even suggested that the relationships between systems in the system of systems could be both cooperation and competition simultaneously.
- **The importance of making strategic choices to adapt or even survive in such changing ecosystems.**
  In the business world these strategic choices may be influenced by the introduction of new competitors, downsizing or mergers and acquisitions. In many cases, the interaction among systems within a system of systems could lead to unpredictable results [41].

It is our experience in working with groups of industry practitioners trying to develop a representation of system of systems that particular interpretations of the ISO 42010 standard were needed in a system of systems context, and that requisite capability requirements were not considered in the ISO standard. Combining use of the standard with a business model view helps overcome these shortcomings.

In order to support decisions on business opportunities, the system of systems modelling approach should have the following characteristics:

(a) **Metrics are available to measure the system's performance**

The system of systems will be operated in parallel with the complex engineering system. Service is qualitatively different to the familiar product-based approach where hard artefacts are delivered to the asset owner. Service is a negotiated exchange with the asset owner (and operator) to provide intangible outputs that are usually co-produced with the asset owner. A service is usually consumed proximate to the time of production, if not coincident with it. Services cannot be transferred to other asset owners in the same way that products can. Hence, the development of appropriate performance metrics is essential and most of these are supported by advanced information and computational technologies.

(b) **System design based on proven system architectures**

A system of systems incorporates system design knowledge that draws upon principles derived from a wide range of engineering disciplines including systems engineering, logistics engineering, project management, information systems and many others. The knowledge helps the system support engineer to take into account as many constraints as possible during the system design phase. These constraints are imposed by the environment in which the complex system and the business are operating.

(c) **Sustainability capability built in the system design**

The performance based services are characterised by the need to create value for both asset owner and the service provider. As such both sides are treated as co-innovators in the design of the service support solution. Many decisions are made based on incomplete data rather than fully analysed data set. There are a lot of risks, both from the point of view of data availability, as well as subjective human judgement and communication.

4.5 The 3PE Model

Extending from traditional enterprise modelling methodology, Chattopadhyay et al. [42] developed a business model for virtual manufacturing with particularly emphasis on the need for intense collaborative network for a variable-variety, variable-volume and manufacture-to-order situation with provisions for recycling and reverse logistics. The concept was further developed as an aggregated model resembling nature's atomic and molecular interaction after studying the supply chain in China [43]. These new attempts to incorporate human participation in modern global enterprises have highlighted the effect of new information and communication technologies in bringing the human dimension in enterprise architecture to a dominated position.

Figure 4.5 shows that the physical element of a system as the "product" that is built from fundament engineering sciences. This is the common view of most users and society in general. The "product" is the tangible element that can give the "touch-and-feel". An example is the ear implant system that enables patients who have lost

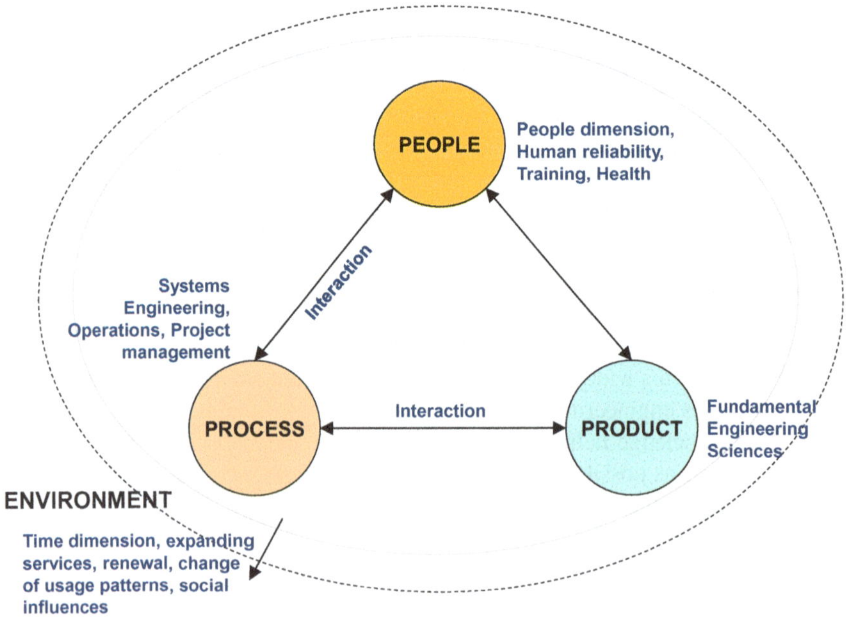

Fig. 4.5 Elements of a system within an environment

hearing capability to be able to hear sound again. This case is obvious because the implant is the "product". Even in software system, the "product" is the program that sits in the computing device. In commercial sense, this is what the customer feels that he/she pays for.

Product is the element that is normally regarded by engineers as the "system". It consists of the hardware and software, supposed to be integrated together as a working system. However, the common system concept does not include the other two important elements. Without trained operators (people), the "product" will not work by itself. Even if the "product" is already highly automated, it still needs someone to set up and plan for its actions before it is going alone. In any case, there are always customers and the public involved in its operation.

The element "people" in the system's point of view does not limit to the user. It includes all human participants involved to enable successful operation of the system. In ear implant example, the user is obviously the patient. However, who puts the device into the patient's body? Who provides regular checks and training, or even system upgrades?

To use the "product" properly, a set of procedures, i.e. "process," should be defined and followed. A defined set of procedures not only allows the "people" (remember there could be many people) to synchronize with the reactions of the system at different inputs during operation, but also ensures the system to be used safely, reliably, smoothly and continuously.

Needless to say, these elements are interacting among themselves as shown by the double arrows. Without these interactions, the "product" is not used by "people", the "people" do not follow the "process", the reaction of the "product' is unpredictable without a defined "process".

The "process" element governs the proper use of the "product" by "people", and proper management of the whole operation within the boundary of the overarching "environment". Within the boundary of the "environment", the three elements are interacting among themselves in various ways to achieve the goal of the system, i.e. expected performance.

Hence, from the above point of view, a functional system is a collection of elements "product", "people" and "process" interacting among themselves as a system within a defined "environment". The elements can be arranged within the system's boundary as a hierarchy of functional elements. Modelling of these interactions can be done with standard system modelling tools such as function model, functional flow diagram, data flow diagram, process model, data model, etc.

The world is always changing. Changes affect the environment in which the system operates. This is represented in Fig. 4.5 as the expansion (theoretically, this change can be contracting too) of the environment. The change in "environment" affects the system both in terms of the size of the elements and the interactions among them. In other words, some or all of the elements have to be changed to adapt to the new "environment". If nothing is done to the system while the "environment" has changed, the system can become out-of-date and obsolete.

The 3PE model represents a logical way of describing system relationships with elements in the system. By carefully analysing the evolution and interlinks between

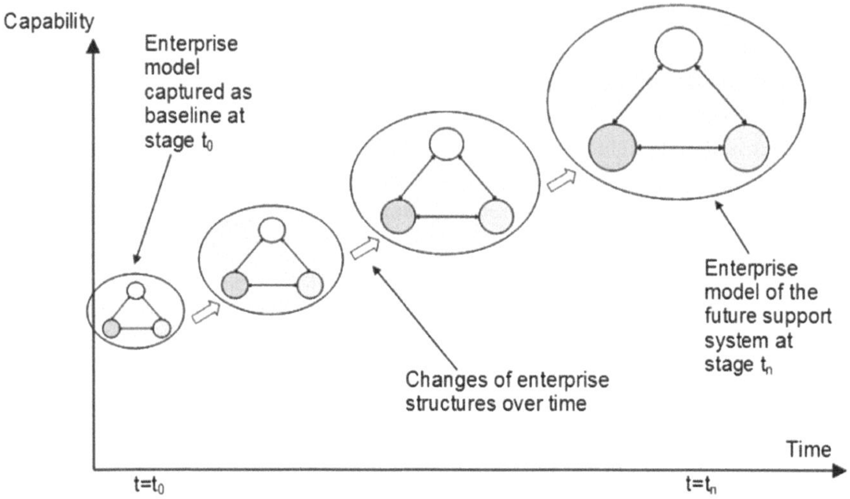

Fig. 4.6 3PE models in a transition trajectory

different functions, data and processes, a development continuum could be mapped out to form a trajectory as shown in Fig. 4.6. The new (future) architecture covers the additional "changing" aspect of service system by integrating the concepts of product, process, people to changes in environment over time.

To manage the changes in the system evolution process, system modelling and representation should be done in a way that can be traced throughout the system's lifecycle consistently, for benchmarking and knowledge management purposes. The 3PE framework can be applied to transition of a system over time.

4.6 System of Systems Modelling Using 3PE

It was noted that in network-centric operations actors may be viewed as part of a continuously adapting ecosystem. The extensive nature of this interaction is illustrated in Table 4.3, which shows a different representation of the ISO 42010 model. The highlighted areas indicate where interactions take place and numbers can be used to link with descriptions of the nature of such interactions.

When these interactions are implemented to the participating organisations, the resulting system of systems can become a chaotic structure [44]. Figure 4.7, which is a representation of complex mining equipment support considerations mapped onto the top half of ISO 42010.

Two case studies are described and their key features are highlighted. The cases are earlier forms of service and support system representing various degree of success in creating new system of systems. The ad hoc systems of systems established at the time of these cases provide good examples for benchmarking current thinking of the

Table 4.3 Example interaction matrix

From \ To	Stakeholders	Architectural description	Rationale	Mission	Environment	System	Architecture	Concerns	Viewpoint	Viewpoint library	Views	Models
Stakeholders	X	1	1			1		1	1			
Architectural description	1	X				1					1	1
Rationale		1	X									
Mission				X		1						
Environment					X	1						
System	1	1		1	1	X	1					
Architecture						1	X					
Concerns		1						X	1			
Viewpoint	1	1							X	1	1	1
Viewpoint library									1	X		
Views		1							1		X	1
Models		1							1		1	X

Fig. 4.7 ISO/IEC 42010 mapping of interactions in a particular organisation

design and implementation of this type of systems. These cases are chosen because the parties in the cases have tried to apply a defined enterprise infrastructure that links different parts of the service system working in conjunction with the product. Subsequently, the service system has to be designed and tailored to characteristics of the product or the enterprise.

The products in the cases are complex engineering systems. Case 1 is a computer controlled plasma cutting machine that can cut steel plates up to 50 mm thick. The machine has been sold over the world. Case 2 is a chemical plant that is designed and built by a Japanese engineering company. In order to support the customer with minimum costs, the design of the service support system used the Internet, which was evolving at the time when the project was done.

4.6.1 Case Study 1: Signal Based Condition Monitoring System

System health monitoring plays a critical role in preventative maintenance and product quality control of modern complex engineering products. The effectiveness of management can directly impact their efficiency and cost-effectiveness. A condition monitoring system monitors the products using various classical methods of signal analysis such as spectrum or state-space analyses [45]. Maintenance decisions are then made according to the prediction of system performance.

Using time based signals available from normal machine sensing mechanisms, a CNC machine manufacturer in Australia developed a remote condition monitoring system for plasma CNC cutting systems with the aim of servicing the customer anywhere in the world via the Internet. Figure 4.8 shows the network structure of

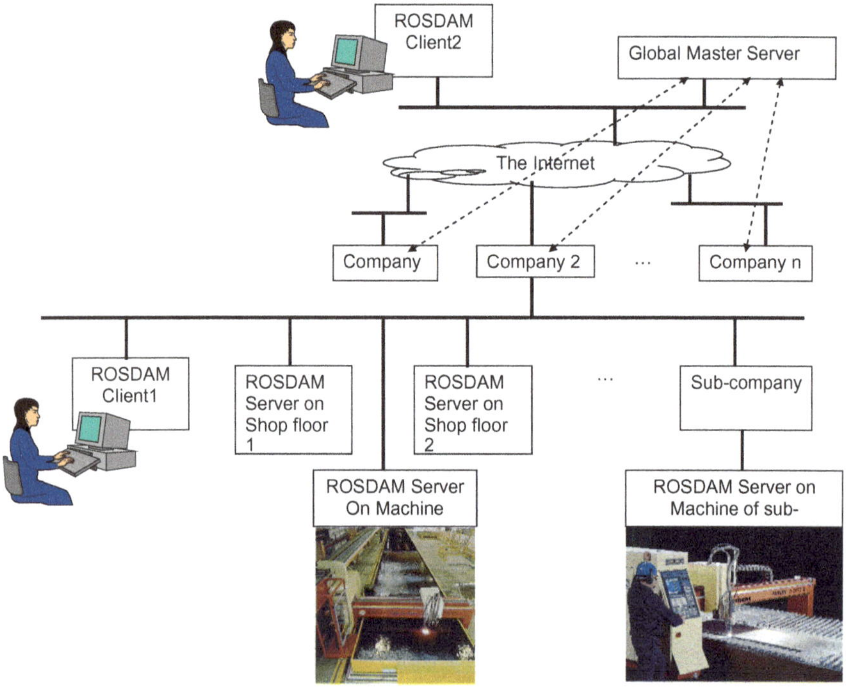

Fig. 4.8 Signal based condition monitoring service system network structure

the system known as ROSDAM [46]. The nature of metal plate cutting varies from very large pieces like bridge and buildings to very small pieces such as mechanical parts casing. All ROSDAM enabled machines were configured as servers that had functionality communicating with the global master server. Information about the operation of the machines was captured through individual companies' database. The significantly improved sources of information enabled the product manufacturer to decide the best option that supported operation and maintenance of the plasma cutting machine from a distance.

In this case study, the 3PE inter-enterprise interactions could be identified and modelled. These elements are mapped to the 3PE model as shown in Table 4.4.

The new service system significantly increased the efficiency of service and support of the new machines and reduce the cost of operation of the OEM.

4.6.2 Case Study 2: Global Operation Support System

Complex assets are normally built from a large number of components and involving a large number of engineers and contractors. In the past, customers as plant owners usually maintain their own service department. However, the increasing complexity of the plant and operating conditions such as environmental considerations require service personnel to have a higher level of analysis and judgment capability.

In managing the design and manufacture of a chemical plant for their customer, Kamio et al. [47] established a service virtual enterprise (SVE) with several partner companies around the world providing after-sales services to a customer (Fig. 4.9).

Each partner in Fig. 4.9 was an independent entity that had its own unique capabilities and competencies, assuming responsibility to perform the allocated work. The SVE was designed as a "hosting service" which had a broad range of services including plant monitoring, preventive maintenance, trouble-shooting, performance simulation and evaluation, operator training, knowledge management and risk assessment. Participants of the virtual enterprise had well-defined roles and responsibilities.

An essential element in the design of a service enterprise is to develop efficient system architecture and provide the right resources to the right service tasks. By synchronising organisational activities, sharing information and reciprocating one another's the technologies and tools, each partner in the service enterprise will be able to provide services that would have been impossible by individual effort. The support solution therefore requires properly designed systems to support the use of technology in the provision of support services to customers.

It should be noted that the engineering product remained the same as it was designed initially. There was no noticeable engineering change required on the product itself in order to implement the support service offered by SVE.

This case has four types of companies participating in the network: OEM, Service Provider, Supplier and Customer. Table 4.5 shows the interaction matrix among all four types of enterprises. It should be noted that the diagonal interactions occur

Table 4.4 Mapping of service elements to 3PE in case 1—two companies network model, OEM to client

		OEM		
		People	Process	Product
Customer	People	Peer to peer relations: Knowledge sharing/transfer—transform customer data to knowledge New data processing algorithms were developed as software modules that were required to process data on machine to knowledge useful to enhance operational efficiency	Adaptation: Engineering information integrated for supporting more effective customer service Engineering information such as bill of material, machine configuration management, parts inventory and resources planning were integrated from different sources including CAD, MRP and various manufacturing sources to create seamless operation database for the machine	Training: Communication networks and IT systems based on client–server model The controller of the machine was significantly changed from a normal standalone operating system to one that can act as a server in a network environment
	Process		Negotiation: The interaction in this case is primarily a procurement process interaction. The customer tries to negotiate for a value-for-money purchase to get the most suitable machine, whereas the OEM tries to maximize the opportunity to receive a favourable purchase order	Exception handling/Customisation: Customised database for customer's machine The new system design requires upgrade of field products Field upgrade for machines that were already installed at customers' location was progressively rolled out according to contracted maintenance schedules
	Product			Engineering change/New R&D: New signal based processing theory to support on machine signal based diagnostics capability A new diagnostics software module based on chaotic theory and digital signal processing was developed to assist identification of faults

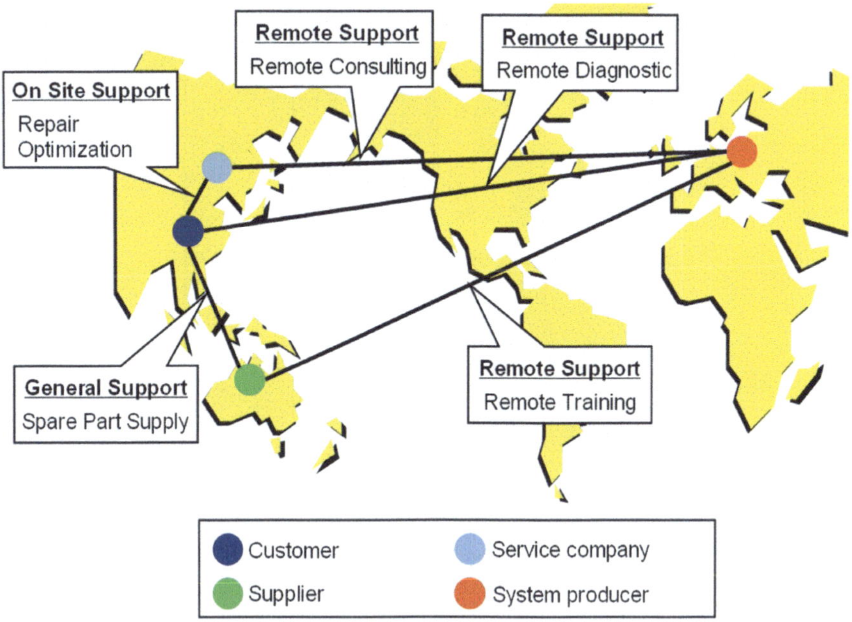

Fig. 4.9 A globally service virtual enterprise

within companies and the lower half triangular cells are repeating the upper half. Cells with hatching do not have interaction.

Interactions in Table 4.5 are:

[A] Peer to peer relations:
The relationship within a SVE was definitely different from a totally authoritative company structure. A much more flexible human organisation structure was established. Peer-to-peer relationships between OEM and service providers are mostly cooperative and knowledge sharing, while there is little interaction between OEM and Suppliers, OEM and Customers. There are interactions between service providers and customers because the service providers are in fact front end of the type 2 contracts.

[B] Adaptation:
New IT and communication systems were installed to enable inter-company exchange of information as well as personal interaction. The new SVE is basically a specially developed communication infrastructure for supporting the type 2 contracts. The service providers had to adapt their system to interact appropriately with the OEM's IT system in order to obtain latest service information of the products. The SVE had a customer interface to support the customers' operations. This interface was also used by the service providers. The suppliers made use of the SVE to manage spare parts and consumables for the service providers and customers.

Table 4.5 Participant interaction chart

		OEM			Service providers			Suppliers			Customers		
		People (P1)	Process (P2)	Product (P3)	People (P1)	Process (P2)	Product (P3)	People (P1)	Process (P2)	Product (P3)	People (P1)	Process (P2)	Product (P3)
OEM	P1				[A]	[B]			[B]			[B]	
	P2					[C]						[C]	[D]
	P3												
Service providers	P1								[B]		[A]	[B]	
	P2											[C]	[D]
	P3												
Suppliers	P1												
	P2												
	P3												
Customers	P1												
	P2												
	P3												

[C] Negotiation:
The SVE was implemented on the Internet allowing global access by customers. The SVE helps the customer to negotiate the most relevant service and support packages to form the contracts. Negotiation on different combination of service provisions and warranty terms could be carried out on SVE among the OEM, service providers and customers.

[D] Customization:
Access to the product at customer location was enabled by SVE for the OEM and service providers. Work items were analysed individually so that the link from individual level to group level can be streamlined ensuring minimum duplication of work and conflicts.

It can be seen from the two case studies that the inter-enterprise interactions have certain generic characteristics that form the basis of modelling a system of systems. These characteristics have been labeled in the corresponding cells in Table 4.5.

4.7 Conclusion

The design, manufacturing of modern engineering systems such as an aircraft or a frigate can be handled effectively using traditional systems engineering approach which focuses on single system and single product development lifecycle modelling. For through-life services and support, many more enterprises and organizational units are involved and the "system" becomes complex, multifaceted and may change over time. These engineering systems are working in an environment that has multiple individual users, complicated supply chain and affecting their performance by many other stakeholders. In essence, these systems are working as a system of with many autonomous systems that are governed by their own individual set of rules and possibly operate with different architectures and system processes.

Engineers trying to apply the theory of systems engineering to "design" a system of systems find the outcome often unpredictable and uncontrollable, as the linked systems operate with high degree of independence. The system of systems approach is evolved from the demand of many large engineering system owners that require services and support contract, which is identified as type 2 contracts in this chapter. These contracts can't be handled by systems engineering methodology alone.

System of systems concept assumes multiple systems working in a network-centric arrangement. System operations are embedded in business networks that are evolving and changing all the time. Since the system agents voluntarily participate in the network, they can come and go at any time without warning. This highly uncertain relationship requires a different modelling approach. This chapter addresses the modelling requirements by extending the people-process-product in environment (3PE) modelling methodology. The 3PE can represent networks in a more logical and orderly way allowing easier analysis of the interactions. This new modelling method for system of systems is demonstrated by two case studies in this chapter.

References

1. Department of Defense (2008) Systems engineering guide for system of systems, version 1.0
2. Dahlmann J, Lane JA, Rebovich G, Lowry R (2010) Systems of systems test and evaluation challenges. In: 5th international conference on system of systems engineering. https://doi.org/10.1109/sysose.2010.5543979
3. Bouziat T, Combettes S, Camps V, Boes J (2017) Cooperative multi-agent approach for computational systems of systems architecting. In: 9th international conference on agents and artificial intelligence (ICAART 2017), 24 Feb 2017–26 Feb 2017, Porto. https://doi.org/10.5220/0006190101740181
4. Defence Materiel Organisation (2011) ASDEFCON (support) v3.0 performance management framework, Version 3.0, Department of Defence, Australia. Accessible from http://www.defence.gov.au/dmo/gc/asdefcon/asdefcon_support/vers_3/PPBC_Framework.pdf
5. Olwell DH, Henry D, Pyster A, Hutchison N, Enck S, Anthony JF Jr (2013) Analysis of the references from the guide to the systems engineering body of knowledge (SEBoK). Procedia Comput Sci 16:1000–1006
6. Honour E (2013) Designing for adaptability and evolution in system of systems engineering. Presented at National Defense Industry Association systems engineering conference, Arlington, Va. Accessed 21 Nov 2017
7. Dahmann J, Rebovich G, Lane J, Lowry R, Baldwin K (2011) An implementers' view of systems engineering for systems of systems. In: Proceedings of the IEEE systems conference, Montreal, QC, Canada, 4–7 Apr
8. Sherwin DJA (2000) A review of overall models for maintenance management. J Qual Maint Eng 6(3):138–164
9. Barabadi A, Barabady J, Markeset T (2005) Maintainability analysis considering time-dependent and time-independent covariates. Reliab Eng Syst Saf 96(1):210–217
10. Chan FTS, Lau HCW, Ip RWL, Chan HK, Kong S (2005) Implementation of total productive maintenance: a case study. Int J Prod Econ 95(1):71–94
11. Tam ASB, Chan WM, Price JWH (2006) Optimal maintenance intervals for a multi-component system. Prod Plan Control 17(8):769–779
12. Colombo S, Demichela M (2008) The systematic integration of human factors into safety analyses: an integrated engineering approach. Reliab Eng Syst Saf 93(12):1911–1921
13. Mo JPT (2012) Product services systems and their risks. In: Asset management conference ICOMS2012, 4–8 June 2012, Paper No. 944, Hobart, Australia
14. Chesbrough H (2010) Business model innovation: opportunities and barriers. Long Range Plan 43:354–363
15. Teece DJ (2010) Business models, business strategy and innovation. Long Range Plan 43(2):172–194
16. Johansson P, Olhager J (2006) Linking product–process matrices for manufacturing and industrial service operations. Int J Prod Econ 104(2):615–624
17. Neely AD (2009) Exploring the financial consequences of the servitization of manufacturing. Oper Manage Res 2(1):103–118
18. Shen ZJM, Daskin MS (2005) Trade-offs between customer service and cost in integrated supply chain design. Manuf Serv Oper Manage 7(3):188–207
19. IfM, IBM (2007) Succeeding through service innovation: a discussion paper. University of Cambridge, UK, 33 p. ISBN: 978-1-902546-59-8
20. Tukker A (2004) Eight types of product-service system: eight ways to sustainability? Experiences from SusProNET. Bus Strategy Environ 13(4):246–260
21. Spohrer J, Maglio PP, Bailey J, Gruhl D (2007) Steps toward a science of service systems. Computer 40(1):71–77
22. Ng ICL, Parry G, McFarlane D, Tasker P (2011) Towards a core integrative framework for complex engineering service systems. In: Ng ICL, Wild P, Parry G, McFarlane D, Tasker P (eds) Complex service systems: concepts and research. Springer, UK. ISBN: 0857291882

23. Doucet G, Gøtze J, Saha P, Bernard S (2008) Coherency management: using enterprise architecture for alignment, agility, and assurance. J Enterp Archit 4(2):9–20
24. Veneziano V, Jones S, Britton C (1999) Adding a systemic view to the requirements engineering processes. In: Proceedings, tenth international workshop on database and expert systems applications, 1–3 Sept, Florence, Italy, pp 321–325
25. Bernus P, Nemes L (1996) A framework to define a generic enterprise reference architecture and methodology. Comput Integr Manuf Syst 9(3):179–191
26. Chattopadhyay S, Mo JPT (2010) Modelling a global EPCM (engineering, procurement and construction management) enterprise. Int J Eng Bus Manage 2(1):1–8
27. Mo JPT, Nemes L (2010) Issues using EA for merger and acquisition. In: Doucet G, Gøtze J, Saha P, Bernard S (eds) Coherency management: architecting the enterprise for alignment, agility, and assurance, chapter 9. AuthorHouse, pp 235–262. ISBN 978-143899-60783
28. Bessant J, Tidd J (2007) Innovation and entrepreneurship. Wiley, Chichester
29. Miles I (2013) Interactive impacts – foresight as a product, service and coproduction process. In: Meissner D, Gokhberg L, Sokolov A (eds) Science, technology and innovation policy for the future. Springer, Berlin, Heidelberg, pp 63–81. ISBN: 978-3-642-31826-9
30. Peruzzini M, Marilungo E, Germani M (2015) Structured requirements elicitation for product-service system. Int J Agile Syst Manage 8(3/4):189–218
31. Teece DJ, Pisano G, Shuen A (1997) Dynamic capabilities and strategic management. Strateg Manage J 18(7):509–533
32. Lewis G, Morris E, Place P, Simanta S, Smith D, Wrage L (2008) Engineering systems of systems. In: Systems conference, 2008 2nd annual IEEE, Montreal, QC, Canada, 07 Apr–10 Apr 2008, pp 1–6
33. Gezgin T, Etzien C, Henkler S, Rettberg A (2012) Towards a rigorous modeling formalism for systems of systems. In: IEEE 15th international symposium on object/component/service-oriented real-time distributed computing workshops, 11 Apr 2012, Shenzhen, Guangdong China, pp 204–211
34. Håkansson H, Snehota I (eds) (1995) Developing relationships in business networks. Routledge, London
35. Bondar S, Hsu JC, Stjepandić J (2015) Network-centric operations during transition in global enterprise. Int J Agile Syst Manage 8(3/4):355–373
36. Maier MW (1996) Architecting principles for systems-of-systems. In: INCOSE international symposium, vol 6, no 1, pp 565–573
37. De Laurentis DA (2005) Understanding transportation as a system-of-systems design, problem. In: Proceedings of the AIAA aerospace science meeting exhibit, Article AIAA-2005-123
38. Möller K, Arto R (2007) Rise of strategic nets—new modes of value creation. Ind Mark Manage 36(7):895–908
39. Osborne S (2002) Public-private partnerships: theory and practice in international perspective. Routledge, Abingdon
40. Bengtsson M, Kock S (2000) "Coopetition" in business networks—to cooperate and compete simultaneously. Ind Mark Manage 29(5):411–426
41. Vaniya N, Noran O, Bernus P (2014) Merger and acquisition preparedness building: an enterprise architecture perspective. In: José Escalona M et al (eds) Improving enterprise communication. Proceedings of the 22nd international conference on information systems development. Springer, Berlin, pp 171–183
42. Chattopadhyay S, Chan DSK, Mo JPT (2010) Business model for virtual manufacturing – a human-centered and eco-friendly approach. Int J Enterp Netw Manage 4(1):39–58
43. Chattopadhyay S, Chan DSK, Mo JPT (2012) Modelling the disaggregated value chain – the new trend in China. Int J Value Chain Manage 6(1):47–60
44. Browning TR (2001) Applying the design structure matrix to system decomposition and integration problems: a review and new directions. IEEE Trans Eng Manage 48(3):292–306
45. Yang S, Sammut M, Kearney T, Mo JPT (2005) Engine condition monitoring with ignition signals. In: Intelligent vehicle & road infrastructure conference, 16–17 Feb 2005, Melbourne, Australia

46. Mo JPT (2003) Case study - farley remote operations support system, chapter 21. In: Bernus P, Nemes L, Schmidt G (eds) Enterprise integration handbook. Springer, Berlin, pp 739–756. ISBN 3-540-00343-6
47. Kamio Y, Kasai F, Kimura T, Fukuda Y, Hartel I, Zhou M (2002) Providing remote plant maintenance support through a service virtual enterprise. In: VTT symposium 224, global engineering and manufacturing in enterprise networks, 9–10 Dec 2002, Helsinki, Finland, pp 195–206

Chapter 5
Traceability in Engineer-to-Order Businesses

Fredrik Elgh and Joel Johansson

Abstract A rapidly growing strategy in product design and manufacture, with great potential to improve customer value, is mass-customization. The main idea is to divide the product into modules that can be shared among different product variants. This will support a wide range of options for the end customer to select among, while an internal efficiency, similar to mass-production, can be achieved. This has been a success for many companies acting on the consumer market. However, many manufacturing companies are engineer-to-order (ETO) oriented, such as original equipment suppliers (OES). They design a unique solution, often in close collaboration with other companies. The solution can then be manufactured in different quantities depending on the client's need. For these companies, there is a strategic need for developing high quality engineering support to further utilize and exploit the information and knowledge produced during product development and to succeed with a strategy influenced by the principles of mass-customization. This has to include the implementation and management of systems enabling highly custom-engineered products to be efficiently designed and manufactured. One challenge when introducing such flexible support is to enable traceability of decisions taken, tasks executed, knowledge used and artefacts developed throughout the whole life-cycle of an individual product. In this chapter, it is shown that traceability can be achieved by introducing support for capturing, structuring and mapping between decisions and resulting outputs, such as geometrical building blocks, knowledge implemented as rules, and the argumentation for the selection, design and specification of these. Three examples are presented where the concept Design Description has been modelled based on an item-oriented, a task-oriented, and a decision-oriented perspective which show the generality of the Design Description concept. The three examples demonstrate how to use the Design Description to enable traceability in platform design, product design, and manufacturing development processes.

F. Elgh (✉) · J. Johansson
School of Engineering, Jönköping University, Jönköping, Sweden
e-mail: fredrik.elgh@ju.se

J. Johansson
e-mail: joel.johansson@ju.se

J. Stjepandić et al. (eds.), *Systems Engineering in Research and Industrial Practice*,
https://doi.org/10.1007/978-3-030-33312-6_5

Keywords Customization · Engineer-to-order · Product platform · Traceability · Design rationale

5.1 Introduction

A rapidly growing strategy in product design and manufacture, with great potential to improve customer value, is mass-customization [1]. The idea is to strive for a broad offer of products and at the same time ensure mass production efficiency. To succeed with such a strategy, it is necessary to make effective and efficient use of information and knowledge-rich systems (e.g. configuration systems, KBE-systems and design automation systems) supporting the whole product realization process from customer specification through design and manufacture of individual customized products. From a scientific viewpoint, most research concerning these systems has focused on the functionality and, to some extent, system development methods, whereas questions concerning what characterizes and how to support efficient implementation and management of these engineering support systems are less well understood. The problems and challenges of bringing a principle technical system solution into operations and support its management have been clearly stated by original equipment suppliers (OES) as key issues [2].

In addition, highly custom-engineered products require the adoption of an engineer-to-order approach in development, quotation preparation and order processing. This allows products to be adapted to large variations in the customers' specifications, which bring more value to the customer and profit to the company by efficient utilization of material and manufacturing resources. However, to quickly go from answering a request for quotation, engineer the product and move it into production, while maintaining the most competitive pricing, is based on the exercising of a very rich and diverse knowledge base of the products, their production and the required resources for design and manufacture. This requires the utilization of systems for efficient design of product variants with associated specifications for automated manufacturing and entails a significant investment in time and money for system development, implementation and maintenance. The complexity and scope of these systems can vary from applications to be used as a support in the design process, to fully automated systems for the design and production preparation.

Regardless of the complexity and scope, the experience is that problems often arise when a system is to be implemented in current operations as well as in its management. Of central importance are issues relating to methods of generating and managing documentation such as engineering calculations and simulations combined with the principles of traceability from the product to the underlying knowledge and vice versa, and versioning of rules, models and systems. Many OESs are engaged in development of complex products and their systems are sub-systems of an overall system. A specific sub-system interacts, affects, and is affected by, other sub-systems developed by other suppliers. An OES can deliver solutions to different original equipment manufacturers (OEM) for a wide set of different products.

A methodology to master development projects of complex products is Systems Engineering (SE). A thorough description of SE can be found in [3], among others. Traceability, from requirements to detailed design and throughout the whole product lifecycle, is important in SE [3]. It is well known that it is a challenge to efficiently enable traceability, while different approaches are required for a multidisciplinary system across stakeholders throughout its whole lifespan [4]. Support for a standardised approach to enable traceability in traditional product development has been discussed to support knowledge sharing in dispersed organisations [5]. However, the specific circumstances in which engineer-to-order-oriented companies operate are not addressed.

This chapter presents results from two research projects focusing on how engineer-to-order businesses can work both strategically and practically to increase traceability. The chapter is organized as follows. First, the business environment in which engineer-to-order companies act is described. The state-of-art in related areas is then presented and followed by a theoretical foundation for working with traceability. Three case studies, focusing on enhancing traceability in three businesses, are then described, evaluated, and compared. The chapter ends with a summary including conclusions and areas for further research.

5.2 Business Environment for Highly Customized Products

There is an increasing interested in industry to support customization and individualization. It is common that customization is achieved by combinations of standard parts and a limited number of unique parts. The authors of this chapter have more than 25 years of experience in improving engineering processes in this area, and have worked in close collaboration with several companies throughout the years. One of the companies, a large international supplier of tooling solutions and know-how to the metalworking industry, can serve as an illustrative example of an engineer-to-order business that has automated their engineering processes to an extreme level. The company has worked with rule-based design for about 30 years of which ten years in collaboration with the authors. The organization for development at the company includes departments across the globe for product design, design space development, automated design, automated process planning, CAD-method development, and development of special IT-support systems. The company is a pioneer when it comes to total automation and the business processes require advanced support. At the company, executable product platforms are developed which contrasts with the more common approaches targeting single one-of products or the utilisation of a modular system from which alternative solutions can be derived. Even if the product individuals generated from the executable product platforms do not include many parts, they are not predefined and are all unique (in fact even if two identical orders are place they will be treated as if they were not identical). The company can represent the target condition for other companies as a major challenge in customization concerns the efficiency in design and manufacture of the unique parts. The

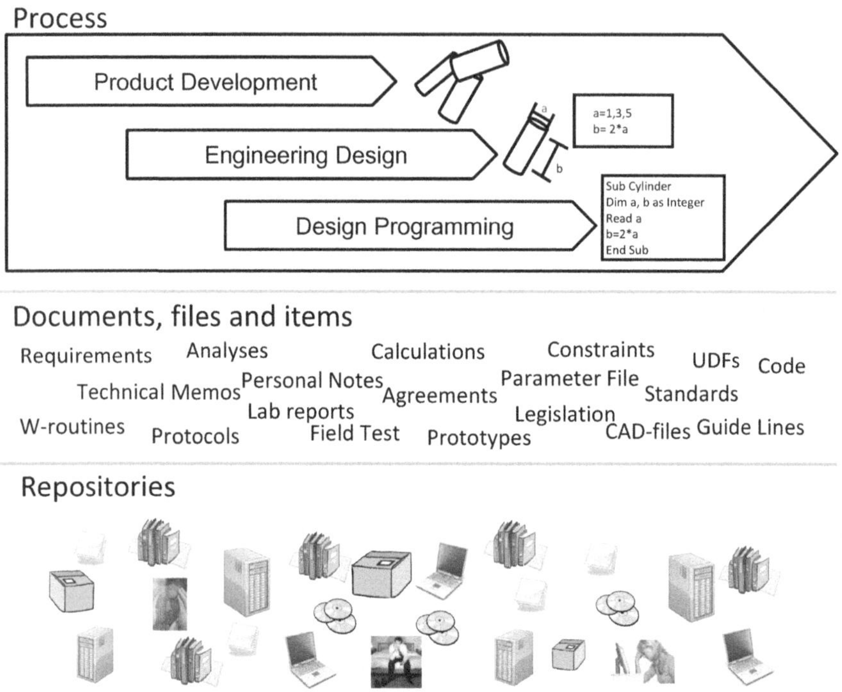

Fig. 5.1 The development process with information carriers and repositories at a large international company

need for improved traceability in a complex environment is shared with many other companies adopting a product platform strategy, regardless of level of automation.

The development process at the company includes three steps: product development, engineering design, and design programming (see Fig. 5.1). The product development process is the process of converting identified market needs to a set of *requirement ranges*. The requirement ranges include all identified needs of the customers and can be of size and performance dimensions. The product development continues in a traditional way with the main difference that instead of developing one single product, a set of product instances are developed. The instances are selected to be in the extremes of the requirement ranges. In the second step of the process, engineering design, the product instances from the first step are used to define a continuous and complete design space that will make up a product platform that includes not only the initial product instances but also any instance that can be derived from requirements in the requirement ranges. Finally, the design space is computerized with specially developed programming tools so that it is possible to automatically create cost calculations, generate product documentation and manufacturing data for any point in the design space. The result of the process is a product platform defined as a space with implicit solutions, which are generated based on individual customer requirement. Theoretically, it is an infinite design space. A single unique solution can be automatically generated, manufactured and shipped to a customer within hours after an order has been received.

In Fig. 5.1 the process is modelled at the top. In the mid-section, the different documents and files produced through the process, including amongst other lists of requirements, CAD-models and test reports are illustrated. At the bottom of Fig. 5.1 it is schematically shown that these files are stored in various repositories, including databases, individual computers, post-it notes, binders, and in the mind of individuals.

5.2.1 *General Development Model*

The development process at the company described above can be generalized as shown in Fig. 5.2. The process starts with the identification of needs and demands on the market, which are transferred into product requirements. During that step there are many decisions to be made based on the knowledge of the market experts. They need to make good decisions when requirements are to be presented as ranges, defining minimum and maximum values for the different requirements. Based on the requirements ranges, many decisions are made based on the knowledge of the engineers. A number of tests are done to gain the knowledge required to make correct decisions. The result is a platform from which individualized product instances can be derived, manually or automatically depending on the company strategy.

When a customer submits an enquiry to the company, a sales person uses its knowledge to make decisions regarding what products in the product portfolio is suitable for this specific customer and converts the needs of the customer into a list of requirements. Based on engineering design knowledge, new decisions are taken during the process of deriving a unique design from the platform that fulfils the specific customer requirements. Subsequently, manufacturing knowledge is used when making decisions on how to manufacture the individualized product.

As seen in Fig. 5.2 there are five types of decisions that are based on five sets of knowledge. The possibility and benefit to fully formalize and completely automate

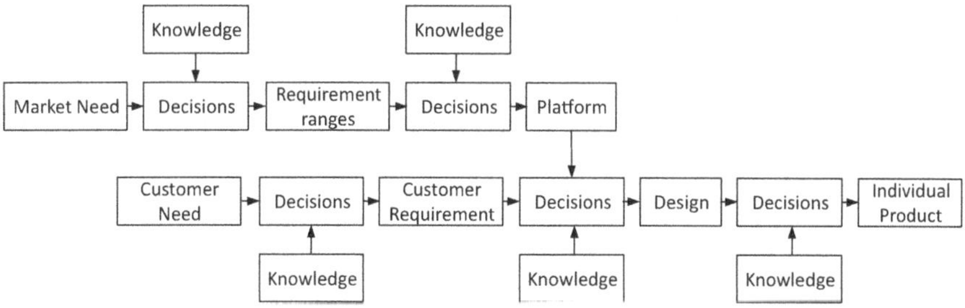

Fig. 5.2 Generalized development process

the knowledge and decisions differs from product to product and for the business case. In some cases, a combination of manual work and automated task is to be preferred.

5.2.2 Traceability in Practice

From an industrial perspective, the implementation is a critical process and of high importance for the actual use and consequently the benefits achieved and future return on investment. User acceptance is of high importance and strongly related to the access and understanding of the underlying knowledge, which requires a high level of system transparency and traceability. In addition, the long-term management is of major importance for system longevity. Two life-cycle perspectives must be considered when addressing management, a knowledge perspective (on the executable product platform description) and a product perspective (for every single delivered product variant). Management concerning knowledge includes the adaptation of rules and models to changes in production technology, new product knowledge, new markets, changes in legal requirements, etc. Issues related to flexibility, stability, quality assurance, traceability and documentation of a system's different constituting parts and underlying knowledge can be critical unless adequate measures have been taken in the development phase. Management concerning the product focuses mainly on documentation, traceability, and version control. As the governing framework and models are updated and refined due to shifting prerequisites, the system and hence the solutions generated for a single specification will change over time. This affects product management and the ability to meet legislation and customers' requirements regarding documentation and traceability, as well as the company's ability to provide services, maintenance and supply spare parts.

The implementation and management of systems enabling highly custom-engineered products is challenging. A key enabler is traceability, i.e. the ability to describe and follow the life of a conceptual or physical artefact [6], including decisions taken, tasks executed, knowledge used, and artefacts developed throughout the whole lifecycle of an individual product. There are several reasons why an engineer-to-order business would benefit from this ability, such as:

- Controlling spare parts (what parts were used in a specific product individual, and what were the manufacturing circumstances)
- Tracking which parts are in use by customers and in which product individuals
- Analyse change propagation to see what parts and customers are affected due to changes in technology
- Persistently understand why certain relations, parameters and values are used in the platform
- Maintain products
- Maintain knowledge
- Support the work to adapt products to include new technologies
- Knowledge and solutions reuse.

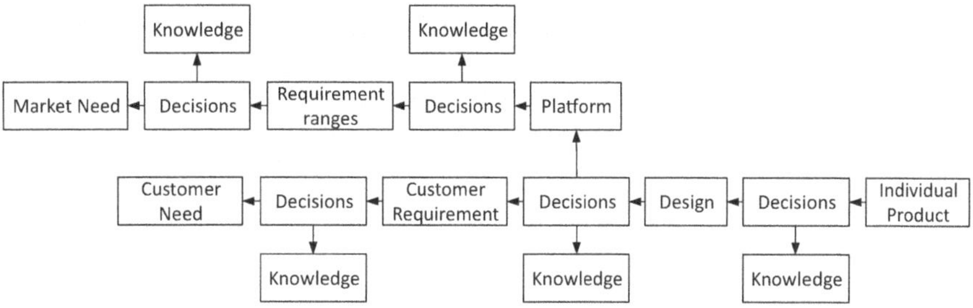

Fig. 5.3 Traceability in an engineer-to-order business

Traceability in engineer-to-order businesses includes the ability to trace all the decisions and their underlying knowledge for production, design, customer requirements and needs, platform development, requirement ranges and market needs for an individual product instance. In an engineer-to-order business it is hard to achieve such functionality on a very fine level as it implies that links are created in and between every part in Fig. 5.2. These links will enable backtracking, as shown in Fig. 5.3, which proactively requires measures to be taken in each and every step on a fine level of granularity. The big challenge is the vast number of variants, versions, file formats, and data storage repositories together with a project-based working method (commonly engineer-to-order companies close the project databases when a project is finished and it can be hard to get access to historical information stored within them).

5.2.3 The Need for Design Rationale

What is it that must be done proactively to enable traceability? When studying the documentation of products and platforms in industry it can be concluded that it is mainly directed towards describing the final results of the different activities, answering questions what the product is and how it is to be manufactured. These questions are answered as the development progresses, as in Fig. 5.2. When moving upstream to trace the decisions and underlying knowledge, as in Fig. 5.3, the questions to be answered are instead why the product is constituted and produced the way it is, and what other solutions were tested but rejected. The answers of these questions are easy (often too obvious) to give during the development project but are not persistently stored and soon forgotten by those making the decisions. Even if persistently stored, the vast amount of information regarding why the product is constituted the way it is will soon become overwhelmingly large and hard to navigate [7].

In one of the case studies (see Sect. 5.5) the identified documentation was classified as either process related or product related. *Process-related documentation* is associated to a specific product development project including documents for project management, meeting protocols and other documents used for sharing information between project members. Process documentation is stored and managed using a project database. *Product documentation*, on the other hand, includes *object documentation* describing the result of an activity (e.g., a description of a parametric CAD model), and *object process documentation,* which describes the work related to the object (e.g. considerations, tests, analyses, decisions, assumptions etc.), and guidelines regarding the product design considering some specific aspect (e.g., manufacturing and environment). *Object documentation and guidelines* are stored, managed and published in a company internal portal. The design engineers provide the material regarding the product design that is to be published on this portal. However, no central system for storing and managing *object process documents* existed. Some individuals make notes in documents or in programming code for personal use or to be used by other group members. An overall summary of in-depth interviews with decision-making personnel pointed out that [8]:

- The purpose of documentation in general and project documentation in specific is not seen by all company employees.
- The quality of the documentation is quite varying.
- The corporate project database is used for finding work prerequisites and to learn from earlier projects.
- It is perceived hard, by the respondents, to find project documents for non-project members.
- The information in the corporate project database is coarse and not easily accessible for non-project members, especially when the project has been closed. The system is mainly used to find specific individuals for consultation regarding, by example, reuse of product descriptions.
- The documents are weakly connected to the different product families.
- Specific geometries, CAD models, are reused to some extent but design rules and principles are seldom reused.
- It is difficult for individuals who have developed good solutions to share these solutions. The reason given for this is that there is no present system for such documentation.
- The access to information is seen as most difficult by design engineers and design programmers.
- A general view is that reuse could be augmented at the company and improved documentation could support this. However, it is important that documentation can be easily done.

These issues and needs call for the development of methods for increased traceability making it possible to get answer to questions not only on how an individual product instance is constituted or how it was manufactured but also why, i.e., the design rationale. Traceability, in the meaning to navigate and search for decisions, is not enough; the rationale of decisions are required if changes or reuse is to be supported. In the next section, research and development concerning traceability and design rationale from a product perspective (answering questions of what and how), and from a knowledge perspective (answering question of why) are introduced and discussed.

5.3 Fundamentals and Existing Support

The previous sections have introduced the business environment for companies that deliver unique solutions for each customer. Furthermore, the need to trace decisions and knowledge for these companies to support development, maintenance, re-use and management of individual products has been described. This section presents and discusses fundamental concepts and existing support that could be considered as candidate means in a framework that would aid companies in the development of solutions and practices that tackle the challenges of traceability. This includes the two concepts of traceability and design rationale as well as methods for development of means for customization together with methods and tools for knowledge modelling. Finally, to what extent the methods and tools can provide support for traceability is discussed.

5.3.1 Traceability and Design Rationale

The development of a support system for customization of products is preferably a part of, or integrated with, the development of the product platform that it will represent, such as the example in Fig. 5.1. Four major outputs can be identified within such a development process: the product design, the design space, the system-adapted definition of the design space, and the system implementation. Traceability, defined as the ability to describe and follow the life of a conceptual or physical artefact [6], across and within these outputs, is essential. The main artefacts of concern in this context are product platform and the support system for customization where the product platform is described and can be executed for creating a new product variant. A new support system is designed for every new product platform. The system encapsulates product knowledge that has been expanded and transformed into different levels of completeness and generalization throughout the four sub-processes. Traceability, both forward and backward, across different knowledge levels would support the identification of affected objects when changes occur in the premise of a design or support the redesign of an existing solution to be used under new circumstances by identification of the original design decisions; i.e., knowledge traceability, defined as the ability to follow the life of a knowledge component (i.e. a fragment of knowledge from a specific source) from its origins to its use and vice versa [6], is required.

Design rationale is the set of reasons behind the decisions made during the design of an artefact (e.g., a product or an application system). The access to a design rationale can support the development of new artefacts, modification of existing artefacts (design changes) or the reuse of an existing solution in a new context. The realization of a design rationale system includes methods and tools to capture, structure, manage and share information across organizations, processes, systems and products. The requirements concerning the scope and the granularity of a design rationale to be captured depend on future needs. These needs can be difficult to predict. A limitation must be set as is not feasible to capture every action taken during the design process. Two different approaches to represent design rationale are argumentation-based and template-based [9]. Argumentation-based representation uses nodes and links whilst Template-based representation makes use of predefined standard templates. The selection of an approach affects the scope, the granularity, and the structure of the captured design rationale. However, the key factor for successful implementation of a design-rationale recording tool is the simplicity of using it [10].

5.3.2 Methods for Development of Customization Support

There are numerous examples in industry where systems supporting the ability to provide custom-engineered products have been applied with varying methods and degree of sophistication. The most common applications in industry are various forms of configuration systems [1, 11], but there are also many applications regarding parametric component design [12, 13], structural analysis tasks [14] and systems of a more generative nature [15, 16]. There are mainly three classes of systems, either based on the *configuration* of a set of predefined product modules and attributes (i.e., retrieval of valid combinations of discrete sets), *design automation* executing different engineering tasks generating a product definition, or *knowledge based engineering,* (*KBE*) where set of rules representing engineering knowledge (combining discrete and continuous domains) are integrated with a pre-defined parametric geometry model. The two latter are best suited to support engineer-to-order processes and the first is best suited for configure-to-order processes. Despite all examples of successful implementations in industry and prototypes in research, a limited number of methods for systematic development have been reported.

In the area of *configuration*, Claesson has introduced and developed the concept of configurable components [11]. The concept is built on function-means modelling and focus on as-is modelling of a (existing) product platform. A more extensive work is by Hvam et al. [1]. They describe a complete and detailed methodology for constructing configurable product platforms supported by a configuration system in manufacturing companies. They suggest an iterative process including the following activities: analysis of product portfolio, object-oriented modelling, object-oriented design and programming, among others. Every activity results in a description of the problem domain with different levels of abstraction and formalisation. The analysis of a product portfolio results in a Product Variant Master (PVM) and Class Relationship Collaboration (CRC) cards.

A general method for how to plan a *design automation* system is described in [17]. A top-down approach is suggested starting from the specification of system requirements and a description of the problem characteristics followed by a mapping to appropriate methods for system realisation. A set of criteria of system characteristics is defined in [18] including transparency, knowledge accessibility, flexibility, ease of use and longevity. Most likely, these characteristics affect system implementation and management. The criteria are to be considered and weighted in the planning of a design automation system.

Stokes describes a methodology called MOKA (Methodology and software tools Oriented to Knowledge Based Engineering Applications) [19] for the development of *knowledge based engineering* applications. La Rocca et al. have developed the Design and Engineering Engine, DEE, approach [20–22]. This approach partly builds on the principles of MOKA and consists of three major elements: The first element is concerned with the design process, which includes multidisciplinary optimisation. The second major element is the Multi-Model Generator (MMG) that uses the product model parameter values in combination with formalised domain knowledge to generate product models. Report Files are generated and fed to the third major element, the detailed analysis modules. These modules calculate the design implications. Finally, the loop is closed by analysing the data files using convergence and evaluation checks. Curran et al. [23] extends the DEE approach to the Knowledge Nurture for Optimal Multidisciplinary Analysis and Design, KNOMAD, methodology. The KNOMAD acronym highlights method process of: (K)nowledge capture; (N)ormalisation; (O)rganisation; (M)odeling; (A)nalysis; and (D)elivery. These implementation steps are taken and repeated as part of the knowledge life cycle and in this context.

5.3.3 Methods and Tools for Knowledge Modelling

The development methods for customization support focus mainly on specifying the requirements of a system implementation, different activities for conducting the work and the representation of the final solution, both the product platform constructs and the system realisation. They provide little support when it comes to capturing and structuring knowledge. However, there are methods and tools that could be used for this. One *method for knowledge modelling*, applicable in the domain of design automation systems, is the Systems Modelling Language (SysML) [24]. SysML is a general-purpose modelling language for systems engineering applications. It supports the specification, analysis, design, verification, and validation of a broad range of systems and systems of systems. These systems can include hardware, software, information, processes, and facilities. The language provides graphical representations with a semantic foundation for modelling system requirements, behaviour, structure, and parameters, which are used to integrate with other engineering analysis models. CommonKADS is a method to document and manage engineering knowledge [25]. It acts as a baseline for system development and research projects. CommonKADS originates from the need to support development of industry-quality knowledge systems on a large scale, in a structured, controllable, and repeatable way. CommonKADS has a predefined set of models (organisation, task, agent, knowledge, communication, and design), each of them focusing on a limited aspect, that together provides a comprehensive view. Product Variant Master (PVM) is an operational tool to model and visualize a product platform [1]. In general, a product platform in PVM can be modelled as a Part-of structure, which shows the components included in the product, and a Kind-of structure that shows the variants available.

Several readily available *tools supporting knowledge modelling* (of which some are free and open source) exist with functionality suitable for the purpose of capturing and structuring a design rationale. PCPACK includes different functions that provide user-friendly graphical interfaces to structure knowledge [26]. By example, categories, sub-assemblies, and sub-components can be represented and visualized in PCPACK. Another use of this software is for defining and presenting relationships and properties associated to pieces of knowledge. Ten tools are defined to make the knowledge modelling more easy and flexible; five acquisition and modelling tools and five specialized tools. In order to support re-use of knowledge, PCPACK uses XML, which is fully compatible with modern web technologies such as the semantic web and provides a formal machine-readable content. Another application is Design Rationale Editor (DRed) [10]. DRed allows designers to record their design rationale at the time of its generation and deliberation. The design rationale is displayed in a document as a graph of nodes linked with directed arcs. The user creates the nodes by choosing from a predefined set of element types. The functionality is based on four main applications for: diagnosing a problem (problem understanding), designing a solution (solution synthesis), completing a standard checklist template, and communicating the final design and its rationale. A third application is Product Model Manager (PMM). PMM is a tool built upon the principles of Haug et al. [27]. PMM has a user-friendly graphical interface to model a product structure including its parts, assemblies, interchangeable modules, variables, and rules. The main purpose of the tool is to support modelling activities and documentation of configurable products. Finally, there is Semantic MediaWiki (SMW) [28], which is a free open-source extension to MediaWiki that enables querying data within a wiki's pages. The purpose of SMW is to allow users to improve the structure and organization of the knowledge in a wiki by adding simple, machine-readable information to wiki articles. With this additional information, searching, browsing, and sharing the wiki's knowledge can be improved, both within the wiki's pages and from external computer programs.

5.3.4 To What Extent Can Traceability Be Supported?

The previous two sections describe methods for development of means for customization and methods, as well as dedicated tools, for knowledge modelling. In this section, the means, and to what extent these provide support, for traceability are discussed.

The configurable component concept [11] includes a function-means model to provide design rationale for the encapsulated design solutions, which could support the understanding of the system and thereby support system implementation and maintenance, however, this is not described in detailed or exemplified. In the work using PVM and CRC-cards [1] it is suggested that the maintenance is to be organised by introducing Model managers. The Model managers are responsible for the delegation, coordination, collection, and documentation of domain-expert knowledge. The programmers then use this documentation to update the system. Haug et al. [29]

have developed a prototype system for the documentation of configuration systems founded on the PVM concept that can support the Model managers. The documentation system is separated from the implemented product configuration system.

The general method for how to plan a design automation system [17] does not include aspects such as user-friendliness, maintainability, or documentation despite the author's statement that they are of significant importance for success in industrial praxis. The authors argue that implementation and management issues are to be considered only when the fundamentals of the problem at hand have been solved. The criteria for system characteristics [18] do not give concrete answers to implementation and management issues as stated in the reference. However, a possible means to support the management of a systems' incapsulated knowledge base is to strive for an implementation that allows for continuous revision and documentation.

Two central parts of the MOKA methodology for knowledge-based engineering development are the Informal and Formal models [19]. The Informal model is used to document and structure knowledge elicited from experts, handbooks, protocols, literature etc. The Informal model can be regarded as paper-based with text and illustrations. The Formal model is derived from the Informal model with the purpose to model and structure the knowledge in a fashion suitable for system specification and programming. The Formal model is described by an object-oriented annotation, MML, based on the UML standard. The importance of maintenance is stressed but detailed practical support is missing [30]. KNOMAD [23] is argued to support the whole Knowledge Management across the product lifecycle. It includes an approach for multidisciplinary design (optimization) and for knowledge capture, formalization, delivery, and lifecycle nurture. Exactly how this is to be achieved is not described in detail.

SysML provides support to model and visualise the rationale, requirements, constraints and rules by using the concept of block diagrams [24]. In CommonKADS [25], all information from design to delivery can be included and clearly visualized. Storing experience, geometry and data that are related to a product and present them within different classes and views are supported by MOKA [19]. Regarding the three specific applications, it can be concluded that PCPACK [26] provides an integrated suite of ten knowledge tools designed to support the acquisition and use of knowledge. Support in analysing knowledge from text documents and structuring knowledge using various knowledge models makes PCPACK an extensive system. DRed [10] is a software tool that allows engineering designers to record their rationale during the execution of design process. It supports the capture of issues addressed, options considered, plus associated pro and con arguments (arguments for or against an answer), in the form of a directed graph of dependencies. PMM [27] is a tool considered to be easy to learn with an intuitive structure and graphical notation. However, support for advanced queries, revisions, and authorization are not included. Improved data structures by using categories and access to information according to user's specific queries are the advantages of SMW [28]. Support for revisions and authorization are also supported by SMW.

When it comes to reducing costs, risks and lead-time in a KBE project as well as providing a way of developing and maintaining applications, MOKA provides the most sufficient support. Product variant master (PVM) gives a general overview of the product according to sub or super parts with relations between different components, which all can be seen on a big piece of paper.

Despite the methods described above, and the numerous applications describe in scientific publications, a number of issues and challenges to be addressed still exists. For example, the major shortcomings of KBE have been identified in an extensive review by Verhagen et al. [31]. Four of these, that most likely impact implementation and management, are: system transparency, knowledge sourcing and re-use, semantics of knowledge models and traceability. The authors' experience is that this is also the case when it comes to configuration and design automation systems.

In summary, it can be concluded that development methods and associated models mainly focus on the development phase and the delivery of product/process models that can be executed based on different requirements to generate a specific solution. The ability to maintain, expand or reuse constructs of these models is not in focus. This is commonly pointed out as important but little support and guidance can be found. In some cases, it seems to be taken for granted that support for maintenance, expansion and reuse are inherently integrated parts of the proposed methods whilst others acknowledge the challenges to achieve this in an industrial setting but lack in advice or support of counter measures. Documentation is an important enabler for efficient management, but there is no support enabling back-tracking throughout the development process. It must be possible to navigate among, across and within different collections of information. The specific decisions, knowledge and requirements supporting an individual product's design as well as the decisions, knowledge and requirements forming the product platform from which the product variant design was derived must be traceable. It is essential that the collected information is of value. Otherwise, it is a complete waste of resources.

5.4 A Foundation for Traceability in Engineer-to-Order Businesses

The tools and methods presented in the previous sections are either general modelling languages or tools developed to describe the product and its related knowledge. It is difficult to make use of these tools and methods without an overall framework. Such a framework should be based on a holistic view for the work with traceability in engineer-to-order companies. The framework should make it possible to (adapted from [8]):

- Facilitate capturing and structuring decisions, design constructs (e.g., geometry and rules), their related knowledge and essential design rationale
- Enable persistent storage in a way that easy retrieval of the relations and content is possible
- Support visualization

- Enable versioning control
- Include authorization control
- Preferably reside upon existing available tools.

A foundation for such framework is presented in the following sections where the key concepts are defined based on a case study at the company introduced in Sect. 5.2 (more details, see Sect. 5.5.1).

5.4.1 Key Concepts

When investigating the geometrical building blocks of a product platform and especially ways in which rules regarding these building blocks relate to the concept of knowledge, lead to the conclusion that rules implement a kind of knowledge. In this case, knowledge is an intentionally defined element that systematically transforms input to output. Rules can implement computations, actions, consequences, and relations but they do not encapsulate the argumentation for their existence or the reasons behind their constitution. The definition process of a product platform includes decisions, which constitute another kind of knowledge that, if captured, provides a deeper understanding of the knowledge implemented through the rules. Such knowledge answers to questions: why the rule is defined the way it is, when to use it, valid ranges of input/output of the rule, the origin of the rule and its supporting theories, and what simplifications it is based on. Since that knowledge constitutes knowledge about knowledge it is referred to as meta-knowledge.

To support reuse, expansion and maintenance of the building blocks and connected knowledge, it is required that the focus in the product development process is not limited to the definition of geometry and rules exclusively, but also includes the definition and collection of associated meta-knowledge. Potential and relevant meta-knowledge can appear in different contexts (e.g. meetings, coffee table discussions, or directed thinking) stored in different formats (e.g. text-documents, CAD-models, or hand-written notes) in different repositories (servers, e-mails, memory sticks, white boards), see Fig. 5.1. These pieces of information, that we label Meta-Knowledge Carriers, occur throughout the development process while defining the geometrical building blocks and connected knowledge. But, the focus during the development activities is not on the Meta-Knowledge Carriers. No mapping between used Meta-Knowledge Carriers, the resulting building blocks and other supporting knowledge sources is done. No descriptions are added that provides the reason for the decisions, the context, and the meaning. As there is no mapping, traceability is not supported. To enable traceability, the concept of Description is introduced [8]. A Description is an object intended to carry both knowledge and meta-knowledge that typically occur during the development process in an engineer-to-order business. The concepts of Design Definition and Design Rationale are also introduced [8]. The main focus of the Design Definition is the construction and the function of process output objects implicitly defining the design space by a set of rules to be executed to generate the specifications for a product variant. These process output objects form a set of Knowledge Objects [32–35] that transform a set of input to output using formalized engineering principles. The definitions of rules can be based on heuristics, physical laws, simulations or testing [33]. The main focus of the Design Rationale is the argumentation and supporting descriptions unfolding and justifying the object's design (meta-knowledge). Both the Design Definition and the Design Rationale provides essential meta-knowledge about the process output object and together they constitute the foundation for the Design Description. The Design Descriptions serve as containers with the following main function and properties:

- Support capturing of Design Definition and Design Rationale
- Links to supporting documents, models and items
- Links to preceding Descriptions
- Written for a clearly defined purpose and potential users
- Based upon templates with predefined headings, keywords and fields
- Simple and visual
- Continuously updated
- Versioning control
- Authorization functions
- Has an owner.

The intention with Descriptions is to facilitate the work of documenting and to support high quality documentation. The content of a Description includes, by example, an explanation of the overall product, its building blocks at different levels (e.g. product, assemblies, parts, features and geometrical entities), relations between building blocks (e.g. functional structure and assembly sequence), parameters (input, internal and output), and rules describing the design space. This will constitute the Design Definition of a Description. By adding information and links concerning aspects such as calculations, analyses, field test, underlying principles for design, assumptions, constraints, context, valid ranges of parameters and aspects for validity of rules, together with statements regarding what to consider when changing, ideas not yet implemented and workarounds, the Design Rational of a Description is completed. Means for information representation include tree models, text, illustrations, pictures, tables, formulas, links and meta-data. Process output objects (e.g. Knowledge Objects), different Design Descriptions, Knowledge Carriers (e.g. project documents, models and items), meta-data and links are stored in a database managed by a Database Management System. System functionality includes means to enter, structure, map, store, retrieve, search and visualize information, together with versioning and authorization control and of essential importance is the underlying information model. The Description concept has been used in the three real case examples presented in the following section.

5.5 Real Case Examples

Three case examples from industry will show the potential of the design description approach. The first case adopts an item-oriented approach, the second one a task-oriented approach, and the third one a decision-based approach. These three perspectives affect the information models derived from the design description approach. The case specific information models together with short descriptions of the case companies and screen shots from the implemented systems are presented in the following subsections.

5.5.1 Case 1—An Item-Oriented Approach

Company A is a world-leading supplier of tools, tooling solutions and know-how to the metalworking industry. The company is active in an internationally very competitive market and needs to constantly cut development lead-time by seeking means to improve their processes and system maintenance. The company has a longstanding tradition in automation of quotation and order processes and has adopted an engineer-to-order business model supported by systems for automated design and production preparation of customized product. A request for quotation of a custom engineered product is replied within hours including detailed design drawings and a final price. All the necessary documents and manufacturing programs are automatically generated when the bid has been accepted by the customer.

The representation of knowledge, incorporating both the design definition and the design rationale, in the Design descriptions repository is based on the information model depicted in Fig. 5.4, which also acts as a template. Of central importance is the Rationale class that connects to all other classes except the Product Family[1] Description (PFD) class, individually or in combinations. The Rationale class also enables specification of relations to Rationale classes in the Product Instance Description (PID) and Design Module Description (DMD) domains and relations to supporting documentation. The central concept of the information model as seen in Fig. 5.4 is Item. An item can be any representation of a physical artefact or parts of it but also rules regarding it.

A system, labelled Design descriptions repository, founded on the presented framework for modelling and management of product knowledge together with the functionality provided by Semantic Media Wiki (SMW) [28] was developed. A previously developed product platform was selected for setting up a PFD, which is a type of design description. When setting up the PFD, the concept of classes has been used and the product platform is explained according to these classes. By example in Fig. 5.5, the PFD is described by linking to seven articles. For each article a wiki page is created. The documentation of knowledge relevant for the class is placed

[1]The company used the term product family for their product platform and in this case description these two terms have the same meaning.

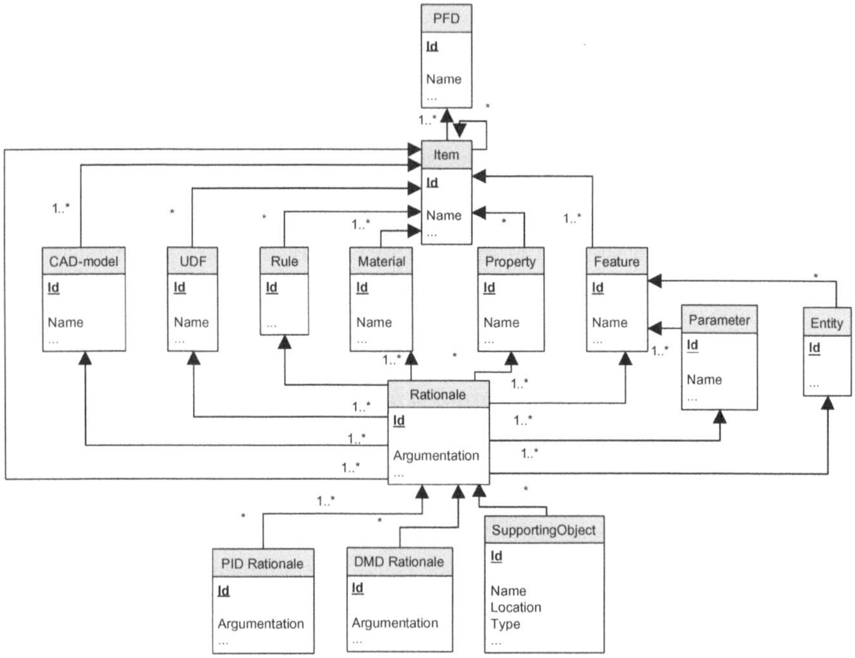

Fig. 5.4 Item-oriented design descriptions for product families [36]

Fig. 5.5 Main page of the design descriptions repository [36]

within that page. An article can contain supporting documents such as Microsoft Excel workbooks and Microsoft Word documents. These can be added to a page by uploading the specific files and then create links to them. Current documentation at the company of product families focuses solely on the design definition and in order to set up a PFD, design rationale should be recorded as well. During several meetings and discussions with the designers, the rationale behind every rule and the knowledge applied were discussed and documented. The design rationale was recorded and then entered for storage in SMW. The information and knowledge was described by using text, figures, tables, rules, schemas and tree structures. A page describing a component contains both the design definition and the design rationale to form a complete description. The text describes different parameters that are used to design the component. It also includes the principle for designing, the function of the component in the product, the rules and their validity for the product platform. It is important to prevent multiple records of the same information and knowledge. For example, in the documentation of the test product platform, some information, tables or values are general for a range of parts and have previously been stored for each of those parts separately. In order to prevent duplication, documentation can be done in two categories; (1) a general category, containing general information which is valid for a range of parts; (2) a specific category for the knowledge which is valid just for the specific part.

5.5.2 Case 2—A Task-Oriented Approach

Company B is a global supplier of products to the automotive industry. The company acts in the business areas Interior, Driveline, Fluid Transfer and Driver Control. Mostly all products have to be adapted to the specific requirements of the OEMs for different car models and variants. One example is a seat heater system that has to be adapted to the specific performance requirements as well as the geometry of the seat where it is integrated and the systems it interacts with. The company has the technology and a concept solution. However, the requirement specification is not complete when an OEM calls for quotations. If a contract is signed, the company joins a development project of a new car model, including its different variants in seat options. The project can last for years and the requirement specification is frequently changed. Is very important to be able to quickly assess what a change implies; if there is a solution, if any trade-offs have to be made affecting other systems and the implications on cost, lead-time, quality, risk etc.

The structure of the information and knowledge entered into the system for documentation and knowledge management is based upon the principle information model presented in Fig. 5.6, which is a derivate of the design description approach. The main page of the prototype system is depicted in Fig. 5.7. The principle of structuring the knowledge and information was to sub-divide the process into different tasks and functions at different levels to be able to support both a contextual meaning and the access to detailed descriptions. The Rationale class can be used to describe

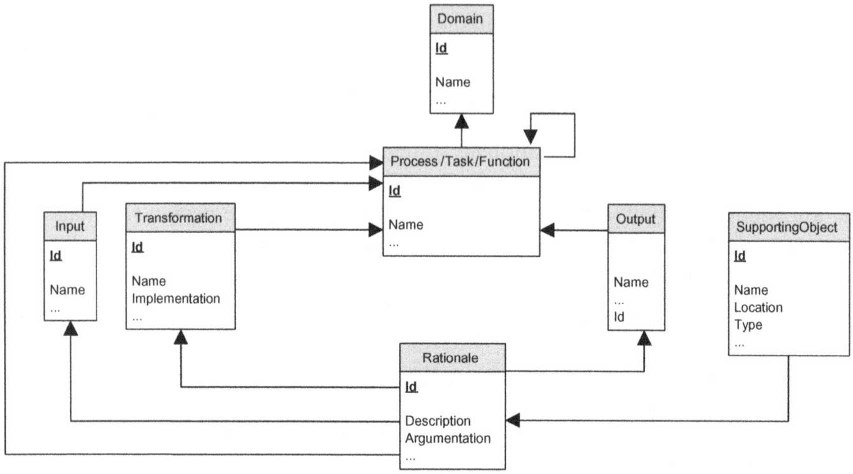

Fig. 5.6 Task-oriented design descriptions [37]

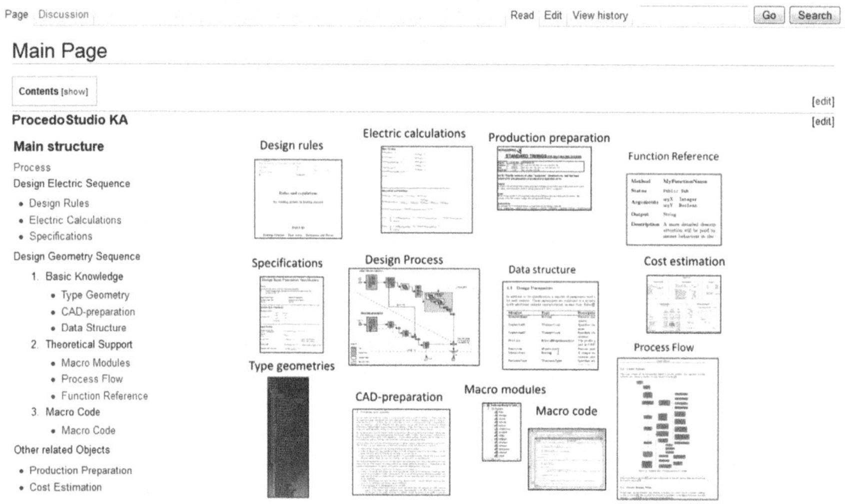

Fig. 5.7 Main page of the system [38]

why a Process/Task/Function exists or in detail describe the set of Input, the set of Output, and the transformation associated with a specific Process/Task/Function. The Supporting Object enables traceability to reports, protocols, guide-lines, standards, legalizations etc. by introducing relations between concepts. The information model also specifies the content of the wiki pages, i.e. it defines a template.

Wiki pages for the electric calculations, process planning and cost estimation were developed together with documentation of a macro generating wire lay-outs.

The documentation of the macro required a lot of effort due to its size and the amount of internal relations. To sub-divide the design process into design tasks and use existing applications to define executable files that automate each and one of these tasks was one fundamental principle of the system for automatic design of seat heaters. The applications used should preferably provide means to enter text and illustration for the purpose of, in natural language, describing the principles of the defined algorithms and rules. The macro was however programmed in CATIA VBA with no support to sub-divide the code into separate files and no support to attach illustrations to the code. The macro was, however, divided into modules that were copied into separate wiki pages and annotated with descriptions and figures.

The front page of the system, showing its architecture, is depicted in Fig. 5.7. On top is the entry page describing the principle design process from which the individual pages for each task/function can be reached. All input and output parameters managed by the design automation system are listed on two separate pages. A relation between output and input represents the use of an output parameter from one task as an input parameter in the execution of another task. One section is used to describe the principal transformation of input to output as implemented in the knowledge database together with a link to the file for its implementation. A page for a task/function also includes a section describing its rationale with references to more detailed documents stored in, for example, a database for documentation of development projects. Links are one of the most powerful tools in SMW and they were extensively used to create relations between different pages allowing for mapping between concepts. The search facilities also provide means to find and track knowledge and information in the domain which supports both detailed selection and aggregation of information.

Traceability is in focus, especially targeting the design automation system and it is suggested that a design automation system is to be founded on the principle that tasks or functions drive a parametric product model and that the associated knowledge is structured according to a process view. The reason for selecting this approach is that it allows for grouping of relations and statements, which operate across the product structure, at different levels. If those relations and statements were to be structured according to product items (i.e. parts or assemblies) they have to be placed on a level where the affected items are all included. For some design problems, this would lead to a very coarse subdivision where portions of the knowledge would be put on assembly levels and very high up in the product structure. This would counteract the objective of fine granularity and context. However, if traceability between reasoning (i.e. tasks and functions) in the knowledge base and product constructs (i.e. items and features) are required, it can be included. To achieve this, the Wiki database is suggested to be expanded with pages for each product item including its features that are affected by the Knowledge database. Traceability is then supported by declaring relations between individual Task/Function pages and the Item pages representing the product constructs that the task/function operates upon.

5.5.3 Case 3—A Decision-Oriented Approach

Company C is a global manufacturer of a wide assortment of products for transporting equipment by car; including roof racks, bike and water sport carriers, and roof boxes. There is a strategic need to considerably cut time and cost in development and manufacture of roof racks for cars. Every car model requires an individual adapted attachment consisting of a footpad and bracket. The ability to quickly launch a roof rack for a new model is considered as very important as it is common that a roof rack with accessories mounted is included as additional equipment when a new car is bought. The company is currently working on the development of a system that enables reuse of existing attachments. Parts of the system are used in operation globally at the company and have proven to reduce the development lead time, reduced the number of new designs and significantly reduced both the development and the tooling cost. The development has up till now focused on system development and technical aspects. Now, when the system is about to be fully integrated into the development process and become a strategic important tool for business success new needs have emerged concerning maintenance, expansion and traceability. In this setting the design descriptions were modelled as shown in Fig. 5.8.

The Design Rationale and the Decision classes are central in the model. Design Rationale objects carry general information (text and picture based descriptions) regarding a concept or an idea and is connected to a set of Decision objects. The

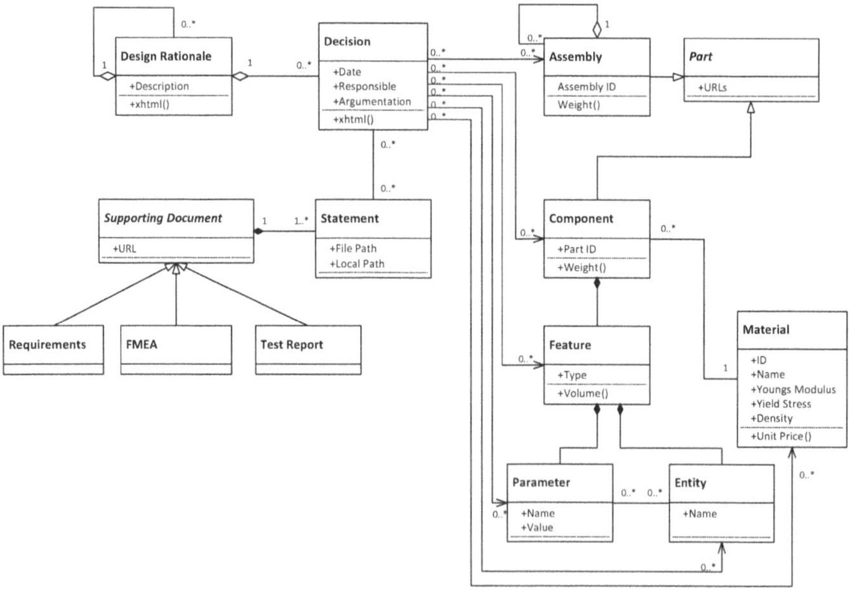

Fig. 5.8 Decision centred design descriptions. Adapted from [39]

Decision objects carry information regarding decisions taken during the design process such as date, who took the decision and, importantly, the argumentation behind the decision. The decision object also contains pointers to the information the decision and its argumentation was based upon. Since that information is scattered, these pointers have to be very specific, yet since that information is of many different types the pointers has to be very general. In the information model, pointers to documents that are not related to CAD-models are called Statements. Statement objects capture information typically found in test-reports, lists of requirements and FMEA documents and are not necessarily connected to items as presented in the first case example but can also regard processes as in the second case example. (Note that the pointers target sub-sets of the supporting documents, not entire files.) Since decisions made during the product development process affects the geometry of the product to a large extent it is possible to make the Decision objects point to Assembly, Part, Feature, Parameter, and Entity objects in CAD-models. A decision may also affect the material selection of components of the product or the tooling and in such cases, there are pointers to Material objects.

Decisions can be made throughout the entire product life-cycle and Design rationale should be captured at its origin using the proposed class diagram. This can be supported by providing an integrated digital environment. In such an environment, the tools for capturing design rationale as well as representing it are integrated to software already used by the engineers. The designers can perform design tasks in different software and applications and concurrently capture design rationale which is a great advantage of such an environment. Design rationale can be captured and represented in formats that the designers are already familiar and would prefer to work with. A prototype system was developed using the information model above and is based on the integration of SolidWorks, Microsoft Word, Microsoft Excel, and wiki pages. The reason of choosing this software was to that design rationale and design definitions are represented in different formats. SolidWorks was chosen as representative for 3D modelling, Microsoft Excel for rules definition and drawing tables, and Microsoft Word for specifications and textual content were selected. Besides the software, wiki pages delegating the explicit description as the fourth part of the system, was chosen. The system has been tested in product development and on tooling design. Figures 5.9 and 5.10 show how the design rationale system was integrated with SolidWorks and Microsoft Word.

5.5.4 Evaluation and Comparison

The three different cases presented in the previous sections have different scopes (see Fig. 5.11). Cases one and two focus on the documentation of a product platform consisting of different assets from which derivate solutions are generated. Traceability and access to design rationale are required to maintain and develop the platform over time. Case three, on the other hand, focuses on documentation of individual products. Traceability and access to design rationale supports re-use of solutions and change management throughout the product lifecycle.

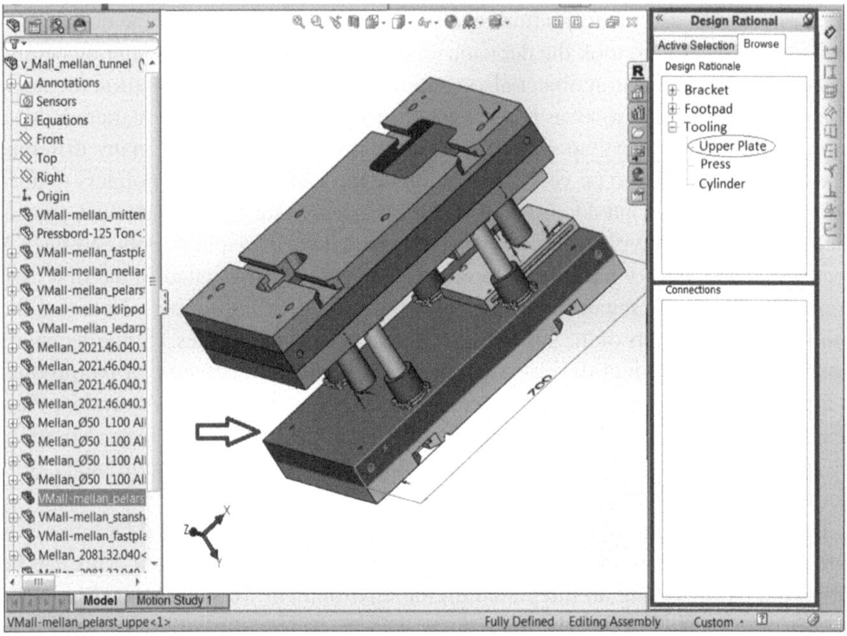

Fig. 5.9 A tooling in a CAD-software together with the design rationale management system as an add-in visible to the right [39]

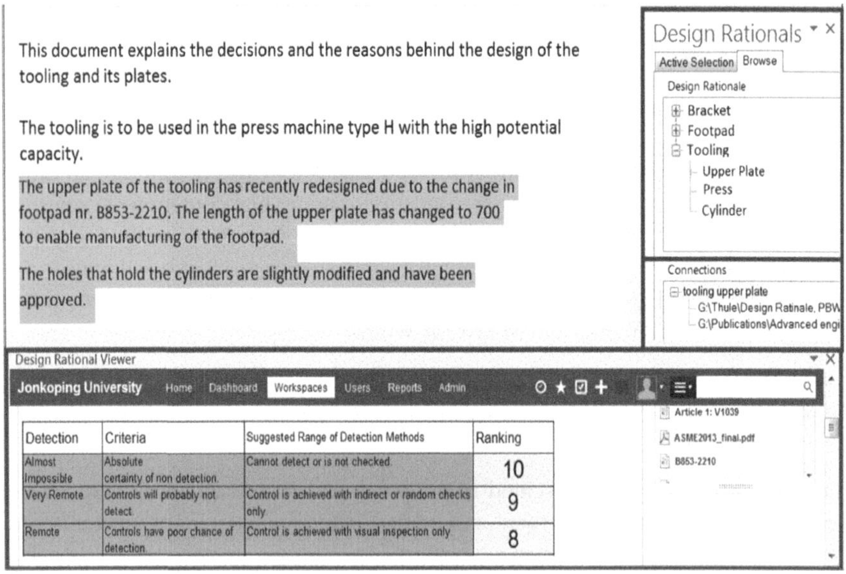

Fig. 5.10 A document in commercial software and wiki page addressing the design decisions made and failure effect ranking. A view of the design rationale management system as an add-in is visible to the right [40]

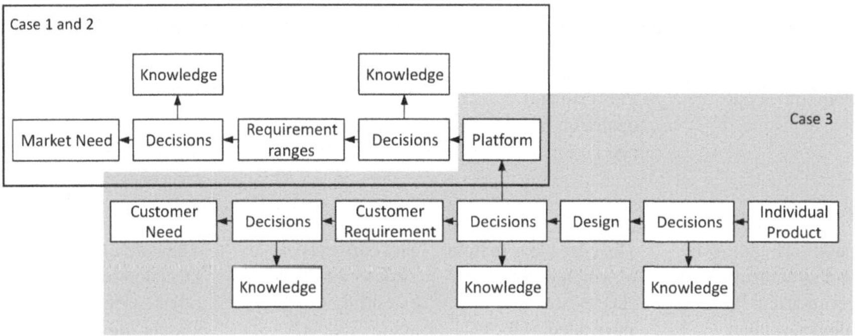

Fig. 5.11 The scope of the tree different industrial cases

Evaluations by practitioners have been done for case one and three, see [36, 39] for detailed information. Regarding case one, the evaluation was done qualitatively using a questionnaire with open-ended questions. Four stakeholders participated in the evaluation; the design automation manager, a project leader, an engineering designer, and a system developer. In summary, both the design automation manager and the project leader were convinced that the Description concept will be used for product families at the company. The system developer considered the concept as essential for traceability, knowledge reuse and a streamlined parametric and rule-based process. The use of the Description concept for product families was judged, by the engineering designer, to be more efficient than the existing support for documentation. Concerning the semantic part of SMW, there were different opinions regarding the cost-benefit. In a further investigation, the design automation manager wanted a focus on the way of documenting and storing information using the Description concept for product families while the project leader preferred to investigations on how the designers can manage such information in an efficient way.

Case three was evaluated through an evaluation session. Three stakeholders participated in the evaluation session; a product development manager, a chief engineer, and, a production engineer. The session started with a presentation of the system after which open-ended questions were individually answered by the participants. In summary, the method of linking the related information (statements) across software was judge as important and essential for tracing the effected knowledge when updates across the product documentation in the company are required. They all were optimistic that the benefits would outbalance the efforts. However, the prototype system has to be tested on a broader scale and then fine-tuned to turn it into an operational tool. Finally, the issue of maintenance of an additional system where pointed out as an important aspect not to be overlook.

In Sect. 5.4, a set of criteria has been stated. These criteria have guided the work of the overall framework that supports the development and implementation of tools and methods enabling engineer-to-order companies to work systematically with traceability. The three different cases are evaluated based on the criteria and individually compared in Table 5.1.

Table 5.1 Evaluation and comparison of the three cases

Case	1	2	3
Potential scope	Platform and upstream to market need	Platform and upstream to market need	Individual product and upstream to platform and customer need
Criteria			
Facilitate capturing and structuring of geometrical building blocks, related knowledge, and design rationale	The core class is Item to which other classes are associated. All classes have a set of attributes. These classes, attributes and associations act as templates that facilitates capturing and structuring of information and knowledge	The core class is Task to which other essential classes are associated. All classes have a set of attributes. These classes, attributes and associations act as templates that facilitates both capturing and structuring of information and knowledge	The core class is Decision to which other essential classes are associated. All classes have a set of attributes. The information model can capture a vast number of information formats, on different levels of granularity
Persistent storage and easy retrieval of knowledge	SMW is based on a server acting as a central storage container. Content can be browsed or searched for by queries	SMW is based on a server acting as a central storage container. Content can be browsed or searched for by queries	XML files were used in the prototype but can be migrated to database
Support visualization	Figures are supported by SMW	Figures are supported by SMW	Captured information is interactively highlighted in its resident software
Versioning control	Supported by SMW	Supported by SMW	If implemented in a database
Authorization control	Supported by SMW	Supported by SMW	If implemented in a database
Reside upon existing available tools	SMW is an extension to MediaWiki. Both are free and open-source. Links to files with supporting content can be included	SMW is an extension to MediaWiki. Both are free and open-source. Links to files with supporting content can be included	XML is a standardized way of structuring data. The underlying information model can be implemented in PostgreSQL, a wide-spread open-source database system

One main difference between the cases is constituted by the principles of capturing and structuring information and knowledge to support traceability and design rational management. The origin of the different principles are the characteristics of the product, the problem domain, and the design process. However, independently of the main approach used, the others can be supported by introducing additional concepts, if required. Another difference is the scope of the methods. Neither of them covers the whole meta-domain from a single product to individual needs and platform assets to market need. However, as they share the concept of platform assets it would be possible to merge them. A major part of the criteria depends on the foundation on which a system is implemented. A database has some essential functionality that can be used for structuring data, secure persistent storage, and manage versioning and authorisation. Case three has limited support for versioning and authorization, however, an information model exists which supports an implementation using a database.

5.6 Conclusions and Further Work

Traceability enables engineer-to-order companies to adapt their products to changing customer requirements, regional legacy, and new technologies. It also makes it possible to maintain and evolve the corporate product and process knowledge. To achieve traceability, i.e. the ability to follow the life of a knowledge component from its origins to its use, proactive work is required in each step of the product realization process. The work has to be expanded to not only focus on the product and manufacturing definition but also on the capturing of rationale. This proactive work includes answering questions regarding why the definitions are constituted the way they are, and what solutions were rejected and why.

5.6.1 Conclusions

A set of conclusions can be drawn upon the work presented in this chapter. One is that the product realization process in ETO oriented businesses commonly includes some kind of platform model with different assets that are used in the development of a single solution. However, this platform model may not be explicit, coherent, or systematically managed. The process also includes a number of stages where decisions are taken to deliver a perceived output based on available input and knowledge. Traceability is needed for many purposes, e.g. maintain products, provide unique spare parts, and tracking of parts. Decision is a core concept enabling links to be created between sets in different outputs. Traceability will then be supported and it would be possible to track all the decisions and their supporting knowledge of production, design, customer requirements, platform development, and market needs for an individual product instance. It can also be concluded that traceability must include design rationale to enable the understanding of why things where designed the way they are. When it comes to support in the existing development methods, the methods for knowledge modelling and the specific tools for knowledge management, the support is limited for traceability covering the scope that from a single individual customized product trace the specific decisions and knowledge used in the development process. The Design Description concept described in this chapter can be used to support the work of capturing and structuring the design rationale. Three examples were presented where design descriptions were modelled based on an item oriented, a task oriented and a decision-oriented perspective which show the generality of the design description concept. The three examples demonstrate how to use the design descriptions to enable traceability in platform, product design, and manufacturing development processes. Neither of three examples covers the whole meta-domain. Based on the fact that they share the concept of platform assets, the conclusion is that they can be merge and thereby create support for traceability from a single product to individual needs, and through platform assets to market need.

5.6.2 Further Work

Further work in this domain is required and an area of significant importance is methods that enable decision makers to easily capture design rationale in their daily work and map preceding decisions and knowledge sources independently of their digital environment. Another challenge is to facilitate the work of capturing and structuring of design rationale so it won't hamper the creative work and cause extension of lead-time or require additional resources. Design is of experimental character and a process of synthesis-analysis where solutions are rejected from one day to another. How to minimize the effort in the solution seeking and at the same time ensure that essential detailed information is not lost? Complex products are not the result of a single engineer. Many people are involved from different disciplines, using different methods and tools, on different sites and in different organisations. How to support everyone in capturing essential decisions and enable cross-disciplinary mapping? To some extent, free-text input is probably required as it is difficult to set up a system that is prepared for everything. But the question that follows is how quality then can be assured? In addition, the question of granularity is of importance. To create a link between two large sets of information is valuable to a limited extent if specific relations cannot be easily identified. How can mapping on a sufficient level be supported and guaranteed? On top of this comes security aspects. Companies are very restricted concerning access to project databases and product knowledge as it is a valuable company asset. Especially during development when it concerns products for the future business. How can the need for a complete map and detailed design rationale be combined with very limited access in the individual case? This touches upon the need to involve practitioners in an industrial setting even more in future activities. At this stage, the focus has been on the more technical side of the problem were issues concerning organisational structures, company processes, legacy systems, upscaling, and maintenance have been considered but not completely investigated, and this has to be done to ensure applicability and usefulness in practice.

Finally, to efficiently support traceability is balancing act between upstream efforts and downstream return on investment. However, it's hard to know what can be of value in the future and creativity should not be hampered by administrative work. On the other hand, building and sharing knowledge across individuals will support the relay race for new innovations and preventing the same mistakes to be repeated. This might be of even more importance with the increased deployment of value driven design that will lead to optimized individual solutions, the change in business models from providing artefacts to functions, and the increased focus on sustainability values and the utilisation of scares resources that undoubtedly will increase refurbishment, upgrading, and remanufacturing of products.

References

1. Hvam L, Mortensen NH, Riis J (2008) Product customization. Springer, London
2. Hjertberg T, Stolt R, Poorkiany M, Johansson J, Elgh F (2015) Implementation and management of design systems for highly customized products—state of practice and future research. In: Curran R, Wognum N, Borsato M, Stjepandić J, Verhagen WJC (eds) Transdisciplinary lifecycle analysis of systems. Advances in transdisciplinary engineering, vol 2. IOS Press, Amsterdam, pp 165–174
3. Biahmou A (2015) Systems engineering. In: Stjepandić J, Wognum N, Verhagen WJC (eds) Concurrent engineering in the 21st century: foundations, developments and challenges. Springer International Publishing, Cham, pp 221–254
4. Königs SF, Beier G, Figge A, Stark R (2012) Traceability in systems engineering—review of industrial practices, state-of-the-art technologies and new research solutions. Adv Eng Inform 26:924–940
5. Ouertani MZ, Baina S, Gzara L, Morel G (2011) Traceability and management of dispersed product knowledge during design and manufacturing. Comput Aided Des 43:546–562
6. Mohan K, Rames B (2007) Traceability-based knowledge integration in group decision and negotiation activities. Decis Support Syst 43(3):968–989
7. Tang A, Babar MA, Gorton I, Han J (2006) A survey of architecture design rationale. J Syst Softw 79(12):1792–1804
8. Elgh F (2011) Modeling and management of product knowledge in an engineer-to-order business model. In: Culley SJ, Hicks BJ, McAloone TC, Howard TJ, Chen W (eds) Proceedings of the 18th international conference on engineering design (ICED11). Design Society, Glasgow, pp 86–95
9. Tang A, Jin Y, Han J (2007) A rationale-based architecture model for design traceability and reasoning. J Syst Softw 80(6):918–934
10. Bracewell R, Wallace K, Moss M, Knott D (2009) Capturing design rationale. Comput Aided Des 41(3):173–186
11. Claesson A (2006) A configurable component framework supporting platform-based product development. Dissertation, Chalmers University of Technology
12. Amadori K, Tarkia M, Ölvander J, Krus P (2012) Flexible and robust CAD models for design automation. Adv Eng Inform 26(2):180–195
13. Cederfeldt M, Sunnersjö S (2003) Solid modelling with dimensional and topological variability. In: Folkeson A, Gralen K, Norell M, Sellgren U (eds) Proceedings of the 14th international conference on engineering design (ICED11). Design Society, Glasgow

14. Boart P (2007) The enabling of product information in the conceptual phase. Dissertation, Luleå University of Technology
15. Elgh F (2012) Decision support in the quotation process of engineered-to-order products. Adv Eng Inform 26(1):66–79
16. Johansson J (2011) How to build flexible design automation systems for manufacturability analysis of the draw bending of aluminum profiles. J Manuf Sci Eng 133(6)
17. Sunnersjö S (2012) Planning design automation systems for product families—a coherent, top down approach. In: Dorian M, Mario S, Neven P, Nenad B (eds) Proceedings of DESIGN 2012, the 12th international design conference, Dubrovnik, Croatia, pp 123–132
18. Cederfeldt M (2007) Planning design automation: a structured method and supporting tools. Dissertation, Chalmers University of Technology
19. Stokes M (2001) Managing engineering knowledge—MOKA: methodology for knowledge—based engineering applications. Professional Engineering Publishing Limited, London
20. Lisandrin P, van Tooren M, (2002) Generic volume element meshing for optimization applications. In: 9th AIAA/ISSMO symposium on multidisciplinary analysis and optimization, Atlanta, 4–6 Sept 2002
21. La Rocca G, Krakers L, van Tooren M (202) Development of an ICAD generative model for blended wing-body aircraft design. In: 9th AIAA/ISSMO symposium on multidisciplinary analysis and optimization, Atlanta, 4–6 Sept 2002
22. van Tooren M, La Rocca G, Krakers L, Beukers A (2003) Design and technology in aerospace. Parametric modeling of complex structure systems including active components. In: 13th international conference on composite materials, San Diego 14–18 July 2003
23. Curran R, Verhagen WJC, van Tooren M, van der Laan TH (2010) A multidisciplinary implementation methodology for knowledge-based engineering: KNOMAD. Expert Syst Appl 37(11):7336–7350
24. Friedenthal S, Moore A, Steiner R (2012) A practical guide to SysML. Elsevier, Amsterdam
25. Schreiber GT, Akkermans H (2000) Knowledge engineering and management: the CommonKADS methodology. MIT Press, Cambridge
26. Epistemics (2014) PCPACK. http://www.epistemics.co.uk/Notes/55-0-0.htm. Accessed 24 June 2014
27. Haug A, Hvam L, Mortensen NH (2009) Implementation of conceptual product models into configurators: from months to minutes. In: Proceedings of MCPC—the world conference on mass customization and personalization, Helsinki, 4–8 Oct 2009
28. SemanticMediaWiki (2016) Semantic MediaWiki. https://www.semantic-mediawiki.org/wiki/Semantic_MediaWiki. Accessed 6 Dec 2016
29. Haug A, Degn A, Poulsen B, Hvam L (2007) A prototype of a documentation system that supports the development and maintenance of product configuration systems. WSEAS Trans Inf Sci Appl 4(5):1048–1055
30. Stjepandić J, Verhagen WJC, Liese H, Bermell-Garcia P (2015) Knowledge-based engineering. In: Stjepandić J, Wognum N, Verhagen WJC (eds) Concurrent engineering in the 21st century: foundations, developments and challenges. Springer International Publishing, Cham, pp 255–286
31. Verhagen WJC, Bermell-Garcia P, van Dijk REC, Curran R (2012) A critical review of knowledge-based engineering: an identification of research challenges. Adv Eng Inform 26(1):5–15
32. Elgh F, Cederfeldt M (2007) Concurrent cost estimation as a tool for enhanced producibility—system development and applicability for producibility studies. J Prod Econ 109(1–2):12–26
33. Johansson J (2007) A flexible design automation system for toolsets for the rotary draw bending of aluminium tubes. In: ASME 2007 international design engineering technical conferences and computers and information in engineering conference, Las Vegas, 4–7 Sept 2007, pp 861–870
34. Elgh F, Johansson J (2014) Knowledge object—a concept for task modelling supporting design automation. In: Cha J, Chou SY, Stjepandić J, Curran R, Xu W (eds) Moving integrated product development to service clouds in the global economy. Advances in transdisciplinary engineering, vol 1. IOS Press, Amsterdam, pp 192–203

35. Johansson J (2015) Howtomation© suite: a novel tool for flexible design automation. In: Curran R, Wognum N, Borsato M, Stjepandić J, Verhagen WJC (eds) Transdisciplinary lifecycle analysis of systems. Advances in transdisciplinary engineering, vol 2. IOS Press, Amsterdam, pp 327–336
36. Elgh F, Poorkiany M (2012) Supporting traceability of design rationale in an automated engineer-to-order business model. In: Dorian M, Mario S, Neven P, Nenad B (eds) Proceedings of DESIGN 2012, the 12th international design conference, Dubrovnik, Croatia, pp 23–132
37. Elgh F (2014) Automated engineer-to-order systems: a task-oriented approach to enable traceability of design rationale. Int J Agile Syst Manag 7(3–4):324–347
38. Nan J, Li Q, (2012) Design automation system: supporting documentation and management. Master thesis, Jönköping University
39. Poorkiany M, Johansson J, Elgh F (2016) Capturing, structuring and accessing design rationale in integrated product design and manufacturing processes. Adv Eng Inform 30(3):522–536
40. Poorkiany M, Johansson J, Elgh F (2014) Supporting tooling design of customized products by instant access to design rationale. In: Stahre J, Johansson B, Björkman M (eds) The 6th international swedish production symposium, Gothenburg, 16–18 Sept 2014

Chapter 6
Decision Analysis and Interface Management in Systems Engineering

John C. Hsu

Abstract The crosscutting technical management process facilitates both the systems design and product realization processes. The eight sub-processes within the crosscutting technical management process are: technical planning, requirements management, interface management, technical risk management, configuration management, technical data management, technical assessment, and decision analysis. The technical management processes make the link between project management and technical team. Subsequently, individual members and tasks are integrated into a functioning system that meets cost and schedule pre-requisites. The crosscutting functions serve to execute project control on the apportioned tasks. In this chapter, we will put our focus to explain decision analysis and interface management. Decision analysis is the process of making decisions based on research and systematic modeling of tradeoffs. The objective of a decision analysis is to discover the most advantageous alternative under the circumstances. Decision analysis may also require human judgement and is not necessarily completely machine driven. In detail, we show how to conduct the trade study. While interfaces are connection points between parties or elements, interface management provide a systematic methodology to handle with multiple parties or technical elements. Implementing an interface management process on a project identifies critical interfaces, streamlines communication, and monitors ongoing work progress while mitigating risks. Under interface management, we provide interface definition, identification and interface management tools.

J. C. Hsu (✉)
AIAA, Reston, USA
e-mail: john.hsu@csulb.edu

INCOSE ESEP, San Diego, USA

The Boeing Company, Long Beach, USA

California State University Long Beach, Long Beach, USA

System Management and Engineering Consulting Services, Cypress, USA

UK Royal Academy of Engineering, London, UK

Queens University, Belfast, UK

J. Stjepandić et al. (eds.), *Systems Engineering in Research and Industrial Practice*,
https://doi.org/10.1007/978-3-030-33312-6_6

Keywords Systems engineering · Functional analysis and allocation · Trade studies · Interface management

6.1 Introduction

The technical management processes are used to establish and evolve technical plans for the project, to manage communication across interfaces, to assess progress against the plans and requirements for the system products or services, to control technical execution of the project through to completion, and to aid in the decision-making process [1]. The technical management processes are the bridges between project management and the technical team. In this portion of system engineering, eight crosscutting processes provide the integration of the crosscutting functions that allow the design solution to be realized. Every member of the technical team relies on technical planning; management of requirements, interfaces, technical risk, configuration, and technical data; technical assessment; and decision analysis to meet the project's objectives. In this chapter, we will put our focus to explain decision analysis and interface management.

6.1.1 Backround of Decision Analysis

The term decision analysis was coined in 1964 by Ronald A. Howard, professor of Management Science and Engineering at Stanford University. Decision analysis (DA) is an systematic, quantitative and visual approach to supporting decision makers conduct rational decisions [2]. In opposite to descriptive view of decision-making, it is the normative field, called also decision engineering, and looks like a calculator to make a decision [3]. It incorporates systems engineering and decision theory, enhanced by the ability of modern computation to build value-based models and accomplish computations needed to deal with complexity in the number of factors, uncertainties, and dynamics [4]. It also includes the processes for reaching good decisions with real people and organizations, and gaining their commitment to carry them out. Furthermore, it comprises a systematic procedure for transforming opaque decision problems into transparent decision problems by a sequence of transparent steps.

DA is not a single approach, but a discipline with underlying principles and procedures that are adapted to diverse situations. The decision problem is structured by identifying alternatives, one of which must be decided upon; possible events, one of which occurs thereafter; and outcomes, each of which results from a combination of decision and event [1]. DA uses a variety of tools like decision theory, influence diagrams, system dynamics, game theory, to generate good decisions in new product development, business strategy, space system safety, etc. [4]. Such tools aim to evaluate all relevant information to aid in the decision making process and incorporates aspects of psychology, management techniques and training, and economics. DA is often used to assess decisions that are made in the context of multiple variables and which have many possible outcomes or objectives. It can be used by individuals or groups attempting to make a decision related to risk management, capital investments and strategic business decisions. A graphical representation of alternatives and possible solutions, as well as challenges and uncertainties, can be created on a decision tree or influence diagram. To better understand the components of decision analysis, it is important to recall the six elements of decision quality [3].

The core of the decision analysis process is the elicitation or synthesis of the decision basis. The basis has three parts: the choices or alternatives the decision-maker faces, the information that is relevant, and the preferences of the decision-maker (Fig. 6.1) [2]. The alternatives may be readily apparent or may be generated as a major activity of the formulation using tools. By information, we mean any models, relationships or probability assignments that may be important in characterizing the connection between decisions and outcomes. The models could be complex and dynamic, or very simple. The uncertainty that remains would be characterized by probability assignments. The preferences of the decision-maker would be represented in at least three dimensions. The decision-maker would have values on one outcome as opposed to another, and time preference considerations on outcomes now versus outcomes later. Finally, the decision-maker would have a risk preference governing outcomes with different degrees of certainty.

As a typical transdisciplinary methodology [5, 6], decision analysis combines social, engineering, and natural science knowledge to support complex real world decision processes in almost all areas of the human life [7]. The frequent application of DA can be found in the following areas: decision analysis networks [8], engineering management [9], life cycle sustainability assessment [10], critical infrastructure vulnerability [11], and interactive visualization for group working processes [12].

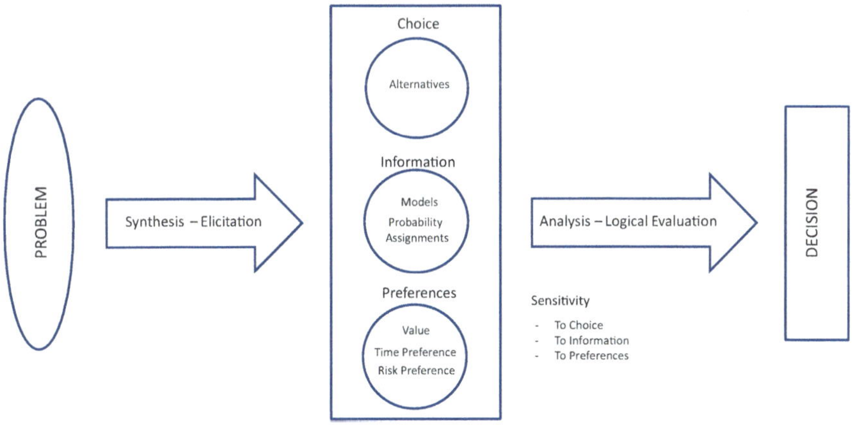

Fig. 6.1 Elicitation and evaluation of the decision basis [1]

6.1.2 Backround of Interface Management

The management and control of interfaces is crucial to successful programs or projects. Interface management (IM) is a process to assist in controlling product development when efforts are divided among parties (e.g., Government, contractors, geographically diverse technical teams, etc.) and/or to define and maintain compliance among the products that must interoperate [1]. During product integration, interface management activities would support the review of integration and assembly procedures to ensure interfaces are properly marked and compatible with specifications and interface control documents. The interface management process has a close relationship to verification and validation.

Interface management has been developed in 1970ties as a discipline within the the project management: planning, coordinating, and controlling the work of others at project interfaces [13]. Today's systems include various types of interfaces, such as hardware interfaces, software interfaces, signaling interfaces, service interfaces, data interfaces, and application program interfaces. Interfaces become more and more critical and meanwhile connect geographically distributed systems. The importance of interfaces requires special attention and appropriate approaches to design, manage, and operate interfaces within a configuration management process. So a Interface Modelling Communication Language (ICML) has been developed for space applications [14].

Extremely large, complex projects with numerous geographically distributed stakeholders discover substantial risks related to the interfaces among stakeholders [15]. This is particularly true during project definition and design, where despite discrete deliverables across the interfaces, decisions require some iteration. Managing interfaces has become feasible with the advent of internet and electronic product and process-management systems to the extent that full-time interface-management positions exist in practice. However, IM lacks formal structure, integration with critical-path-method scheduling, and methods to identify high-risk interface points (IPs) [15].

Specific approaches are necessary on how an integrated approach creates cross-discipline data relationships that provide teams with better information for decision making and progress tracking, and thus helps to improve project performance. In particular, specific attention is paid to interface management and its impact on the following well-established practices [16]. Poor interface management can lead to time overruns and negatively impact the relationship between time, cost, scope, quality, and resources. The trend of streamlined organizations with reduced staffing often requires companies to rely on multiple parties to complete a project [17]. IM problems lead to scope creep in which the balance between project scope, time, cost, quality, and resources is upset.

Currently, the terms interface and interface management in the context of interorganisational relationships between suppliers and customers are under intensive discussion. Central issues are the emphasis placed upon the coordination of activities at business interfaces, and the conviction that the concept can make business operations more efficient and thus more effective. That interface management in supplier–customer relationships is discussed in various academic disciplines and characterized by various thematic approaches makes this field complex. A study has investigated the state of interface research, identifying select topics that promise further significant research in the field of interface management in supplier–customer relationships [18].

From the product management perspective, there are multiple interfaces viewed from the perspective of the interaction between brand and stakeholders and each interface has conflicts and other problems which affect the brand relationships. The importance of stakeholders in the brand relationships is analyzed and the interfaces between major stakeholders are regarded as the most influential to the brand [19]. The reasons for the conflicts are also analyzed and some brand relationship interface management methods or strategies based on interaction of brand and stakeholders are established. One of the best opportunities to deliver sustainable principles occurs during the product development process. To be considered truly sustainable, a product must be designed respecting three dimensions: economic, environmental and societal, in a systemic approach. From this perspective, a paper gives the overview of the sustainable product design and interfaces with its supporting processes, taking advantage of the knowledge development process [20]. Interface management is articulated through analyzing product architecture, with a view to interpreting the component interfaces in a more productive way to achieve better modularity [21, 22]. Experience from past single-system integration programs dictates the importance of disciplined and effective interface management in a program's execution of systems engineering process. System-of-systems (SoS) integrations are substantially more challenging than single-systems integrations due to the complexity and sheer number of the interfaces involved. The interfaces are recognized as a major risk area for SoS development programs. Classical systems engineering processes must be modified to mitigate these risks [23]. Advanced systems have common characteristics of complexity as the level of their demanded emergent capability and the resulting interfaces among their components increase. These characteristics make it difficult to manage the interfaces and the failure of the management can lead to the failure of development projects. Model-based systems engineering approach is a successful way to facilitate the interface management [24, 25]. Lessons learned allow a better understanding of the challenge to manage the diversity of interfaces between different actors and different forms of knowledge. Examples are given in area of plant engineering [26], tunnel enigeering [27], rail road engineering [28] and manufacturing networks [29].

The remainder of this chapter is structured as follows. Section 6.2 Decision Analysis will explain the trade study process. Section 6.3 Interface Management leads across in-house disciplines, supplier, and customer. Learn the interfaces, management tools, and open system interface management.

6.2 Decision Analysis

There are several methods and tools to assist the decision analysis. Trade Study is the most common way. Refer to Fig. 2.6. Trade study is one of the Systems Analysis and Management methods to support the Requirements Management, Functional Analysis and Allocation, and Systems Synthesis. Decision Analysis is part of the System Synthesis as shown in the figure.

The purpose of trade studies is to make better and more informed decisions in selecting best alternative solutions and identify and execute trade-offs among requirements, design, schedule, and cost. The results after trade-offs is a compromised solution. It cannot totally satisfy one criterion. Trade studies are required to support decisions throughout the systems engineering process and project life cycle. Trade studies can assist in selecting system concepts, designs and solutions; establishing system, subsystem and component configurations; selecting components, techniques, services and facilities; supporting materials selection and make-or-buy; and selecting proposed changes. Trade studies process is a systemization of thought. Through the process can clarify options, problem structure, and available trade-offs. It can also improve communication of ideas and professional judgment within the organization as well as rationale for action to others. Correct execution of trade studies process will provide confidence that all available information has been accounted for in a decision. A significant contribution of trade studies is to prevent program/project management from committing too early to a design. Chief engineers and senior-level engineers often make this mistake by making a quick decision based on their past years of experience. But the past experience is not applicable to a new project with new technologies trying to satisfy new customer requirements. Another common mistake by these engineers is not making a balanced decision. They often ignored areas, such as training, spares, and maintenance which are the major part of life cycle cost for operators. Trade studies will help make a balanced decision in considerations of design, cost, reliability, maintainability, producibility and supportability. It is a key tool in the development of designs that meet customer requirements. Trade studies are usually carried out by a team, rarely by an individual except it is for personal business. A trade study leader has to be a facilitator with neutral attitude toward the outcome of trade studies. The leader cannot be a stakeholder associated with any of the proposed solutions since the trade study can be easily tailored.

6.2.1 Trade Study Process

There are seven (7) steps:

1. Define evaluation criteria
2. Identify weights for evaluation criteria
3. Identify alternatives
4. Define scoring criteria for each evaluation criterion

5. Score alternatives against evaluation criteria
6. Calculate ratings for alternatives
7. Assess uncertainties.

6.2.1.1 Define Selection Criteria

The selection of evaluation criteria shall be based on the purpose of trade-off studies. For engineering design, the evaluation criteria must reflect all design requirements. Additionally, the criteria must also reflect business objectives. Is it for system's conceptual design or the component level design, such as, a valve, or a pump, etc.? The cost and risk may or may not be considered in trade study since the result from trade study is only a recommendation. If cost and risk are not included, they must be considered in the final decision in conjunction with the trade study recommendation. In reality, most of the final decisions are dominated by political influence.

Each evaluation criterion should be owned by a stakeholder. The stakeholders of the evaluation criteria are trade study team members. All the criteria should be independent and unique of each other. The overlapping criteria or similar criteria will create unbalanced trade-offs since these criteria will have more weights to tip the final score calculations in favour of these overlapping criteria. More importantly, all the team members (stakeholders) should fully understand all the evaluation criteria, not just the stakeholder's own criteria.

6.2.1.2 Identify Weights for Evaluation Criteria

Trade study is a discriminative process, i.e., there are preferences among the alternatives to choose the most appropriate one to meet the purpose of trade studies. If there is no preference, just like when you want to purchase a car and you like more than one car, 2 or 3, etc., equally. There is no need to perform trade studies. You simply buy all the 2 or 3 cars. Assigning weight to each evaluation criterion sets the priority for each criterion. These priorities will help the trade study process to select the most appropriate alternative. Weight assignment for each evaluation criterion should be objective. Weights are determined by team members. They should be objective including their own evaluation criteria. Weight assignments should be agreed unanimously by all team members. Weight assignment to each criterion does not mean the assignment of importance to each criterion. All the evaluation criteria are important; otherwise, they should not be included. But they are assigned based on priority. It is the assignment of precedence, ranking or giving more attention.

6.2.1.3 Identify Alternatives

As presented in Sect. 2.4.3, several system architectures could be established through functional analysis and allocation. These architectures are alternatives.

If there are no pre-determined alternatives, the widest range of different alternatives should be selected from the widest sources including: customer, brainstorming by all team members, experience, and suppliers, etc. One suggestion is to start with the existing (predecessor) system, if there is one, as a baseline. Postulate alternative concepts that replace one or more of the subsystems. If there is no predecessor, generate alternatives by using brainstorming method. Vary the chosen subsystem one at a time or in combination of more than one subsystem. If appropriate, modifying architectures is another way of creating alternatives.

When there are too many alternatives, we need to down select alternatives quickly to only a few. To perform a more accurate trade study, the number of alternatives is considered should not exceed a comfortable limit. The limiting number is most likely to be in the range of 2–8. Some people may prefer a narrower range. If the potential alternatives exceed the limiting number, Kepner-Tregoe method [30] can be employed to reduce a large number of alternatives to a comfortable limit. The evaluation criteria are classified by MUSTs and WANTs. MUSTs are Go/No Go criteria. WANTs are criteria against which the alternatives are evaluated. Screen all alternatives through the MUSTs. A set of MUST requirements needs to be developed that a large number of alternatives can be eliminated quickly. An example is shown in Table 6.1 [31], for common tool selection at The Boeing Company. Each of the potential candidate tools will be evaluated against these MUST requirements. The score will be a "GO" or "NO GO" against each MUST requirement. Only those potential tools that score a "GO" against each MUST requirement will be qualified to be on the final candidate alternative list for further trade study against the detailed evaluation criteria. After a large number of alternatives have been reduced to a comfortable level, then we can apply the evaluation criteria rules discussed in Sects. 2.5.1.1 and 2.5.1.2. For Kepner-Tregoe method, evaluation criteria are WANTs. Compare down-selected alternatives against the WANTs.

Table 6.1 Example for MUST requirements

The tool MUST provide complete traceability from requirements to verification, and lower level specifications
The tool MUST be capable of allowing multiple users on separate platforms to access requirements
The tool MUST be capable of interfacing with other software applications

Table 6.2 Scoring matrix

Evaluation criteria	High score	Medium score	Low score
1	Define score criteria	Define score criteria	Define score criteria
2	Define score criteria	Define score criteria	Define score criteria
3	Define score criteria	Define score criteria	Define score criteria
4	Define score criteria	Define score criteria	Define score criteria
5	Define score criteria	Define score criteria	Define score criteria
More or less pending how may criteria to be evaluated			

6.2.1.4 Define Scoring Criteria for Each Evaluation Criterion

This is the most important step in trade studies. The heart of trade study process is scores that will determine the final outcome of selecting the most appropriate alternative. We need to ensure that the scoring criteria against each alternative are explicit, simple and straight-forward and will not create any ambiguities. Even a clerical person with high school education can understand it clearly and use the scoring criteria to score. It should not need a Ph.D. level expert to run complicated calculations or running a computer simulation model to score. Well-defined scoring criteria can provide consistencies throughout time by different teams with the same kinds of experts. It means that a different team using the same scoring criteria at a later date that could be months or years, can still select the same alternative.

There are several ways to setup the scoring criteria. A common way is to use a scoring matrix, as shown in Table 6.2 [32]. It is a table listing the high, medium, and low scoring criteria for each evaluation criterion.

6.2.1.5 Score Alternatives Against Evaluation Criteria

After finishing the above steps, this is a relatively easy step. Each alternative will be presented by its creator. The creator should present his (or her) alternative clearly with diagrams. We have to ensure that every trade study team member understands all the alternatives. The team members use the scoring criteria to score each alternative against all the evaluation criteria.

6.2.1.6 Calculate Ratings for Alternatives

Use the scoring criteria to score each alternative based on the team members' consensus for high, medium or low score (referring to Table 6.2). The scoring results, calculations, and rating are tabulated in Table 6.3 [33]. The scores obtained from the scoring criteria matrix are called raw scores (RS), as shown in Table 6.3. The

Table 6.3 Trade study template

Evaluation criteria		Opt. #1		Opt. #2		Opt. #3		Opt. #4	
	Wt	RS	WS	RS	WS	RS	WS	RS	WS
1									
2									
3									
4									
5									
6									
Total									
Normalization									

weighted scores (WS) are calculated by assigned weight values multiplying raw scores. When the scoring for all the evaluation criteria is completed, the total score for each alternative can be calculated by adding all the weighted scores (WS). The alternative with the highest total score is the winner (chosen alternative). The scores for alternatives can also be normalized. Normalization is each alternative's total score divided by the perfect score. It will be demonstrated in the example below. The highest (normalization) percentage is the winner. The chosen alternative from the trade studies is only for recommendation. The final decision will be determined by the highest key individual or group with other considerations, such as, cost and risk if not included in the trade studies evaluation criteria, and politics if applicable.

6.2.1.7 Assess Uncertainties

When the scores between alternatives are too close to decide without any doubts, sensitivity analysis may be needed to assess the uncertainties. One way is to change the weight value one at a time to test the impact to outcome until an alternative is obviously the winner. It depends on how the weight values are assigned? If it is one to five and if there are five (5) evaluation criteria, all the possible combinations will be 5 to the fifth, which are 3125. It is not difficult to make 3125 computer runs but analyzing the results will be an enormous task. Another way is to use judgment to vary the sensitive weight(s) for the key evaluation criteria (or criterion) to evaluate the impact. No matter which parametric studies are used, the ultimate goal is to remove the uncertainties to select an alternative beyond any doubts.

6.2.2 Trade Study Example

A trade study example is shown in Table 6.4 [33]. A hydraulic valve was in a location difficult to apply torque wrench. Either over-torque or under-torque will cause hydraulic fluid leakage. Then it needs to be re-installed. To avoid the nuisance of re-installation problem, three (3) alternative designs were proposed. A trade study team was formed with seven (7) members. One is a systems engineer and the other six (6) were the stakeholders of the evaluation criteria. Team members agreed that the seven (7) evaluation criteria were important to be included for trade study evaluation. These criteria were clearly stated. The weight assigned values have to be unanimously agreed among team members. As disused in Sect. 2.5.1.2, they are priorities assignment. It is not easy to reach consensus among team members since each member (stakeholder) wants to have his (or her) criterion higher priority.

The next major task is to determine the scoring criteria for each evaluation criterion. Numerical scoring values need to be determined. The author has three (3) suggestions: First, the scores between High, Medium and Low should have a gap in between to give a bite to avoid the close scoring among alternatives; Secondly, the scores distribution should be linear to avoid double awards to the high score and

Table 6.4 Scoring criteria

Evaluation criteria	5					3					1				
	Wt	RS	WS	RS	WS	RS	WS	RS	WS						
1															
2															
3															
4															
5															
6															

Table 6.5 Trade study example

Evaluation criteria		Opt. #1		Opt. #2		Opt. #3	
	Wt	RS	WS	RS	WS	RS	WS
1. Installation impact	5	5	25	1	5	3	15
2. Design impact	3	1	3	1	3	3	9
3. Spares and interchangability impact	4	1	4	1	4	3	12
4. Weight increase	1	3	3	3	3	3	3
5. Maintainability	3	1	3	1	3	5	15
6. Supplier cost impact (to company)	4	1	4	1	4	3	12
7. Implementation risk	5	3	15	3	15	5	25
Total	125		57		37		91
Normalization			46%		30%		73%

Table 6.6 Mouse trap requirements traceability matrix

Evaluation criteria	5	3	1
1. Installation impact	Better	Same	Worse
2. Design impact	No impact	Design ≤ 100 h	100 h < Design
3. Spares and interchangability impact	No impact	Impacts spares only	Not interchangeable and impact spares
4. Weight increase	0 lb	0.1 lb < Wt ≤ 1 lb	1 lb < Wt ≤ 2 lbs
5. Maintainability	Better	Same	Worse
6. Supplier cost impact (to company)	Zero cost	0 < C ≤ $100 K	$100 K < C
7. Implementation risk	Low	Medium	High

double penalties to the low score; and Thirdly, do not use wide gap between high, medium, and low scores, such as, 20, 10, 1, or even 10, 5, 1. It is seen in the example Table 6.4, the high, medium, and low scores are 5, 3, and 1, respectively.

As discussed in Sect. 2.5.1.4, we need to keep the scoring explicit, clear, and easy to score. Let us discuss each one separately.

No. 1, "Installation Impact"—It is difficult to assign a numerical value to the installation impact. But, it is easy for team members to compare the new designs with the existing design. If it is better than the existing design, high score is 5. If it is the same, the medium score is 3 and 1 if it is worse than the existing design.

No. 2, "Design Impact"—Since this hydraulic valve is contracted to an outside vendor, the high score 5 is for no design impact to the company, i.e., the company engineer(s) does not have to modify the envelope drawing; medium score 3 if the company engineer(s) has to spend 100 h or less to modify the drawing; and low score 1 if the company engineer(s) has to spend more than 100 h.

No. 3, "Spares and Interchangeability Impact:—This scoring criterion covers two areas: spares for the new design parts; and if they are interchangeable between existing design parts and new design parts. Both have no impacts, it is a high score of 5, 3 if only spares impact and 1 if both spares and interchangeability impacts.
No. 4, "Weight Increase"—The system is sensitive to weight; therefore, the less weight increase caused by the new designs the better for the system. This criterion can score numerically. 0 lb. increase is 5; greater than 0 and equal to/less than 1 lb. increase is 3; and greater than 1 and equal to/less than 2 lbs. increase is 1. The team members may be confident that the weight increase for the new designs is less than 2 lbs.; however, this scoring criterion implies that there is a 0 score if the weight increase is greater than 2 lbs.
No. 5, "Maintainability"—This criterion is difficult to score numerically but it has to be explicit and easy to score. It is similar to No. 1 scoring criterion that if the maintainability for new designs is better than the maintainability of the existing design, it is a high score of 5. If the maintainability for new designs is the same as that of the existing design, it is a medium score of 3; and 1 if it is worse than the existing design.
No. 6, "Supplier Cost Impact to the Company"—Since this valve is contracted to an outside vendor for re-design, if the vendor does not charge any money to the company; it is a high score of 5. If the charge is equal to/less than $100,000, it is a medium score of 3 and 1 if the charge is greater than $100,000.
No. 7, "Implementation Risk"—This scoring criterion is like playing golf, the lower the value the better the score. The low risk is a high score of 5, medium risk 3, and high risk 1.

After the three (3) proposed designs have been presented clearly and thoroughly to the team members, the members will use the scoring criteria to score each alternative against each evaluation criterion. The scores are listed as RS (raw scores) and WS (weighted score). WS (weighted score) is calculated by multiplying RS and weight value assigned for each evaluation criterion, for example, under Option 1, No. 1 evaluation criterion: RS is 5; WS is 5 (RS) times 5 (weight value) = 25. No. 2 evaluation criterion: RS is 1; WS is 1 (RS) times 3 (weight value) = 3, etc. The total scores for each proposed design are obtained by adding all the WSs. As it can be seen that the scores for Options 1, 2, and 3 are 57, 37, and 91, respectively. Therefore, Option 3 has the highest score, the winner. It can be normalized by dividing the score of each proposed design by the perfect score of 125 as shown in Table 6.4. The perfect score is to score the highest RS of 5 for each evaluation criterion. The WS for each criterion is the weight value times 5, then add all the WSs. It can also be calculated by adding all the weight values together and then multiplied by 5. The normalization for Options 1, 2, and 3 are 46%, 30%, and 73%, respectively. The highest normalization is, again, the Option 3.

The scores among the proposed designs have wide gap; therefore, parametric studies to assess uncertainties are not needed. The obvious winner is Option 3. Usually, the outcome from trade studies is only a recommendation. Cost, risk and politics will be included for final decision. But in this example, cost and risk have been included as part of the evaluation criteria. There is no political consideration for such a small item as valve. The final decision is to design Option 3.

6.3 Interface Management

Referring to Fig. 2.6, the Systems Engineering Process, interface management is one of the Systems Analysis and Management methods to support the main processes of Requirements Management, Functional Analysis and Allocation, and System Synthesis. The interface subject has been touched in Sect. 2.4.2 elements of functional analysis between two functional blocks with directional sequence and associated information/data definition. It is the product from functional analysis, but needs to implement this identified interface physically or by means of software. Once the interface requirement is identified, it needs to be defined and controlled. Interface can ensure clear communication between Program leadership, IPT's (Integrated Product Team's), subcontractors or other divisions on system interface changes. Interface control can reduce the probability of failure during the integration activity for the program. It starts at the end of the initial concept phase and continues throughout the system's life cycle.

The objective of interfaces is to ensure compatibility between interrelated system elements and provide an authoritative means of controlling the design of interfaces. Interface management supports the system verification and integration processes. The greatest leverage in system architecting is at the interfaces. The greatest dangers are also at interfaces [34]. A good system is the one with fewest number of misfits. Only seamless interfaces between the parts can insure a perfect fabrication and assembly for verification. Successful verification of components and seamless interfaces between components can insure successful product integration.

6.3.1 Interface Definitions

The following definitions are defined.

Interface—A boundary between two system elements.

Physical interface requirement—Requirement pertaining to the physical boundary between two system elements, for example, physical tolerances.

Functional interface requirement—Requirement pertaining to the functional quantity delivered by one system element and received by the other system element. All interfaces have both physical and functional requirements.

Interface Scope Sheet—An agreement between two parties to the interface defining the interface and the scope of the responsibilities pertaining to the interface.

Interface Control Document (ICD)—A technical document defining in detail the functional and physical characteristics of the interface.

6.3.2 *What Are Interfaces?*

Interfaces are places at which independent elements meet and communicate with each other. For examples, Support—two components bolted together, Power—one element supplying power to another, and Command—one element commanding another. There are two types of interfaces. One is Functional Interfaces (What It Does), for example, Thermal Control, Command, and Structural Support. The other one is Physical Interfaces (What Does It), for example, Cooling Water, Electrical Signal, and Fasteners. Physical interfaces (What Does It) fulfil the functional interfaces (What It Does), for example, cooling water fulfils the thermal control, electrical signal fulfils the command, and fasteners fulfil the structural support. There are also external interfaces and internal interfaces. It depends on your reference point. External interfaces are the interactions occur at various boundaries of the system if you are the systems. Internal interfaces are the interactions occur at various boundaries between individual components within your subsystem. Both external and internal interfaces can be functional and physical.

A functional interface is illustrated in Fig. 6.2 [33]. Side A generates the quantity and delivers the function to Side B. Side B receives the quantity and performs a function with the quantity. The interface boundary is between Side A and B. The physical interface is shown in Fig. 6.3 on a scope sheet. Sketch of interface could be engineering drawing, diagram or hand sketch to fulfil the functional interface. It will define and control the features, characteristics, dimensions and tolerances of one design that affect another. It could also include the material properties of the equipment that can affect the functioning of mating equipment. A narrative description of Side A transmits what information/data and Side B receives the information/data to perform a specified function. The responsible individual or group needs to sign on the scope sheet since this serves as an agreement paper.

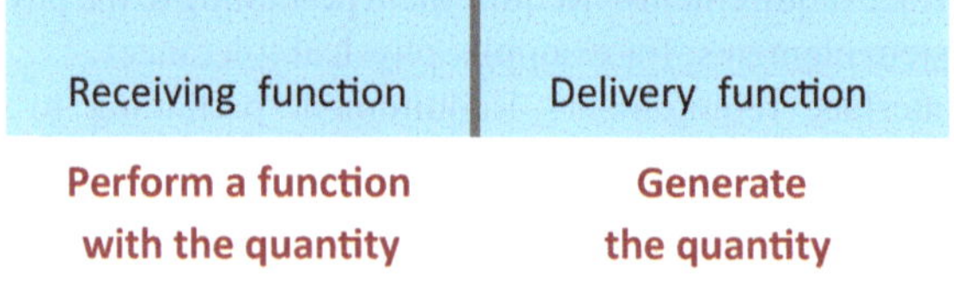

Fig. 6.2 Functional interfaces [33]

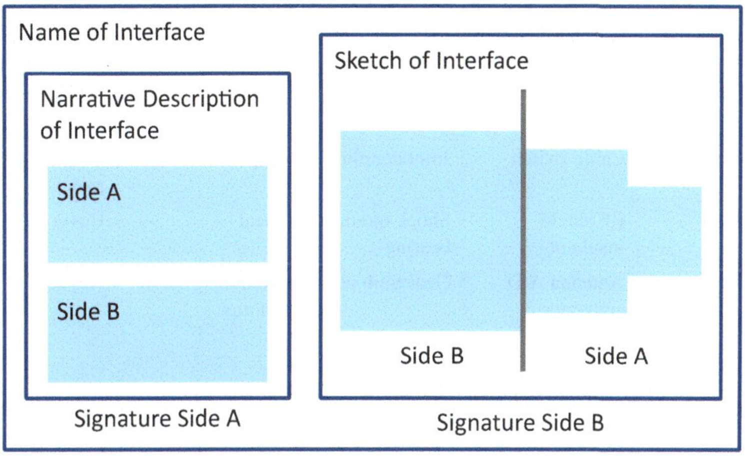

Fig. 6.3 Interface scope sheet

6.3.3 Interface Elements

It is convenient to distinguish three different elements:

Connectors—facilitate the transmission of electricity, fluid, force, etc. between components.
Isolators—inhibit such interactions.
Converters—alter the form of the interaction medium.

Table 6.5 [35] lists a number of common examples of interface elements. For interaction medium of electrical current, the connectors are cable or switch; isolator is RF shield or insulator; and converter is antenna or A/D converter. For interaction medium of hydraulic fluid, the connectors are pipe or valve; isolator is seal; and converter is reducing valve or pump. These must be considered as important design features. The relative simplicity of interface elements belies their critical role in ensuring system performance and reliability. Experience has shown that a large fraction of system failures occurs at interfaces. Assuring interface compatibility and reliability is a particular responsibility of the systems engineer.

Table 6.7 Interface elements examples

Type	Electrical	Mechanical	Hydraulic	Human-machine
Interaction medium	Current	Force	Fluid	Information
Connectors	Cable switch	Joint coupling	Pipe valve	Display control panel
Isolator	RF shield insulator	Shock mount bearing	Seal	Cover window
Converter	Antenna A/D converter	Gear train piston	Reducing valve pump	Keyboard

6.3.4 Interface Identification

From author's experience many organizations do not want to perform functional analysis and allocation due to its time consuming and taking a lot of resources. Sometimes, a legitimate justification is for small sustainment or product improvement project due to limited budget. But skipping this important task is taking short cuts in systems engineering process. Requirements cannot be checked for accuracy and completeness by functional analysis. No functional flow block diagrams show the functional sequences and data/information definitions between functions. The design interface has to be identified through judgement and communications with other designers or intuitively.

Interface identification methods will be presented here under the condition of no identified functional interface. Unfortunately, most of the time this is the case that interface is one of the notorious sources of failures reported. Refer to Fig. 6.4 showing a block with inputs and outputs. This block can represent a component, a subsystem, or a system. If you are the designer of this block, you have been told that there are three inputs and three outputs. If you are a conscious designer, you should be proactively to identify the source(s) of all the inputs from which system, subsystem, or component. If one or two inputs have no sources identified, you should communicate with the design community from which source(s) will provide the input(s). Also, if the inputs are not enough that you need more inputs, you need to communicate with the design community to identify which system, subsystem, or component should provide you what kind of input. Next, you should identify all the outputs feeding to which system, subsystem, or component. It you find out there is no taker of one or two of your outputs; you should communicate with the design community about the needs of this output(s). If not, do not generate this output(s) to save money for

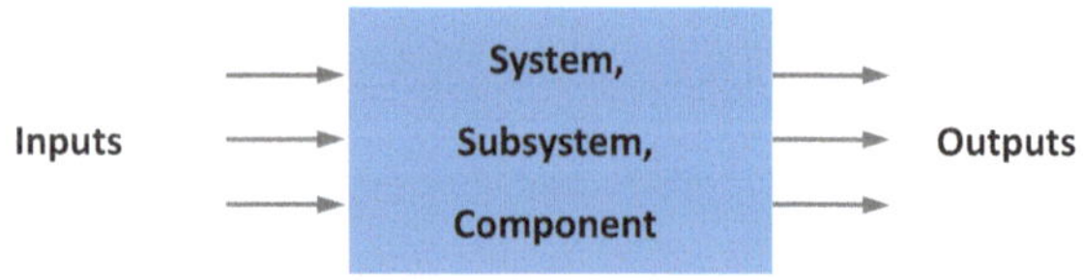

Fig. 6.4 Interface identification

the customer. Further, you could voluntarily inform the design community that you could produce more outputs if there is a need. This design community is the Interface Control Working Group (ICWG) that will be discussed later.

6.3.5 Interface Management Tools

1. N^2 Diagram—It is a matrix method to provide an organized way to identify interfaces.
2. Schematic—Use Block Diagram to depict interfaces.
3. Interface Dictionary—Provides an inventory of all system interfaces in the form of an alpha-numeric listing by Interface ID.
4. Interface Control Documents (ICDs)—It is a document that details the physical and functional interfaces between two system elements.
5. Electronic Development Fixture—Use CAD-3D software to build a 3D system including subsystems and components electronically.

6.3.5.1 N^2 Diagram

Thinking about a Square, the same systems, subsystems, or components are listed on the top and the left-hand side of the Square. It can be used at many levels: functions, processes, and design tools.

If you recall from Sect. 2.4.1 Functional Analysis, the directional sequence with functional data/information flows between functions creating functional interfaces. The N^2 diagram for functional interfaces is shown in Fig. 6.5. The six (6) functional blocks are listed diagonally in N^2 diagram. Functional inputs to functional blocks are designated as in horizontal direction; and the outputs from functional blocks are designated as in vertical direction. This designation is arbitrarily. You could designate input direction vertically and output direction horizontally. But once designated, the direction cannot be changed throughout the entire life cycle of the project. As you can see, F1 (Function 1) has input to F5 (Function 5); F6 (Function 6) has output to F3 (Function 3), F2 (Function 2) has input to F4 (Function 4), and F5 (Function 5) has output to F2 (Function 2), etc. Several N^2 diagrams can be combined to form a compound N^2 diagram as shown in Fig. 6.6. Interface direction convention is still the same as single N^2 diagram except it is from the component, subsystem, or system of a functional block of one N^2 diagram to the component, subsystem, or system of a functional block of another N^2 diagram. N^2 diagrams can only show global view of interfaces for overview or program review. The detailed interface specification is in the interface control document that will be discussed later.

Example 1 The physical interfaces are shown in Fig. 6.7. It is noted that the components are listed horizontally on the top as well as vertically on the left of the N^2 diagram. The physical components can be listed either diagonally as shown in

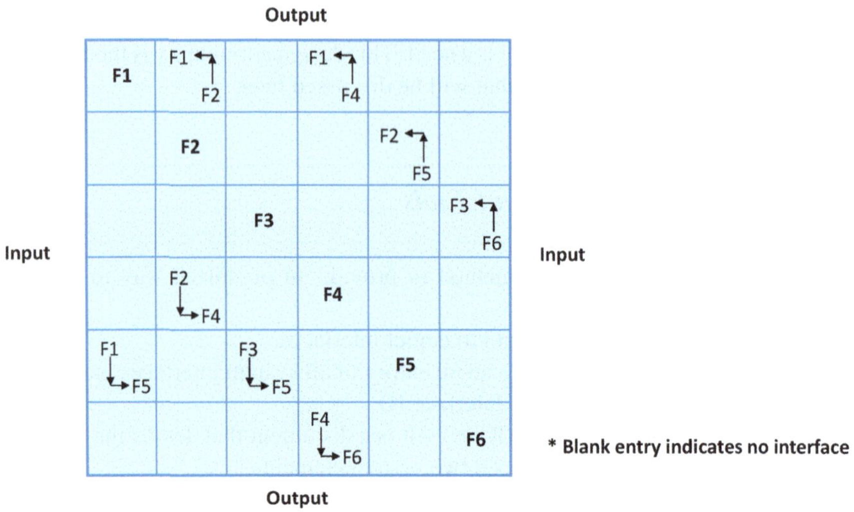

Fig. 6.5 Functional N^2 diagram

Fig. 6.5 or top horizontally and left vertically as shown in Fig. 6.7. However, listed top horizontally and left vertically, the input direction is from horizontal physical components to vertical physical components. Referring to Fig. 6.6, Air Conditioning has input to Room, but Room has no input to Air Conditioning. Electric Motor has input to Air Conditioning, but Air Conditioning has no input to Electric Motor. Generator has input to Electric Motor, but Electric Motor has no input to Generator.

Example 2 A more complete physical N^2 diagram is shown in Figs. 6.8, 6.9 and 6.10 [20]. In Fig. 6.8, the dishwasher subsystems are listed on the top horizontally and on the left vertically of N^2 diagram. The direction of the defined interfaces (blue colour) is from left to top subsystems. Blank blocks indicate no interface. Any defined interfaces can have detailed functional and physical interface requirements. Moving the cursor to any defined interface blocks can give a link to the detailed interface table as shown in Fig. 6.9 for Water Supply/Housing Interface. There are internal and external interfaces. Internal interfaces are the interfaces between washer subsystems while external interfaces are washer subsystems interface with a system, subsystem, or component outside the washer system. Normally, the interfaces are internal. External interface is not necessary only as needed. Water Supply/Housing only has internal interface. Internal or external interface must have both functional and physical interface requirements. This is what shown in Fig. 6.9. It is common to have more than one physical interface requirement to fulfil one functional interface requirement. Water Supply/Housing Interface Scope Sheet is shown in Fig. 6.10. Side A to Side B drawings are accompanied by narrative descriptions of both sides. The scope sheet should be signed by Side A and Side B group or individual as a contractual agreement.

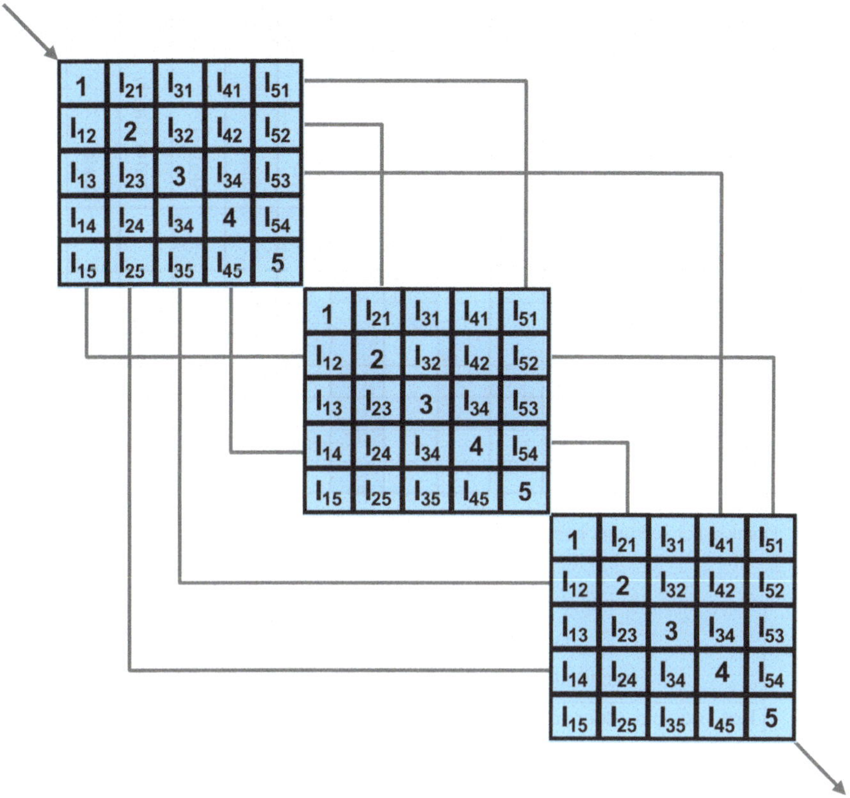

Fig. 6.6 Compound N^2 diagram

Fig. 6.7 Example physical N^2 diagram

	Room	Air Conditioning	Electric Motor	Generator
Room				
Air Conditioning	X			
Electric Motor		X		
Generator			X	

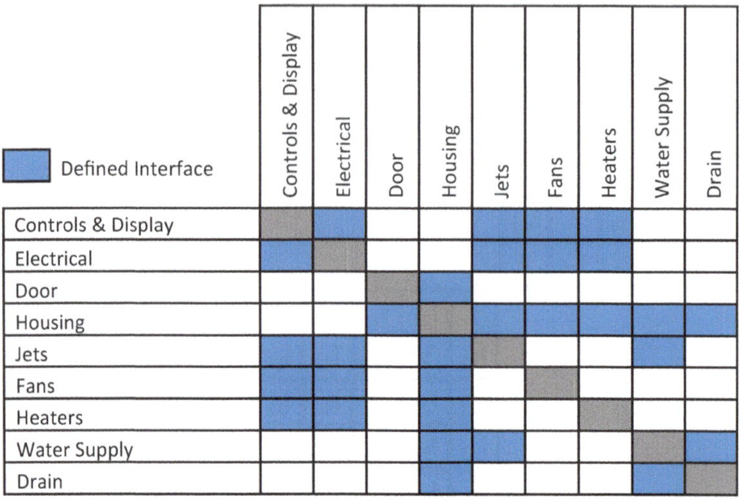

Fig. 6.8 Dishwasher pertinent interfaces

Internal	
Functional	Physical
Hot and cold water supply to housing.	Hot water flows through a gate valve, hose and elbow to washer housing. Cold water flows through a gate valve, hose and elbow to washer housing.

Fig. 6.9 Interface details between water supply subsystem and housing subsystem

6.3.5.2 Schematics

Schematics are mainly block diagrams. It is used to visualize and communicate interface relationship and applicable at all levels; between major mission/system element; between major subsystems within system elements; and between subassemblies within subsystems. It does not substitute for rigorous interface definitions and gives only global view of interfaces between blocks. Most of the time, the block diagrams is used in interface control documents to augment the interface understanding. An example, shown in Fig. 6.11, is the aircraft avionics block diagram. Each block represents an avionics component (box). How these components are connected and interfaced. It also shows the types of cables in compliance with what industrial standards connecting these components.

There are several block diagram types: functional with no distinction between software and hardware required, such as the FFBD and IDEF0 diagrams, presented

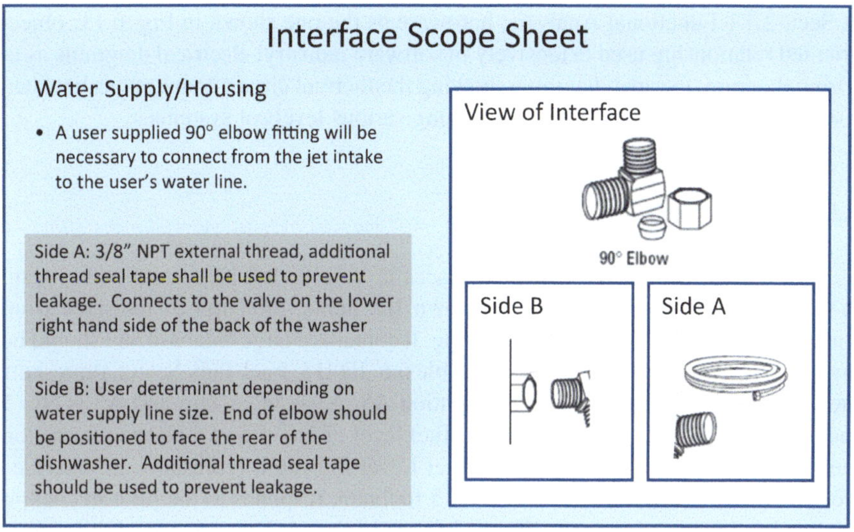

Fig. 6.10 Water supply/housing interface scope sheet

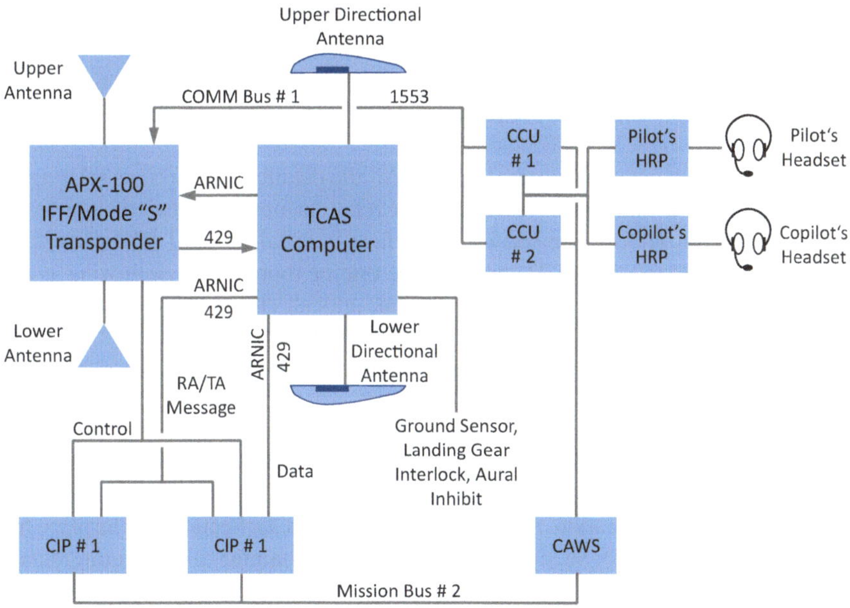

Fig. 6.11 Avionics block diagrams example

in Sect. 2.4.1 Functional Analysis; hardware as the one shown in Fig. 6.11; object oriented relationship used extensively in software industry; electrical diagrams as in wiring diagram; thermal diagrams showing the thermal circuit connections between components; and system diagrams showing various levels of systems.

6.3.5.3 Interface Dictionary

A large system must have many interfaces at various levels, system, subsystem, to component. Each interface has its own ID, name, direction of interface from source to destination, and descriptions, etc. It contains a large database and should be organized systematically, as shown in Table 6.6. ID I15, Fuel Tank Select, represents the interface from Source A1 to Destination A5. A1 is in cockpit and A5 is No. 5 fuel tank. The pilot in cockpit asks the fuel level of No. 5 fuel tank via interaction medium of electrical signal. ID I51, Fuel Level Indicator, represents the interface from Source A5 to Destination A1. No. 5 fuel tank responds to the pilot in cockpit the fuel level via interaction medium of electrical signal. This interface dictionary is only notional suggestions. More interface details can be included in the dictionary pending on the needs for more detailed information. All the interfaces can now be collected and compiled in one place.

6.3.5.4 Interface Control Documents

This is the ultimate interface control details. All the detailed descriptions relating to physical, functional and performance interface relationships are included in interface control document (ICD). For software interface is usually called IDD (Interface Definition Document). An ICD may control one or more of the following types of design requirements, mechanical, electrical, optical, etc., operational sequence, interoperability, installation, envelope, and interconnection, etc. This is a contractual document between two parties. A large volume of ICD between two companies could take several man-years to complete by expertise in different areas. ICD will be thoroughly reviewed, commented, and concurred (signed) by experts as well as managers of highest level of both companies. It could be 20–30 ICDs at various levels for a large project. A typical ICD outlines is shown in Table 6.7. The outlines are similar to system specification. Section 1 is the scope of ICD. Section 2 is the place for references quoted in ICD. Section 3 is the main part containing all the requirements. Section 4 addresses all the verification requirements to verify the requirements in Section 3.

The format and styles of ICD should follow the guidance from customer. If numbering system is used to designate each section at various levels, as shown in Table 6.7, the numbers of digits should not exceed seven (7).

6.3.5.5 Electronic Development Fixtures

Use CAD 3-D models to support system design and integration rather than the more traditional physical mockups or physical development fixtures. The CAD models can be used to share information during the design and integration effort. Just like the physical development fixtures, at the beginning, only system shell is built on 3-D CAD. When each component or subsystem is designed and built on 3-D CAD, a copy of the complete designed component or subsystem 3-D CAD model can be inserted at the correct location inside the system shell. As design goes on more components and subsystems are built and located inside the system shell. The designer can view the interfaces between components or subsystems inside the system electronic development fixture (EDF). For a new design can also reference to the inside of EDF to insure no interference with existing components or subsystems and also a seamless interface between components or subsystems. The EDF can share with customer and contractors. It can provide the 3-D presentation of components or subsystems during the design review. The customer or reviewer can travel through inside the system or subsystem to view components, piping, and wires connection.

The problem is whether the technology can catch up with the demand with 3-D CAD modelling capabilities and data growth. The growth of data volume is rapidly approaching to the point to overwhelm the technology. Every user does not need the same data fidelity or content. The good thing is to save money but the bad part is the efforts spent for software tailoring and management. Technical data is no longer strictly for engineers and can be shared with manufacturing, supplier management, managers and customers. Take maximum advantage of level-of-detail and rendering priority capabilities. Utilize role-specific markups to convey ideas. A reference EDP is shown in Fig. 6.12 [36].

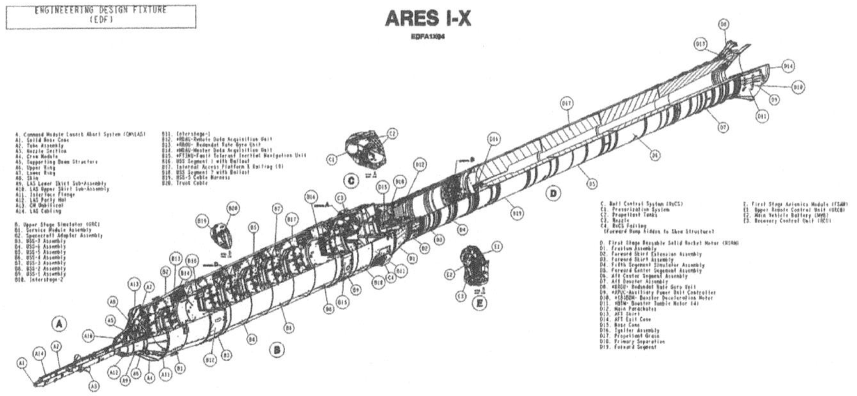

Fig. 6.12 Electronic development fixture for ARES 1-X

6.3.6 Open System Interface Management

Before discussing open system interface management, let us understand what is an open system? Those with engineering background, especially mechanical and aerospace engineering would say that a control volume (or a system) is open to mass flow in-and-out. But in systems engineering, the definition is a system that can be opened to replace inside part(s), whether for obsolescence or improvement, readily plug-and-play without any modifications. A close system is to replace the entire system if any part(s) inside is obsolete or for improvement. Nowadays, electronic components can be quickly getting obsolete. Military and commercial consumers prefer to purchase open systems; therefore, there are guidelines from military on open system interface management. All the parts, regardless manufacturers, have to follow industrial standards, i.e., the parts made by different manufacturers are interchangeable.

Open system interface management emphasizes [37] the correlation of interface requirements between interfacing systems. The interface management effort must control interface design such that interfaces specifically chosen for an open system approach are designed based on the following priority (Note that these are clear priorities, not options): Open standards that allow competitive products; Open interface design that allows installation of competitive products with minimal change; Open interface design that allows minimal change installation of commercial or NDI (Non-Development Items) products currently or planned to be in DoD use; and last, unique design with interfaces designed with upgrade issues considered.

The level at which the interface design should focus on openness is also a consideration. Each system may have several levels of openness depending on the complexity of the system and the differences in the technology within the system. The level chosen to define the open interfaces should be supported by industry and be consistent with program objectives. For example, for most digital electronics that level is the line-replaceable (LRU) and shop-replaceable (SRU) level. On the other hand the Joint Strike Fighter intends to establish openness at a very high subsystem level to achieve a major program objective, development of different planes using common building. The open system approach designed segments of a larger system could have additional openness at a lower level. For example, the Advanced Amphibious Assault Vehicle (AAAV) engine compartment is an open approach design allowing for different engine installation and future upgrade capability. Program objectives (such as inter-operability, upgrade capability, cost-effective support, affordability, and risk reduction) and industry practice (based on market research) drive the choice of the level of openness that will best assure optimum utility and availability of the open system approach.

6.3.7 Interface Control Working Group (ICWG)

Referring to Sect. 2.6.4, during the process of identifying interface inputs and outputs, it has mentioned several times of communicating with the design community from which source(s) providing the input(s) and sending output(s) to which destination(s). The design community is ICWG which is similar to CCB (Configuration Control Board) and could be part of the CCB depending on program/project's execution policy. It consists of representatives from Systems Engineering, applicable Functional Groups, Specialty Engineering, Management, Quality, Subcontractor and Customer as required. The representatives should be assigned permanently to keep the continuity from meeting to meeting since ICWG meeting is held at least once a week or could be 2–3 times a week. The representatives may not know all the answers, but they can delegate to the right expertise in his (or her) organization. The representative can and sometimes should bring the delegate person to the meeting. It is the forum for discussing interface issues. It serves two purposes: to ensure effective and detailed definition of interfaces by all cognizant parties; and to expedite baselining of initial ICDs and subsequent resolution of interface issues.

All ICWG meetings are chaired by the cognizant program organization by performing the following functions:

1. Identify Interfaces—Each IPT (Integrated Product Team) identifies interfaces associated with their respective system elements.
2. Generate Interface Control Plan (ICP)—The ICP is approved by the ICWG and Program Management. Once the initial ICP is approved, this task is completed until ICP is submitted for revision.
3. Initiate ICD/ICN (Interface Change Notice)—The ICWG authorizes the ICD activity to proceed and designates the custodian for each ICD. Internal ICDs within a single company will determine the owner of the ICD.
4. Prepare ICD/ICN—The custodian IPT prepares the ICD/ICN and coordinates with affected IPTs, suppliers, and contractors. The custodian IPT should send prepared ICD/ICN to the affected representatives for their organization's review.
5. Analyze ICD/ICN for Compliance with Interface Requirements—Affected IPTs review ICD/ICNs to assure completeness and accuracy of the interface definition. The advance copies should be sent to the affected IPTs for their review prior to the scheduled meeting. The custodian IPT will maintain a record that the interface has been examined for form, fit and function. A new interface(s) could be identified. If it is justifiable, the new interface(s) will be incorporated in the ICD/ICN. Through the inputs from other IPTs or the Electronic Development Fixture could find the interferences with the proposed ICD/ICN.
6. Approve ICD/ICN—ICWG coordinates approval of completed ICD/ICN through the affected technical and program management organizations. Upon approval, the ICWG authorizes ICD/ICN release.
7. Verify Design Compliance with ICD/ICN—Affected IPTs verify that the design definition is compliant with the ICD/ICNs. If the design is not compliant, redesign will be completed under the System Design Process.

8. Propose ICD Change Interface Change Request (ICR)—Affected IPTs, suppliers, or associate contractors may request changes to the ICD as design definition changes occur or new interface(s) is discovered. The requesting organization will conduct pre-coordination efforts with affected groups.
9. Authorize ICD Change—Following coordination, the Interface Change Request (ICR) is submitted to the ICWG for review of the ICR for technical content, accuracy, completeness, and disposition.
10. Determine Scope of Change Request—The ICWG determines if the authorized ICR is within the scope of the affected contracts. In-Scope ICRs will be returned to the ICR originator and the custodian of the ICD for preparation and release of the ICN. Out-of-scope ICRs are forwarded to program management.
11. Disposition of Change Request—Out-of-scope changes are reviewed by program management to establish next actions. Out-of-scope changes to suppliers and associate contractors must be coordinated with them. Once the Out-of-scope changes are approved, like in-scope changes, ICN processing will proceed as required.

6.4 Summary and Prospective

The technical management processes provide a connection between the project management and the technical team. From eight crosscutting processes, we have presented two in this section: decision analysis and interface management. Further processes belong technical planning, requirements management, risk management, configuration management, technical assessment, and technical data management. Without these crosscutting processes, individual members and tasks cannot be integrated into a functioning system that meets the Concept of Operations (ConOps) within cost and schedule [1].

All systems engineering activities should be conducted in the context of good decision-making. The purpose of decision analysis is to ensure credible, understandable, and timely views to decision makers. If a systems engineering activity cannot point to at least one of the many decisions embedded in a system's lifecycle, one must wonder why the activity is being conducted at all [38]. By using decision analysis as the mathematical foundation, opaque decision situations become transparent. The intelligent decision system provides this functionality to individuals, also to lower-level corporate decision-makers who get the possibility for very rapid decision-making [3].

Both internal and external interfaces between the various systems are traditionally been a source of unreliable operation for complex systems. The system-to-system interfaces cover the whole of the range of areas in a complex system. The interfaces to be controlled have a wide variety of characteristics and features. Effective management of interface requirements, specifications, and designs helps to ensure implemented interfaces are complete and compatible. Finally, interface management needs to become an integral part of the industry's culture. Furthermore, its benefits need to be quantified compared to more direct profit-generating activities. In that sense, interface management definitely has an essential role in minimizing risks of mistakes, rework, slippage and extra cost.

References

1. National Astronautics and Space Administration (NASA), Systems Engineering Handbook, Washington, DC, June 1995
2. Howard RA (1988) Decision analysis: practice and promise. Manage Sci 34(6):679–695
3. Abbas AE (2016) Perspectives on the use of decision analysis in systems engineering: workshop summary. In: 2016 annual ieee systems conference (SysCon). https://doi.org/10.1109/syscon.2016.7490551
4. Matheson JE Decision Analysis = Decision Engineering, INFORMS TutORials in Operations Research, pp 195–212. https://doi.org/10.1287/educ.1053.0015
5. Wognum N, Bil C, Elgh F, Peruzzini M, Stjepandić J, Verhagen WJC (2019) Transdisciplinary systems engineering: implications, challenges and research agenda. Int J Agile Syst Manag 12(1):58–89
6. Sobolewski M (2017) Amorphous transdisciplinary service systems. Int J Agile Syst Manag 10(2):93–115
7. Parnell GS, Driscoll PJ, Henderson DL (2011) Decision making in systems engineering and management, 2nd edn, Wiley, Hoboken, NJ
8. Díez FJ, Luque M, Bermejo I (2018) Decision analysis networks. Int J Approx Reason 96:1–17
9. Yassine A, Chelst K (2018) Opportunities for decision analysis in engineering management. IEEE Eng Manage Rev 46(2):151–161
10. Hannouf M, Assefa G (2018) A life cycle sustainability assessment-based decision-analysis framework. Sustain (Switzerland) 10(11):3863
11. Huff J, Medal H, Griendling K (2019) A model-based systems engineering approach to critical infrastructure vulnerability assessment and decision analysis. Syst Eng 22(2):114–133
12. Bajracharya S, Carenini G, Chamberlain B, Chen K, Klein D, Poole D, Taheri H, Öberg G (2018) Interactive visualization for group decision analysis. Int J Inf Technol Decision Making 17(6):1839–1864
13. Morris PWG (1979) Interface management—an organization theory approach to project management. Project Manag Q 10(2):27–37
14. Gianni D, Schaus V, D'Ambrogio A, Gerndt A, Lisi M, De Simone P (2014) Interface management in concurrent engineering facilities for systems and service systems engineering: a model-based approach, CEUR workshop proceedings, 1300, pp 72–81
15. Shokri S, Haas CT, Haas RCG, Lee SH (2016) Interface-management process for managing risks in complex capital projects. J Constr Eng Manag 142(2):04015069
16. Shokri S, Maloney K (2015) Interface management integration for major capital projects. In: Society of Petroleum Engineers—Abu Dhabi International Petroleum Exhibition and Conference, ADIPEC 2015
17. Crumrine T, Nelson R, Cordeiro C, Loudermilk M, Malbrel CA (2005) Interface management for subsea sand control completions. Offshore, 65(10):88 + 90 + 92
18. Allmayer S, Winkler H (2013) Interface management research in supplier–customer relationships: findings from a citation analysis of international literature. J Bus Econ 83(9):1015–1061
19. Hou L (2009) Brand relationship interface management: Communication, interaction and integration. In: 2009 international conference on information management, innovation management and industrial engineering, ICIII 2009, 4,5370438, pp 246–249
20. De Souza V, Borsato M (2016) Sustainable design and its interfaces: an overview. Int J Agile Syst Manag 9(3):183–211

21. Shamsuzzoha AHM, Kristianto Y, Helo P (2013) Implications of interface management for modularity degree. J Model Manag 8(1):6–24
22. Schuh G, Wentzel D, Rudolf S, Erkin A, Gerlach M, Schaffrath K (2015) Interface management in business-to-business-practice: how companies manage internal and external complexity I [Schnittstellenmanagement in der business-to-business-praxis: Wie unternehmen interne und externe komplexität managen]. ZWF Zeitschrift fuer Wirtschaftlichen Fabrikbetrieb 110(11):694–697
23. Minor RR, Dagli C (2007) An overview of system-of-systems interface management on the future combat systems program. In: IIE annual conference and expo 2007—industrial engineering's critical role in a flat world—conference proceedings, pp 445–450
24. Oh SG, Park P (2015) Building the system interface management environment for the development of complex system. Res J Appl Sci, Eng and Technol 10(8):951–959
25. Di Maio M (2014) Model-driven interface management: a practical approach to systems engineering in large-scale international research projects 8th annual IEEE international systems conference, SysCon 2014—Proceedings, 6819285, pp 383–387
26. Truffer B (2007) Knowledge integration in transdisciplinary research projects—the importance of reflexive interface management I [Wissensintegration in Transdisziplinären Projekten Flexibles Rollenverständnis als Schlüsselkompetenz für das Schnittstellenmanagement]. GAIA 16(1):41–45
27. Piepenburg MA, Louis DA, Magnifico R, Juliano A (2009) Railroad interface management for MTA East side access project tunnels and structures. In: Proceedings—rapid excavation and tunneling conference, pp 814–823
28. Harrison P (2004) An overview of interface management. In: American Society of Mechanical Engineers, Rail Transportation Division (Publication) RTD, 27, pp 89–99
29. Rauch E, Dallasega P, Matt DT (2017) Distributed manufacturing network models of smart and agile mini-factories. Int J Agile Syst Manag 10(3/4):185–205
30. Kepner CH, Tregoe BB (1997) The new rational manager. Kepner-Tregoe Inc., Princeton, NJ
31. Hsu J, Widmann R (2006) A generic tool selection process: achieving the benefits of common tools in a large corporation. In: AIAA aerospace sciences meeting, Reno, NV, USA
32. Hsu J (2014) Decision analysis. AIAA continuing education course, USA
33. Hsu J (2015) Systems engineering lecture notes. California State University Long Beach, AIAA Continuing Education, USA
34. Rechtin E, Maier MW (2009) The art of systems architecturing. CRC Presss, USA
35. Kossiakoff A, Sweet NS (2003) Systems engineering—principles and practice. Wiley, Hoboken Chapter 2
36. Wright R (2010) Ares I-X EDF team large assembly management. NASA
37. Defense Acquisition University (2001) Systems engineering fundamentals. Supplement 17-A
38. Parnell G (ed) (2017) Trade-off analytics. Wiley, Hoboken

Chapter 7
Mechatronic and Cyber-Physical Systems within the Domain of the Internet of Things

Peter Hehenberger, David Bradley, Abbas Dehghani and Patrick Traxler

Abstract There has been a shift in emphasis within systems from hardware-oriented to more software-oriented topics integrated in an overlaying communication framework (e.g., cloud-based services). This chapter presents current research in the field of the interaction between mechatronic and cyber-physical systems. It presents solution basics design methods that are illustrated by some real-world applications. The discussed case studies (Smart Home, Bio-mechatronic Systems, Cyber-Physical Production System, Data-driven analysis) provide illustration of applications involving different functional distributions of activity between the 4 key elements of people, data, mechatronics and cyber-physical system.

Keywords Mechatronics · Cyber-physical system · Design methods · Smart home · Bio-mechatronics · Cyber-physical production system · Data-driven analysis

7.1 Introduction

Recent years have seen the development of the concepts of Cyber-Physical Systems and the Internet of Things, both of which impact in related but different ways on the design, development and implementation of mechatronic components and devices, as well as on larger systems resulting from the combination and integration of these. In the case of Cyber-Physical Systems, individual components, often but not always

---

P. Hehenberger (✉)
School of Engineering, University of Applied Sciences Upper Austria, 4600 Wels, Austria
e-mail: peter.hehenberger@fh-wels.at

D. Bradley
Abertay University, Dundee, UK

A. Dehghani
School of Mechanical Engineering, Institute of Design, Robotics and Optimisation (iDRO), University of Leeds, Leeds, UK

P. Traxler
Software Competence Center Hagenberg, Hagenberg im Mühlkreis, Austria

J. Stjepandić et al. (eds.), *Systems Engineering in Research and Industrial Practice*,
https://doi.org/10.1007/978-3-030-33312-6_7

mechatronic in nature, are combined and integrated through the use of advanced intelligent software. The Internet of Things, then, provides a domain structured around access to and exchange of information, to further aggregate both mechatronic and Cyber-Physical Systems to create new and novel systems focused on users.

Beginning with Mechatronics, itself an interdisciplinary construct within engineering science structured around the integration of mechanical engineering, electrical engineering/electronics and information technology across a range of applications and products, the chapter considers associated design issues and discusses the product development process from both a Mechatronic and a Cyber-Physical System (CPS) point of view, using a model-based approach within the overall domain of the Internet of Things[1] (IoT). In this context, an important requirement in both the modelling and simulation of the interactions between systems at different hierarchical levels is that the product model is consistent across all phases of the design process and across all sub-systems considered.

Where such complex systems are concerned, achieving highly accurate system modelling is often prohibitive both in cost and time. As such, simplified modelling approaches become preferable. Further, an interdisciplinary model-based approach of the system-level models needs the provision of specific methods, languages and tools to support such a multi-view approach, as for instance multi-agent modelling based on an engineering cloud structure.

The chapter begins by looking at the relationships between mechatronics, Cyber-Physical Systems and the Internet of Things, including consideration on the potential impact on associated design methods and strategies in areas such as user-centred design and system modelling. It then uses a series of case studies in Smart Home technologies, Bio-mechatronics, CPS-based production systems and data-driven analysis to provide context for the discussion.

7.2 Mechatronics, Cyber-Physical Systems (CPS) and the Internet of Things (IoT)

While mechatronics has become well established as an engineering discipline, the more recent constructs of Cyber-Physical Systems and the Internet of Things are still evolving both technically and conceptually as well as in respect of their relationships with each other.

In the case of CPS a particular challenge is that of integrating design practice to support the interdisciplinary nature of a CPS, particularly in the allocation of functionality between software and physical elements at a level beyond that generally associated with mechatronics [2, 3].

[1] "The Internet of Things refers to a state where Things (i.e. objects, environments, vehicles and clothing) will have more and more information associated with them and the ability to sense, communicate, network and produce new information, becoming an integral part of the Internet" [1].

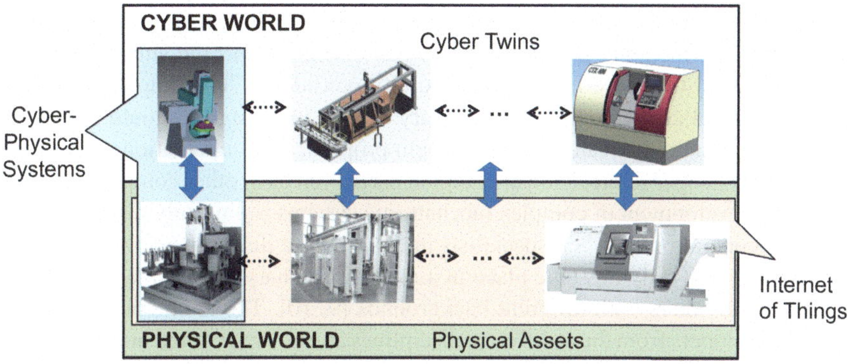

Fig. 7.1 Cyber-physical systems and internet of things (after [3])

An important element in Fig. 7.1 is the "Digital Twin", Digital Twin was brought to the general public for the first time in NASA's integrated technology roadmap and is defined as follows (see [4]):

> A Digital Twin is an integrated multiphysics, multiscale simulation of a vehicle or system that uses the best available physical models, sensor updates, fleet history, etc., to mirror the life of its corresponding flying twin. The Digital Twin is ultra-realistic and may consider one or more important and interdependent vehicle systems, including propulsion/energy storage, avionics, life support, vehicle structure, thermal management, etc. Manufacturing anomalies that may affect the vehicle may also be explicitly considered.

In contrast, the Internet of Things is a much more abstract concept structured around a dynamic network of system artefacts to enable these to collect and exchange data. In 2013 the International Telecommunications Union (ITU) Global Standards Initiative on Internet of Things (IoT-GSI) [5] described the IoT as being "*the infrastructure of the information society*." The IoT thus supports the integration of remote and disparate smart objects as part of a dynamic system configured according to both context and need. In a system structured around the IoT, each individual smart object remains uniquely identifiable while integrating and exchanging data with other such objects through the infrastructure of the internet. From a design perspective, the abstraction of the IoT presents a particular challenge through its ability to incorporate as required both hardware and software elements of defined functionality but of unknown origin given that they can be autonomously selected by the system on the basis of context and need.

7.2.1 Characteristics of Systems

Mechatronic systems integrate sub-systems and components from mechanics, electrical engineering, electronics and software and constitute the basis for the design of products ranging from domestic appliances to machine tools and vehicles [6–8].

In a mechatronic design process the conceptual design phase is especially important. Here, the functional interactions between discipline-specific sub-systems are determined, and therefore have to be carefully considered. This implies that during the phases of conceptual and preliminary design, the designer should be able to quickly and accurately evaluate those system properties and interactions that result from design changes to mechanical components as well as to other components. The successful development of complex mechatronic systems is thus only made possible by close cooperation between specialists in the different disciplines involved. Consequently, design activities take place in a multidisciplinary environment involving engineers and others with different backgrounds [9, 10]. The interactions between product developers from these different disciplines are, however, often hindered by an insufficient understanding of the interactions between the disciplines, and by failing to utilise common platforms for the modelling of complex systems [11]. In respect of Cyber-Physical Systems, these are defined by Lee [12] as resulting from the:

> …. integration of computation and physical processes. Embedded computers and networks monitor and control the physical processes, usually with feedback loops where physical processes affect computations and vice versa.

Thus, a CPS could be considered as being structured around an aggregation and assembly of mechatronic components and devices. Such systems can be found in areas such as aerospace, automotive, energy, healthcare, manufacturing, entertainment and consumer appliances.

For the easy validation of requirements during the overall product development process, methods are needed for the decomposition of high-level system requirements into criteria for design decisions. This will in the future be an increasingly important issue as new system architectures such as those of Cyber-Physical Systems are introduced [13, 14]. As has been seen, the Internet of Things (IoT) then provides access to information, context dependent and otherwise, as well as sourcing a range of software, platforms and infrastructure services and functions. In many cases, these will be autonomously sourced by the system on demand without necessarily any a priori knowledge by the designer or user as to their origins or structure. Thus, individual systems may communicate with each other to establish a network and determine optimum solutions for themselves as well as with other systems.

Referring to Fig. 7.2, within the context of the chapter the relationship between the various system layers of mechatronics, Cyber Physical systems and the Internet of Things can be expressed in terms of increasing levels of abstraction, to which must then be added the user whose primary role becomes that of defining system functionality and the context within which it operates [4].

7.2.2 From Design for All to Design by All

Design for All (DfA) refers to a design philosophy which aims to support access to and use of products, services and systems by individuals without the need for adaptation

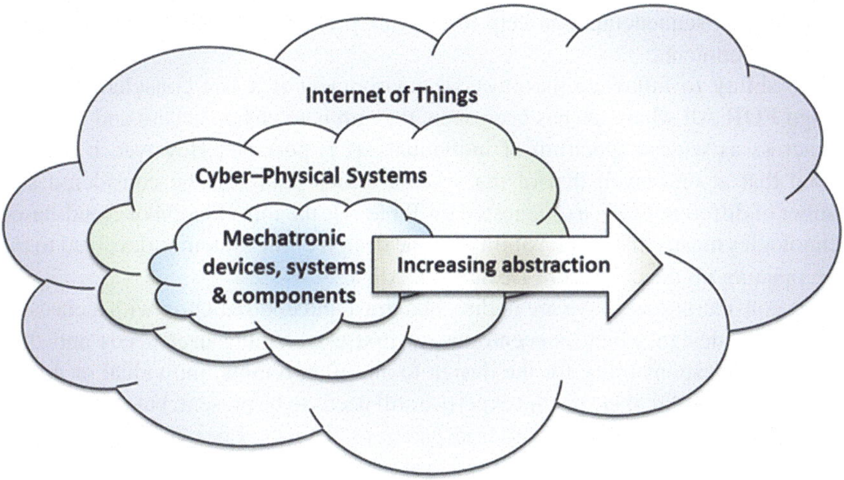

Fig. 7.2 Abstraction of systems [4]

[15]. Further, Design for All is often considered as an aspect of Universal Design (also Inclusive Design and Barrier Free Design), an approach to the physical and aesthetic design of buildings, products and environments accessible to all irrespective of age or physical capability [16–19]. Conceptually, Design for All also forms a part of the Digital Inclusion agenda which aims to ensure equivalent access to information [20].

Conventionally, problems of accessibility have been solved through a combination of adaptation and the use of assistive technologies. Within Digital Inclusion, the concept of Universal Access implies access by anyone, anywhere and at any time. Associated products and services must therefore be capable of accommodating individual user requirements across the full range of contexts of use while being independent of location, hardware or runtime environment, increasingly within an IoT construct [21–25].

In the context of mechatronics, the design process conventionally encompasses a continuum of activities ranging from needs identification and conceptual design to the detailed design of individual components. The advent of cloud-based systems such as those represented by the Internet of Things and its related concepts raises specific design-related issues and questions by increasingly treating information as a commodity whose value to the system will vary according to both context and demand.

A particular aspect of cloud-based systems is the implied integration and transfer between the information and physical domains, and the potential for decisions made in relation to one domain to impact upon, define and influence decisions made in the other. For instance, real-time information on traffic conditions may be assigned a high value by an individual only if it affects his or her intended journey, while that same information, when integrated with traffic flow management options such as traffic light sequencing, can help to mitigate problems of traffic flow within the urban environment.

The ability to influence the physical environment is a key consideration with Design **FOR** All where, as has been seen, the aim is to ensure access and facilitate use across as wide a spectrum of individuals as is possible. However, it may be argued that at one level, that of the system, which itself can be considered at a number of different levels as suggested by Table 7.1, the introduction of cloud-based technologies means that responsibility for the design of the system is devolved to the user, opening up the concept of Design **BY** All.

The considerations above are in direct contrast and contradiction with a classical approach to design which concentrates on first establishing user needs and then transfers the responsibility for the design to an, often remote, individual or group. With Design **BY** All, that specific expertise still needs to be present, but in a form and format which is accessible by, and transparent to, the user. Located in the cloud, as may be much of the system software and management tools, this expertise then serves to underpin decisions and to inform and guide users in relation to their choice based around the selection of suitable components. Specifically, there is a need to support non-expert users in achieving their goals without their being aware of the underlying technologies and its operation. Such transparency of operation in which the user specifies context, requirements and outcomes is increasingly a feature of CPS and IoT based systems such as those of Table 7.1. Thus, a user seeking to control their domestic environment to maximise comfort and minimise energy consumption will focus, with the aid of appropriate smart tools, on the determination and confirmation of outcomes rather than the detail of the control of individual devices.

Structurally, the resulting system is likely to be what is increasingly being referred to as a participatory system such as that shown in Fig. 7.3. The key components of which are:

Table 7.1 System levels

System level	Notes
Mechatronic	The individual components making up the physical structure of the system. In the case of a vehicle, these would range from individual sensors and actuators to smart sub-systems such as engine management, traction control, braking systems and environmental control
CPS	Operating under the control of the intelligent cyber component, the individual vehicle systems are brought together as a Cyber-Physical System. This enables the individual (e.g. mechatronic) components and sub-systems to operate together to facilitate and optimize system behaviour, for instance by linking the engine management, traction control and smart gear boxes to minimize fuel consumption
IoT	Supports information exchange between individual vehicles as well as with other locations. For instance, direct communication between vehicles can be linked to traffic light control to manage traffic flow and, if required, provide clear path routing for emergency vehicles

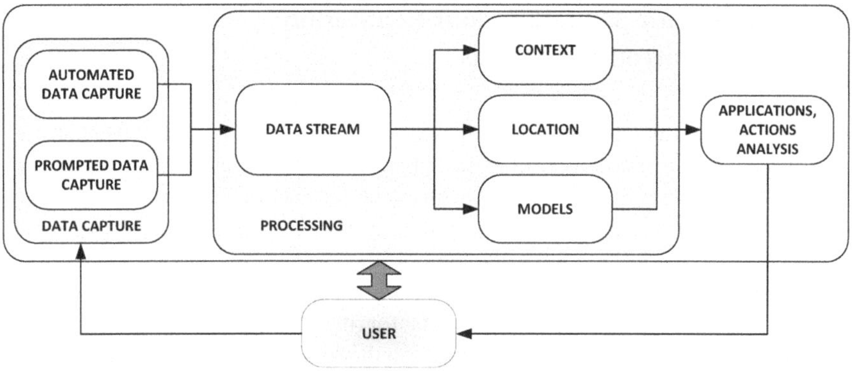

Fig. 7.3 Participatory system

- **Data Capture**: As its name implies, this is associated with the provision of the source data to the system and itself comprises two components as follows:
  - *Automated data* capture implies generic system-level data which is provided on a regular basis from a defined sensor set.
  - *Prompted data* capture then implies data which is not necessarily captured on a regular basis but only in response to a prompt generated by the system user.
- **Processing**: Data is merged into a data stream and evaluated in relation to parameters such as established context and location along with being used as input to relevant system models, as for instance the heat transfer model of a house referred to earlier.
- **Applications, Actions and Analysis**: The outputs from the processing element are taken and used as appropriate in relation to the system outputs, which may themselves be expressed in terms of:
  - *Applications*—As for instance the operation of environmental controls, lighting systems or domestic security.
  - *Actions*—As for instance the turning on or off of a heating system.
  - *Analysis*—As for instance the refinement of the models used in processing the data.
- **User**: Sets and defines the system-level functions including context and constraints, including issues such as the privacy and security of the user and their data.

7.3 Methods for Modelling and Evaluation of (Mechatronic) Systems

The role and importance of modelling in relation to the incorporation of mechatronic components and sub-systems with Cyber-Physical System constructs and their subsequent incorporation as smart objects within the Internet of Things has been introduced earlier in the chapter. This section therefore focuses on two important topics in relation to the modelling and evaluation of mechatronic systems, namely system decomposition and system modelling, including integration with biological systems.

A number of methods currently exists for the modelling and evaluation of systems, including, among others, Design Methodology for Mechatronic Systems [8], Axiomatic Design [26], Property-Driven Development/Design (PDD) [27] and Model Integrated Mechatronics MIM) [28].

7.3.1 Decomposition of Mechatronic Systems

To describe and evaluate complex mechatronic systems, it is necessary to decompose the total system into a hierarchical structure of mechatronic sub-systems and modules. Figure 7.4 provides an example of the hierarchical decomposition of an overall

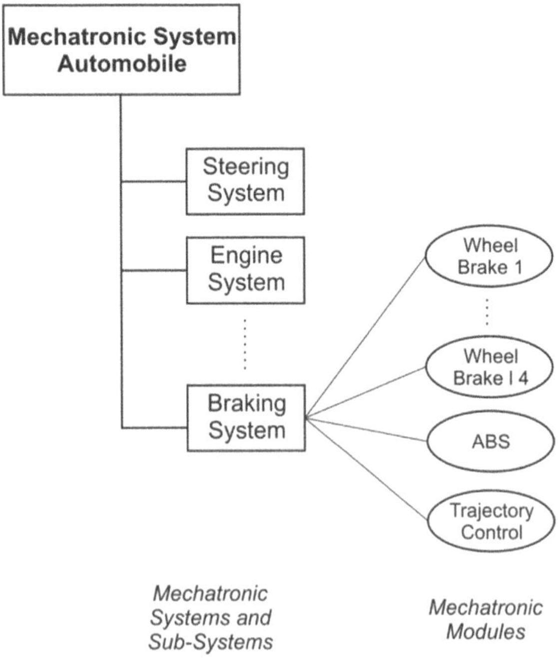

Fig. 7.4 Hierarchical structure of a complex mechatronic system [29]

mechatronic system, in this instance the braking system of a passenger car [29].

In this system, the brake-by-wire mechatronic braking system is structured into four individual mechatronic wheel brake modules. The information from the brake pedal to the modules is transferred electronically and power is transferred to the brake modules via the vehicle's electrical system. As both information and power may be transmitted independently from one another, additional functions as for example, ABS and traction control system can be implemented more flexibly.

The design and optimisation of the mechatronic module 'wheel brake' is a complex task, in part because of the multi-discipline characteristics of mechatronics but also due to the safety-critical and reliability aspects of such systems. The hierarchical structuring supports the recognition and description of internal interactions along with the coupling and integration of all systems at all levels. In this way all relevant interactions, interdependencies and interfaces become transparent.

These characteristics and requirements in combination with cost savings, call for new methods for design, evaluation, optimization and specification of such mechatronic modules building upon and supported by established modelling techniques such as those mentioned in the previous section.

A mechatronic module (according to [29, 30]) invokes several different disciplines, e.g. mechanics, automatic control and sensors, requiring discipline-specific components to be integrated and merged to provide the desired functionality. Thus at the lowest level, a mechatronic module can be decomposed only into discipline-specific (non-mechatronic) components, and not into other mechatronic modules or sub-systems.

A mechatronic module therefore defines the lowest hierarchical level of a mechatronic system and is indivisible within the set of mechatronic sub-systems. With the mechatronic system design model, all couplings between the individual mechatronic disciplines need to be described with the elements or pillars of the model then representing a discipline-specific sub-component or assembly. This in turn is structured into several hierarchical levels corresponding to the degree of detailing as in Fig. 7.5.

This implies that only the first or highest level has an interface to the other disciplines (compare with encapsulated modelling) via the mechatronic coupling level. All couplings between the discipline-specific models (e.g. design parameters and requirement parameters affecting multiple disciplines) are thus captured and described at the mechatronic coupling level.

The model structure has to be adapted if additional couplings between discipline-specific components are detected during a design iteration (design, analysis, integration, performance check, etc.). This is also true if new or additional disciplines come into consideration.

It should also be noted that the Mechatronic System Design Model may also be seen as a representation of a project organization in which each pillar represents a (discipline-)specific group or team. In this context, the Mechatronic Coupling Level may be seen as representing a coordination function within project management. The design of mechatronic systems is an iterative process, as designers often have to move back and forth between steps to redesign or tune what they had previously created. Accordingly, for the process of product modelling of Fig. 7.5, there is a

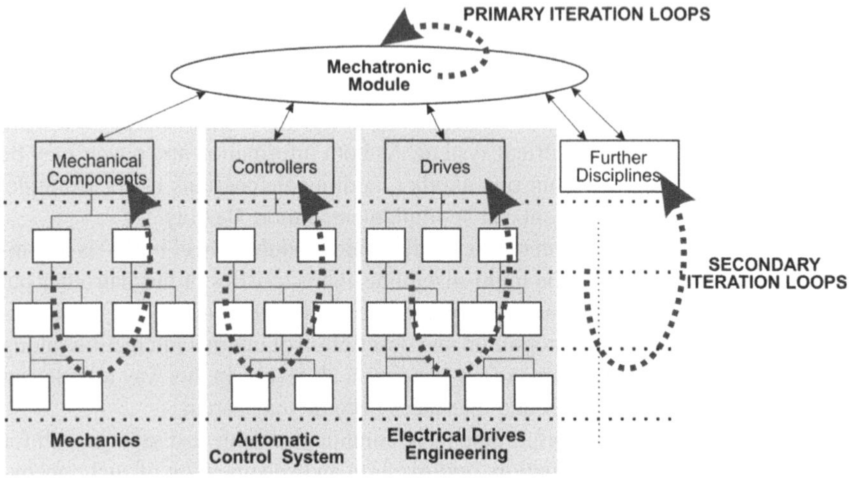

Fig. 7.5 Mechatronic module and design iterations [29]

distinction between two qualitatively different types of iterative loops: one for the different mechatronic disciplines (secondary iteration loops) and one for the couplings between them (primary iteration loops).

Iterations become primary loops if they are characterized mainly by changes in the coupling of the individual sub-systems. Here the requirements and dependencies between the design parameters at the beginning of the development are specified, from which the intersections between the different disciplines are established. It is clear that later changes usually have large effects on the overall operational sequence, thus it should be ensured that subsequent changes are avoided as much as possible, leading to the smallest number of necessary iterations.

The progressive development in the individual partial disciplines is then later characterized by secondary loops. Here, a higher number of iteration runs can be accepted in contrast to the primary loops, as these modifications affect only the requirements and parameters of the specific discipline, and thus have no direct effect on other model pillars as long as the definitions at the coupling level do not need to be changed [31]. The iterative structure of the model allows for a standardized procedure for the development process in subsequent steps.

7.3.2 Bio-mechatronic System Modelling

As has been seen, modelling a mechatronic system is a challenging process. This is due to the fact that such a system comprises mechanical modules (system elements or sub-systems), electronic modules and software to provide the required level of intelligence (including control) for the system operation. The complexity of the modelling is further compounded by the need for the modelling to take into consideration not just the models for individual subsystems/elements of the system, but also interaction between and among them.

A bio-mechatronic system such as has been developed to interact with the human neuromuscular-skeletal system to assist with impaired human motor control [32] thus consists of an integration of biological and mechatronic systems. Biological systems are themselves inherently complex, and as a result the modelling and simulation of bio-mechatronic systems is generally associated with a high level of assumptions and simplifications while taking into consideration the interaction of the two systems.

There is a suggestion for the modelling of such systems using "*a conceptual model for systems design in bio-mechatronics based on the ideas of mathematics and cybernetics originated by systems theory*". It is argued that "*…traditional mathematical models and models of artificial intelligence do not allow describing bio-mechatronic systems being designed on all its levels in one common formal basis*" [33].

Considering a lower limb prosthetic system as a typical example where the ultimate goal is to replace a lost limb of an amputee with an intelligent device to ideally resemble a biological limb both in terms of functionality and aesthetics. Figure 7.6 shows the elements and interactions in this typical example in a very simplified form.

An intelligent prosthetic device as a mechatronic system should have the required sensors, actuators and control to function as expected. In a bio-mechatronic system, as shown in Fig. 7.6, the additional interaction with the biological system which has its own sensors i.e., receptors, actuators or muscles and control (the central nervous system) adds another layer of complexity to the whole system since static and dynamic interactions should also be taken into consideration. This will make the modelling and simulation more challenging. Another example of a bio-mechatronic system is a robotic exoskeleton, which should also interact with the human body but as a separate device. In such a system the interaction should provide some feedback to the system for control purposes. This should also be taken into consideration in the modelling process.

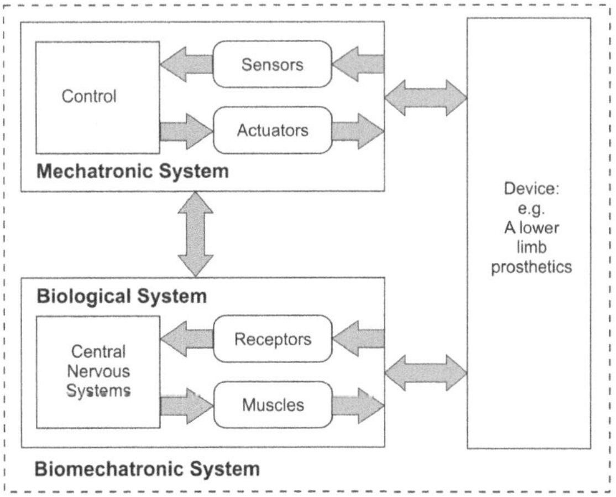

Fig. 7.6 An example of a bio-mechatronic system

The presented methods are only two examples for methods used in mechatronic design processes. In the authors point of view, these play an important role in the first (conceptual) design phases.

7.4 Case Studies

The following case studies provide illustration of applications involving different functional distributions of activity between the 4 key elements of people, data, mechatronics and cyber-physical system. In order to place these case studies into context, each one has been evaluated (on a scale of 0 (minimum) to 5 (maximum)) concerning their design content in respect of these four key elements as follows:

Design content	Assessment criteria
Data	The significance and importance of data in relation to function and operation of the overall system and to the IoT
CPS	The extent to which the system conforms to the general definition of a CPS presented earlier in terms of the balance between the physical and cyber components
People	The expression of the interface between the system and its users, and the importance of that interaction to its operation
Mechatronics	The contribution of mechatronics, generally in the form of components, modules and sub-systems in achieving overall system functionality

Once the individual case studies have been scored by the system designers, the results can be plotted to create the web diagrams of functional distribution of Fig. 7.7 in order to provide a visual representation of the contribution of each of these four elements to overall system functionality. In doing so it needs to be recognised that the result is intended to be informative as to the relative weight of each element at the system level and not definitive in terms of defining its contribution to system operation.

7.4.1 Smart Home

Smart Home technologies and systems for home automation were introduced in the latter part of the 20th century. The advent of the internet then supported new concepts such as the Digital Home, eHome or iHome [34, 35] and saw traditional automation services evolve to include entertainment and communication supported by home networks and residential gateways as suggested by Fig. 7.8 [36–38].

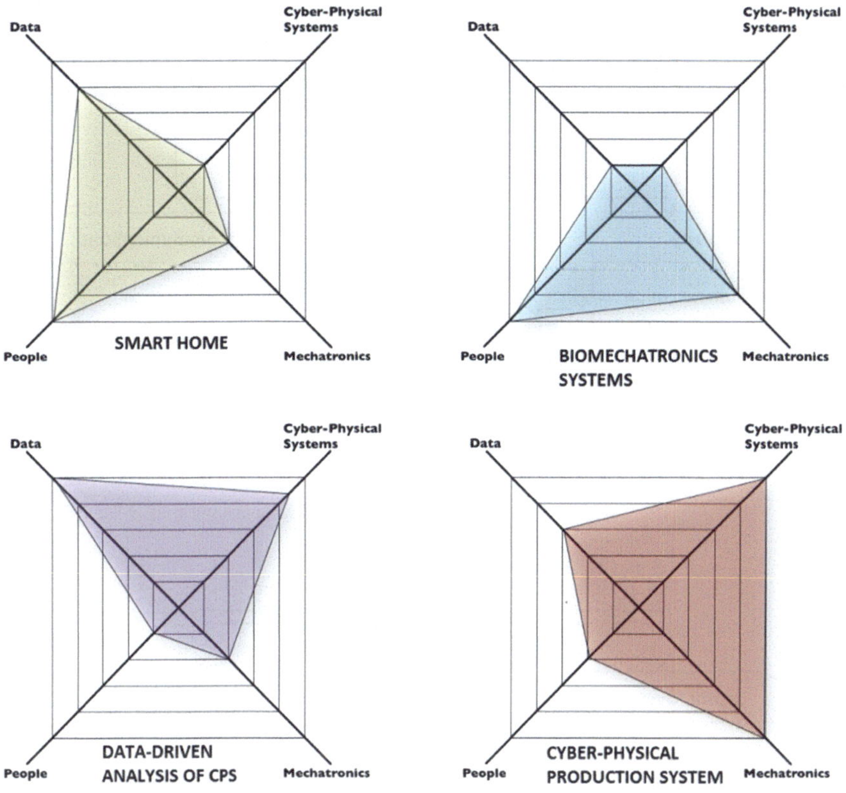

Fig. 7.7 Functional distribution

The 21st century also brought with it new paradigms such as "*ubiquitous computing*" [39] and "*ambient intelligence*". In the case of ambient intelligence, this defines a context in which people will be surrounded by intelligent and intuitive interfaces embedded in everyday objects which recognize and respond to their presence in ways which are context dependent and autonomously and intelligently adapt and respond to the user [40, 41].

This trend of adding intelligence to devices is moving to wider environments as exemplified by the Internet of Things and Smart Cities. The resulting Web of Everything will then integrate Smart Cyber-Physical Systems with the Internet of Things to provide new forms of integrated service [42, 43]. It must also be recognised that this shift brings with it a number of associated risks in relation to matters of individual security and privacy, in particular associated with the collection and use of personal data and the ability to draw inferences about the individual and their personal environment from this. In considering the privacy of the individual it is also necessary to distinguish between the *hard* security issues of Table 7.2 aimed at preventing access to private information, and the *soft* or people oriented aspects of

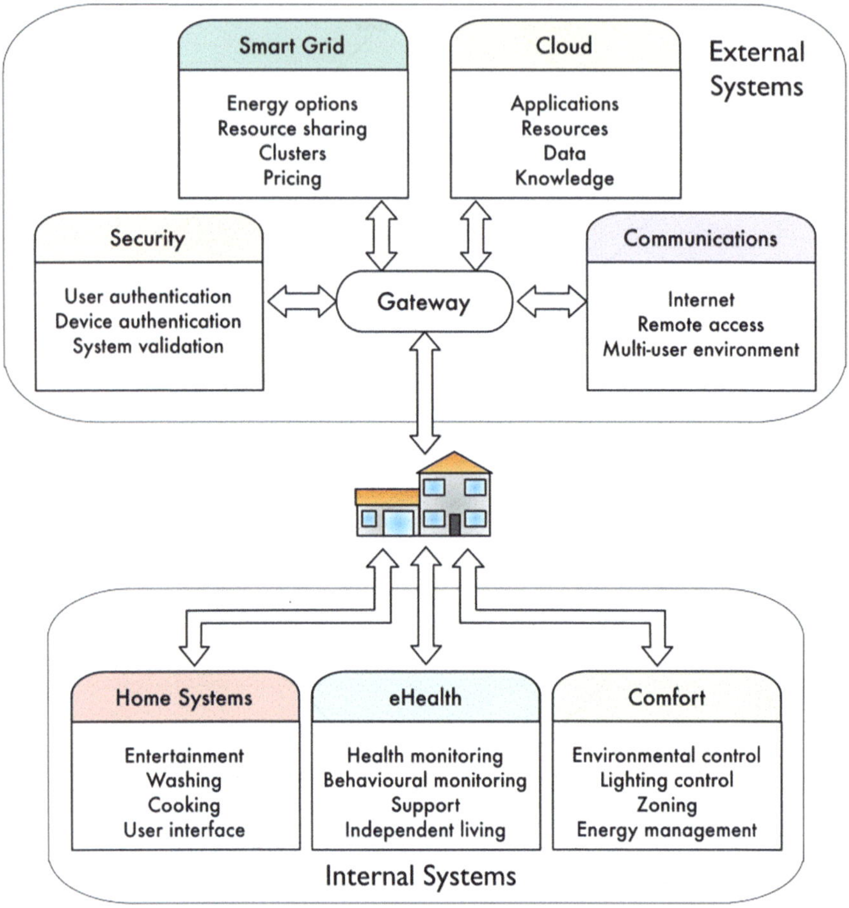

Fig. 7.8 Smart home constructs

Table 7.2 Hard security measures

Measure	Notes
Cryptography	Techniques such as public/private key encryption, digital signatures, steganography, secret sharing, key management, escrow and public and private certificates
Firewalls	Establishes a logical barrier between a trusted and secure network and any external networks
Passwords	Security depends on the passwords being difficult to guess or discover

privacy associated with concepts such as privacy by design [44] aimed at ensuring an individual's right to control over their own data as suggested by Fig. 7.9.

As Information and Communication Technologies (ICTs) develop, individual devices and systems need to communicate with each other in order to support increasingly complex services which involve both co-operation and competition for resources. In this context, the control, multimedia and data functions can be considered according to Table 7.3.

Initially, services were specialized and strongly coupled to the technology that supported them. As the level of embedded intelligence has increased, systems have

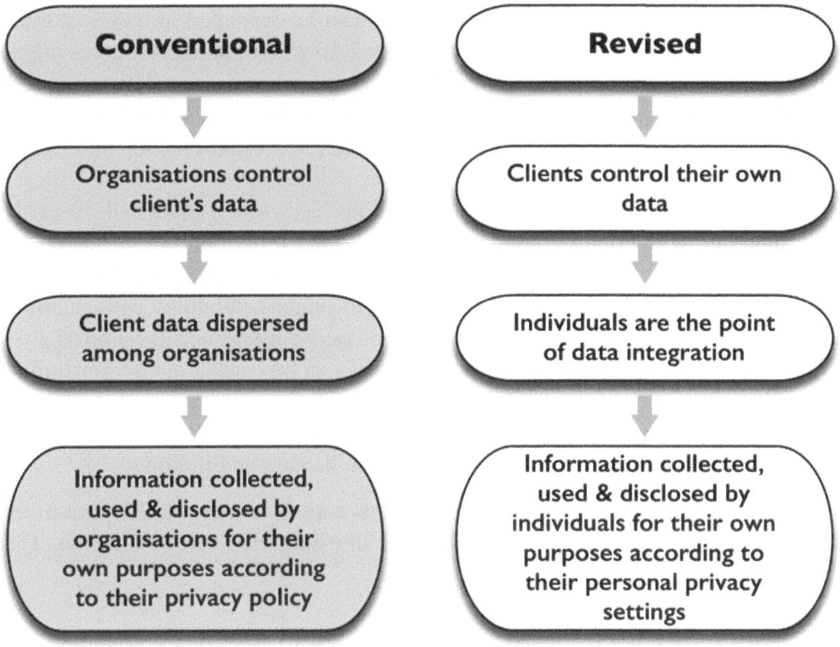

Fig. 7.9 Conventional and revised approaches to client privacy (after [44])

Table 7.3 Input-output transactions table

Control	This provides the infrastructure for the management the home system. Requirements include low cost, ease of installation and reconfiguration, ease of expansion and fault tolerance
Multimedia	This supports the distribution of audio and video media to televisions, HiFi equipment and other media systems. Requirements relate to the volume of data and the quality of service provision required
Data	The data network must provide access to information from anywhere within the home environment for all data sharing devices. Requirements include high bandwidth and low cost

evolved towards a distributed structure in which the gateway supports any device or service with the necessary intelligence [45].

The most important issue is that of enabling individually intelligent devices to communicate and interact with the other devices and actors in the environment. It also needs to take into account the increasing volumes of information home systems are being required to manage, particularly within the context of the Internet of Things where the CONNECT forum has suggested that [46]:

> … the integration of 'Things' as actors in the Internet via massive and innovative sensors, actuators, and real-time reactivity will cause another order-of-magnitude data explosion with challenges that we have yet to understand and deal with.

The underlying concepts of the smart home can be expanded to that of a smart grid environment such as that suggested by Fig. 7.10 where shared resources might, at some future time, encompass local energy systems such as micro-CHP or fuel cells [47, 48]. In such a scenario, the individual houses would be required to negotiate with each other, and with external systems, to balance user requirements, for instance with respect to energy utilization and system level efficiency.

For the user, the requirement is for effective interfaces which adapt to the user and which integrate with context-based approaches to the programming of the system and the requirements for the Design for All and Digital Inclusion agendas introduced in Sect. 7.2 [49–52]. Thus the user interface has to support autonomy while learning to perform tasks for their users and providing proactive assistance as required.

Smart Homes, and their Smart Grid extension, can be considered as constituting a participatory system of the form shown in Fig. 7.3. In order to better understand this relationship, consider the following scenario based around the configuration of a domestic environmental control system for both heating and lighting.

- The first stage of the process would be to first capture the physical layout of the environment and to display this to the user in a suitable form and format. This

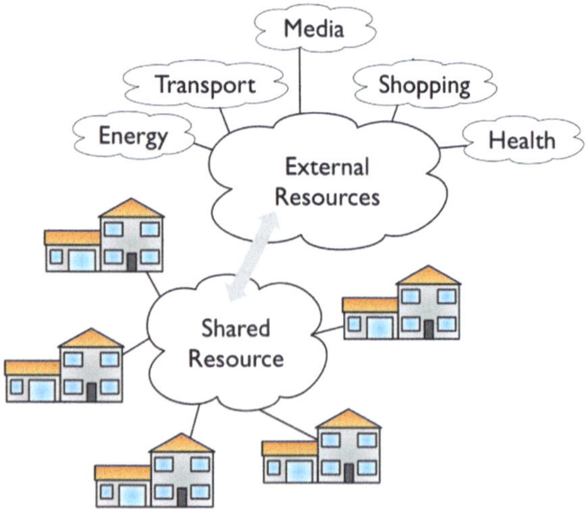

Fig. 7.10 Smart grid

could, but not necessarily, be supported by information on physical properties such as the heat-transfer properties of the walls and ceilings and access to the physical spaces.

- Having defined the environment, then the details of each space can be established along with the use to which that space is put. For instance, do people gather to watch television or listen to music, to eat or to simply sit quietly? This feeds into the modelling of resource utilisation such as heating and lighting, which is in turn reflected into the model of the physical environment.
- Where information such as the heat transfer properties are not available, the system could instead construct over time a model of the profiles for the heating and cooling of the individual spaces, and hence adjust its operation accordingly, taking into account parameters such as the external temperature.

In the above the role of the user is that of data provision and capture, as for instance of the physical dimensions.

- Once the user is satisfied with the needs definition and specification, the design process then devolves to that of component selection structured around user defined criteria such as cost, size, power limitations and so on. This implies an intelligent search function based on some form of recommender system to suggest alternatives to the user from which they can then make a selection.
- Following the selection of components, the system is then required to autonomously generate system-level data such as the optimum location for sensors controlling heating and lighting and the necessary system level functional coding (software) for each device.
- System assembly then evolves to the installation of the components within the environment on a plug, download and play basis. This applies both to individual devices and to the system controller, which may be any combination of smart phone, tablet PC, laptop or panel, and must support external communications for remote operation and diagnostics.

While Smart Home and related technologies are still evolving, it is clear that they present both significant challenges and opportunities across a wide range of technologies from sensors to actuators as well as in relation to context-based approaches to system management structured around the concepts of ubiquitous computing and ambient intelligence. Many of these technologies, and in particular the user interface, are of significance in relation to the requirements of Design for All and Digital Inclusion considered earlier [53–57].

7.4.2 Bio-mechatronic Systems

As introduced in Sect. 7.3.2, bio-mechatronic systems result from the integration of mechatronic and biological systems, usually to assist or support some functionalities of the biological system. Most commonly used examples are prosthetic devices where they are expected to become an integral part of the human body. Robotic exoskeletons, though not becoming part of the human body, integrate and interact with the human body externally to form a bio-mechatronic system. These robotic exoskeletons are traditionally made of hard materials. However, more recently a lot of attention is being given towards soft materials in the design and development of soft robotic systems.

A robotic exoskeleton has a spectrum of applications ranging from individuals with weak body segments or those suffering from a body part pathology, to those involved in heavy duty work such as in the construction industry, to those involved in search and rescue operations after natural and other disasters, and so on. This spectrum of applications can generally be divided into assistive and enhancive robotic exoskeletons.

Taking an overview of the population, and considering normal behaviours and functions, it is possible to identify three core groups of individuals associated with lower than normal physical capacity and capability:

(i) Those with some form of congenital disability from birth;
(ii) Those who due to illness, disease or accident lose some portion of their physical capacity and capability and
(iii) The elderly, a growing proportion of the population who gradually lose capacity and capability as a result of the ageing process.

Recent advances in technology afford opportunities for the design, development and implementation of a range of modular bio-mechatronic systems to address a range of problems associated with all of these categories.

The focus here will be on bio-mechatronic prosthetic systems with particular attention on an above knee lower limb prosthetic device. For the development of a bio-mechatronic prosthetic system, it is important to note that a key feature of human locomotion is its adaptability and robustness to changing situations. For the cases of standing, walking, turning, ascending and descending steps/ramps, and sitting, the lower limb segments and body centre of mass require a sophisticated intelligent sensory-motor-control system to ensure adaptability. This complex system is non-linear with all segments affecting each other [58]. Hence, a sophisticated bio-mechatronic system is needed to be designed and developed to provide full fidelity for a human body segment.

The last decades have seen a technological revolution in the prostheses industry due to technical advances in materials, electronics, sensors and actuators. Current lower limb prostheses are divided into three groups:

(1) Purely passive,
(2) Actively controlled, and
(3) Actively driven or powered prostheses.

Purely passive devices are essentially mechanical systems and require significant voluntary control effort from amputees. The first actively controlled prosthesis was the C-leg, including the first microprocessor knee controlled damping with a hydraulic cylinder. Other prostheses (Smart IP) use pneumatic swing control or the

REHO knee using magnetorheological fluid stance and swing control. More intelligent actively controlled prostheses are now available, such as the *Orion* and *Genium* microprocessor knees. But, these do not provide positive energy for stair ascending and other tasks. Actively driven prostheses are fully actuated and are known as *Power Knees* [59]. These are typically actuated using either brushless dc motors [60] or pneumatic actuators [61]. Although these supply power, they consume more than the human muscles because the muscles of the lower limb are not continuously activated during normal walking. To mimic this behaviour, a hybrid hydraulic lower limb prosthetic based on a semi-active approach was developed [62], but hydraulic systems are inefficient.

In a bio-mechatronic system, dynamic coupling interaction between the segments can be used to produce more efficient and comfortable walking for above knee amputees, with power applied appropriately to the prosthetic knee. The advantage of this is the avoidance of the stiff control of a powered knee, which is often uncomfortable due to the dynamic interaction between the amputee and the prosthetic. Also, using the dynamic natural behaviour of the mechanism provides more energy-efficient walking.

With the user-centred design approach in mind (and also to appreciate the current short-comings in the lower limb prosthetic devices) it can be considered what amputees expect from the next generation of such devices. The following comments are referred to as a sample of lower limb amputees' views:

> adaptability in the foot standing and balance is more difficult than walking; adaptability and sensory feedback in the foot shell; ankle movement is very important; current knees are less adaptable to variable speeds; walking on uneven ground is difficult; turning towards amputated side quite difficult; socket is extremely important; smart socket to relax or firm up according to activity; heat sensitivity; more flexibility at joints particularly ankle joint; shock absorbing mechanism; smart limb alignment; interaction with user feedback is definitely needed (amputees currently use visual feedback, they need to look at the foot); contact feedback: heal, toe, rough ground, driving; do not know where the foot is; comfort of socket fit; adaptable socket for different activities/time of the day; aesthetics are also important; energy expenditure is currently too high; affordable prostheses; a fully integrated limb is needed as part of the body (forgotten limb); powered ankle and knee are essential for some activities; prosthetic leg should adapt to the patient not the patient to the prosthesis; reliable and easy to maintain; to be robust for different environments; learning, training and rehabilitation issues.

Examples of challenges include:

i. In terms of the design and development of bio-mechatronic systems, modelling and simulation is a big issue.
ii. For a lower limb prosthetic bio-mechatronic system to function towards the ideal goal, the interaction at high level is very critical. This means that the user intent should be captured by the system well in advance of an action. Only limited advances have been made in this area.
iii. Static and dynamic coupling between the device and human body is another very important parameter for proper functioning of a prosthetic system. This requires arrays of sensors to be used at appropriate locations in the system.

iv. Currently, there is a bottleneck in a two-way interaction between the human body and the prosthetic device. Feedback to the user is required to allow a natural interaction and hence proper system functionality.
v. Energy efficiency is also crucial for the next generation of such bio-mechatronic systems to allow longer battery life.
vi. Interaction between the bio-mechatronic system and the environment could provide a much better performance. This requires the use of appropriate sensors and more connectivity with the aim of achieving a cyber-physical system. Prevention of falls could be a typical outcome of such systems for example.

In order to design and develop an intelligent lower limb bio-mechatronic device the following processes may be used:

(a) To adopt a user-centered design approach a user community can be set up to allow interaction with the relevant end users.
(b) Detailed biomechanical analysis of human body segment kinetics and kinematics characteristic of common locomotor tasks.
(c) Biomechanical simulation of a human body and lower limb prosthetics, its control and optimization.
(d) Design of optimum prosthetic device and relevant physical experimental work.
(e) Experimental investigation into energy efficiency, recovery and dynamic coupling effect.
(f) Experimental measurement of gait dynamics using appropriate sensors.
(g) Development of algorithms to predict user intent and self-tune to user parameters.
(h) Haptic feedback and its effect on prosthetic performance and consideration of neural connectivity.
(i) Prototype design, development and validation.

Figure 7.11 shows a block diagram of the outcome of such a development. A typical prosthetic system designed and developed using such approach is shown in Fig. 7.12.

Although great advances have already been made in various areas of technology, there are still challenging aspects that require close attention. Two important areas which could greatly advance bio-mechatronic systems including connectivity with the user (neural connectivity) and also with the surrounding environment. Prosthetic devices could be considered as a typical example of such bio-mechatronic systems. User intent is crucially important for proper integration with human body. Interaction with the environment could provide an additional dimension to enhance the performance of such systems as well.

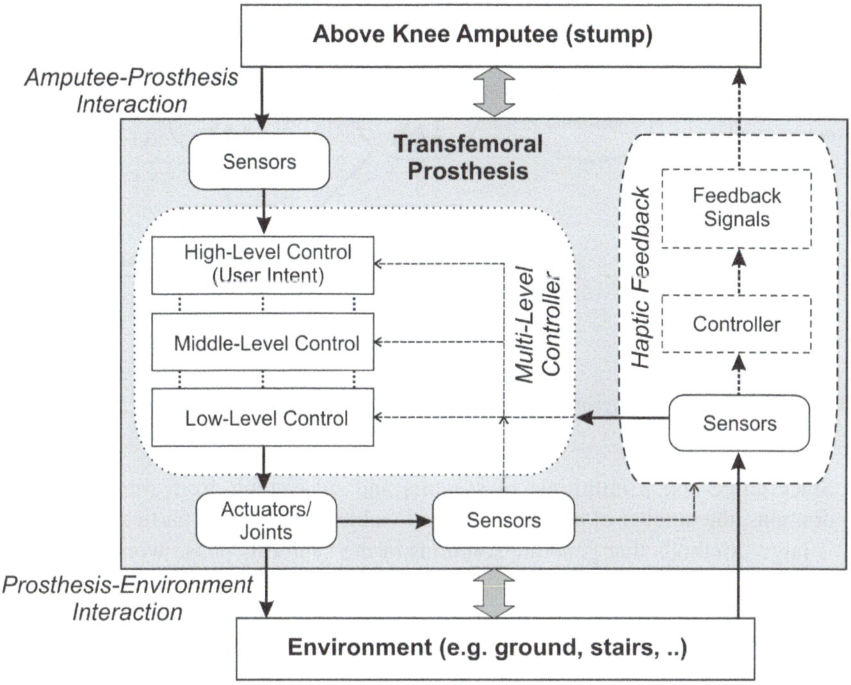

Fig. 7.11 A bio-mechatronic system with user and environment interaction

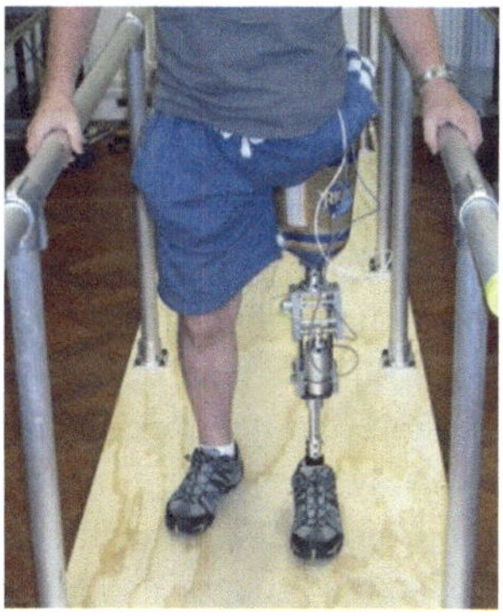

Fig. 7.12 A prosthetic knee during initial user evaluation process

7.4.3 Cyber-Physical Production System (CPPS)

This section discusses improvements to the development of a Cyber-Physical Production System (CPPS) using ontologies. Future research into the efficiency and quality improvement of CPPS engineering faces numerous challenges [63, 64]:

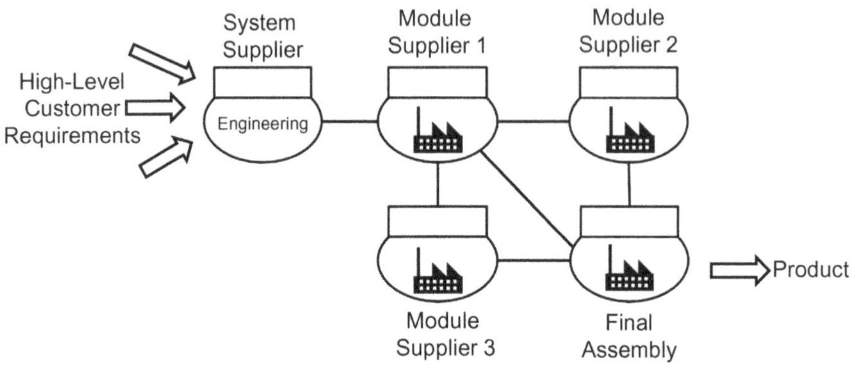

Fig. 7.13 Example of a cyber-physical production system

- Since CPPS use a multitude of systems and subsystems from many different domains, the number of both possible and viable solutions to a particular problem is large. Methods that generate solutions in the synthesis phase would therefore be helpful.
- As there exits a multitude of sustainably solutions to particular design problems, reasonably accurate methods for their evaluation are needed. Including early concept optimization in the methods that support this evaluation saves time.

One solution for tackling these problems is the strict usage of models in different degrees of detail. Since design models can be complex, and their creation and maintenance is often distributed across multiple disciplines, it is difficult to ensure their consistency and correctness.

In the scenario shown in Fig. 7.13, the customers communicate their needs to the system supplier manufacturer (final assembly) and are thus the first link in the supply chain. Depending on the percentage of modules and components manufactured in-house, the process occurs at multiple levels and involves components, component assemblies, mechatronic systems and their integration into the target environment. Based on high-level customer requirements, engineers select and configure standardized mechatronic modules. Building on these modules, the suppliers manage their assembly lines including standardized and user-defined components in the mechatronic module. The manufacturers have to extract internal requirements for their own production process, and must also ensure the consistency of component and module interfaces.

For the optimization of such processes data integration and model consistency play an important role. The involved companies have often problems to guarantee the correctness of data (often embedded in models and their parameters) over time [65]. The aim is now to tackle this challenge by using a semantically-based description of the CPPS to support information and product data exchange.

The knowledge needs to be structured according to a semantic representation. Based on a mechatronic ontology (in [65]) using basic concepts (such as

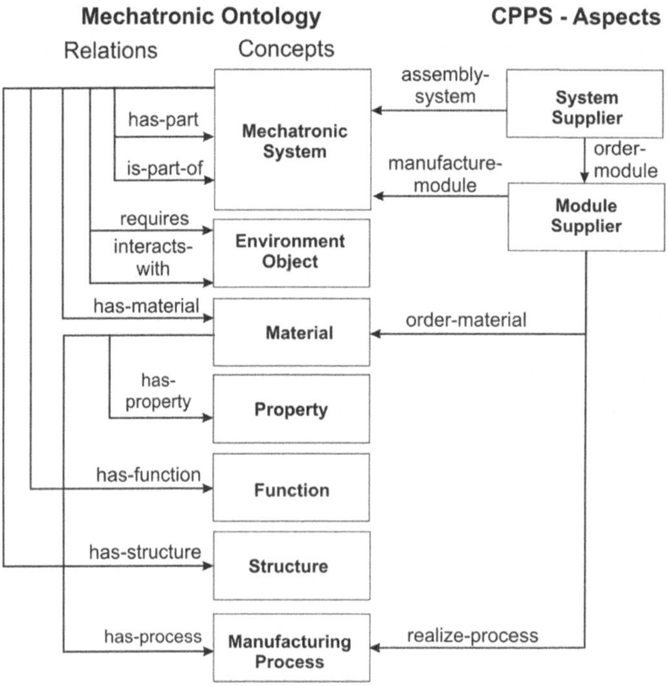

Fig. 7.14 Mechatronic ontology of the CPPS

"*mechatronic system*", "*environment object*", "*material*", "*property*", "*function*", "*structure*" and "*manufacturing process*") and relations (such as "*has-part*", "*is-part-of*", "*requires*", "*interacts-with*", "*has-material*", "*has-property*", "*has-function*", "*has-structure*" and "*has-process*") the relevant knowledge of a CPPS can be presented ("*assembly-system*", "*manufacture-module*", "*order material*" and "*realize-process*"). Figure 7.14 shows the concept of the ontology by means of the above example. This semantic model can be seen as the main element for the horizontal and vertical integration within a CPPS. The wide range of information available to the suppliers can be handled by the described meta-model in a clear and consistent way.

7.4.4 Data-Driven Analysis of Cyber-Physical Systems (CPS)

Asset health and asset performance are essential for efficient and trouble-free operation of CPS. Modern information systems continuously communicate with CPS to analyse and forecast their state of health and performance. In this chapter, a data-driven approach to CPS using the example of the health management of photovoltaic (PV) systems is described.

The scientific foundation of health management applications are the models and algorithms for detecting system faults (fault detection, FD), identifying their possible causes (fault diagnosis, FDD), and the analysis of system performance (performance analysis, PA). Current research assumes the a priori availability of exact models of the target CPS (model-based or knowledge-based diagnosis [66–68]). Manual modelling of systems is however time-consuming in practice and of high costs. Many projects are therefore often not realized.

One goal of data-driven analysis of CPS is thus to automate FD, FDD, and PA and consequently to avoid the need for tedious a priori model building. Robust learning algorithms (RLA) are an important step towards achieving this goal. Robust algorithms learn the necessary model from available observational data. Due to their robustness, they can learn from data of poor quality such as data that itself contains faults. Robust learning and estimation is a current challenge in statistics and machine learning.

The motivation or enabling technology of this research is the Internet of Things (IoT). New technologies for the IoT facilitate the collection and evaluation of data at continuously decreasing costs even for large amounts of data. Technology foundations are cheap and accurate sensors, low-cost internet connections, geo-information systems, cloud-based IoT services. Example applications of learning-based FD, FDD, and PA in the IoT are:

- **Fault message generation and prioritization**. Prioritizing fault messages according to their relevance. This enables the monitoring of a large amount of CPS, e.g. PV systems.
- **Real-time FDD**. Instantaneous identification of possible faults after sudden failures to guide maintenance.
- **Performance analysis**. Analyzing the performance of CPS for use in business planning and in particular planning maintenance.
- **Predictive maintenance**. Forecasting downtimes, system and component failures.

The increasing number of CPS connected to the internet raises new challenges and possibilities. Health management is one of the major benefits the IoT enables. Health management encompasses mere system monitoring and at its core the scientific, data-analytic problems of FD, FDD, and PA. The IoT raises new scientific challenges for data analysis research due to the following reasons:

- **A large number of different system types**—From mobile devices (such as smartphones or tablets) through every kind of vehicle (car or a mobile crusher), household appliances (washing machine or a roof-mounted PV system), sensors for measuring temperature and other environmental parameters and entire facilities for industrial production.
- **A large number of systems**—Besides the large number of different system types, the number of CPS of a particular type itself can be huge. E.g., all cars of a leading manufacturer. Due to the large number of systems, a possible massive amount of data needs to be processed and analyzed (big data).
- **The variety of system configurations**—CPS may exist in different configurations. For example, two PV systems of the same type can have different sets of

sensors. This problem closely relates to the problem of faulty sensors. If a sensor is malfunctioning, its measurements are useless.

- **Incomplete system knowledge**—Knowledge about the system design is often incomplete. Moreover, systems and thus the system models change over time.
- **Unreliable data connections**—Data is often unavailable due to unreliable or slow data connections.

Most of these aspects are new, especially the vast number of system types. Current data analysis algorithms and information systems are often not capable to deal with this new situation without time-consuming manual system modelling. Mastering this situation is the major challenge for scientific disciplines such as machine learning, knowledge discovery, statistics, and fault diagnosis.

Data-driven analysis of CPS relies on methods from machine learning, statistics, and knowledge discovery. The choice of the method depends primary on the CPS, available data, and the analysis task. In the use case of PV systems, the aim is to analyze PV systems to detect and classify faults (FD) [69–72]. Figure 7.15 shows the data-analytic parts of such a health management application. Data is collected from a possible large amount of PV systems. The random process is the conversion of solar energy into electrical energy. Robust learning takes historical data as input and delivers a (linear) model of the PV system to FD. The model relates the system output

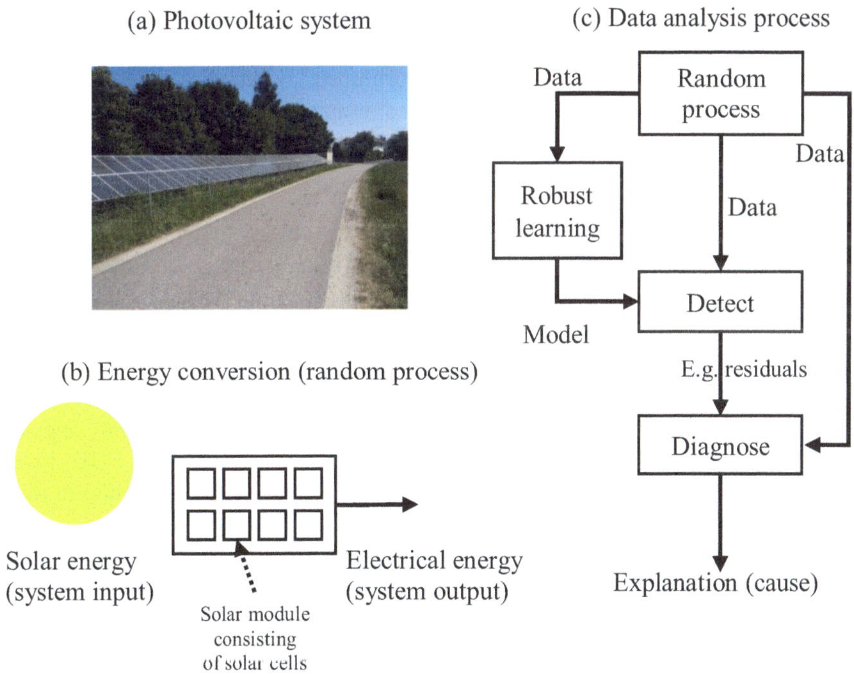

Fig. 7.15 **a** Photo of a mid-sized PV system. **b** Schematic picture of the random process to be analyzed. **c** Overview of the data analysis process based on observations of the random process

(power) to its system input (plane-of-array irradiance) and system state parameters (e.g. module temperature). FD takes current data and the model as input and checks for strong deviations of the observations from the model. FDD tries to identify and explain the faults. One way is to classify faults as melting snow, shading, etc.

The pattern in Fig. 7.15 applies to many other situations in data-driven analysis of CPS. Robust learning assumes that all but k data samples are independent and identically distributed. The remaining k data samples can be arbitrary faults. This notion closely relates to the breakdown point of robust statistics [73]. Any algorithm for computing an estimation with a high breakdown point is a robust learning algorithm (RLA). Such algorithms and heuristics are known for classical statistical problems such as location, dispersion, and linear regression [73], but not known e.g. for decision tree learning or learning artificial neural networks. RLA are necessary if historical data contains faults, which is usually the case in practice.

Fault detection (FD) often reduces to computing residuals, i.e. the model output minus the observed output, and identifying large residuals. Despite some difficulties due to randomness in the data, FD is often a simple task. Fault diagnosis (FDD) however is more involved. The main reason is that learned models are sometimes considerable less accurate than models enriched with expert knowledge. One reason is the existence of critical system components that are not monitored by sensors. In the use case of photovoltaics, a PV system behaves roughly as a linear system. It still does so if a solar module fails. A fault can be detected but its cause cannot be identified if the module is not monitored by a sensor. How to conduct FDD based on learned models is a current challenge. One way is to learn subsystems or cause-effect relations, e.g. via causal discovery algorithms [74]. Another approach tries to include expert knowledge, e.g. learning causal graphical models [75]. Finally, it is possible to concentrate entirely on components, assuming that the necessary sensors are installed. This is done e.g. in [76], which is based on hybrid time automata learning [77].

To meet the challenges of data-driven CPS, several problems had to be solved. There is large variety of PV system configurations due to different solar cell technologies, various common circuit designs, and so on. Thus, linear system models due to their generality were considered [69] noting that they are less accurate than models, which capture the exact non-linear system dynamics.

Because of the large number of PV systems, a fast algorithm especially for model learning for FD [69] was designed and analysed. FDD algorithms are only applied if a fault is detected. This allows to invest more computation time in FDD. FDD algorithms analyse the residuals of robustly learned models. The algorithm in [69] locates faults in residuals, i.e. it estimates the starting time and end time of a fault. Several fault types such as shading or melting snow have typical time patterns. For example, a shading event may happen in the morning over a couple of months. FDD is thus fault classification. In practice, these classifications help to explain the faults.

A common difficulty in data-driven analysis of CPS is the lack of sensors. In the case of PV systems, a plane-of-array irradiance sensor is not present (probable the most critical parameter besides the power output). The idea in [78] is to learn relations (associations) between PV systems that work under similar conditions. The learning method is a robust alternative to correlation networks, a so-called fitness graph. Correlation networks are an important type of associative graphical models. The fitness graph is employed to detect and identify faulty systems. The difficulty here is to learn enough relations to increase the accuracy and reliability of FD.

Most recently, in [79] it is showed how to design fast and parallel RLA for the problem of linear regression, i.e. learning linear models. This method is called Median-of-Means (MoM). MoM is a meta-algorithm following the divide-and-conquer principle. MoM uses the fact that input data is independent and identically (i.i.d.) distributed. It can speedup algorithms. It is robust. It can boost the confidence of algorithms. In addition, it can parallelize sequential algorithms. This last property of MoM in combination with practical algorithms for linear regression is used for analyzing PV systems data.

7.5 Discussion and Conclusions

Mechatronics is an established engineering discipline for which proven and demonstrable design methods and tools are available. However, with the advent of both Cyber-Physical Systems and the Internet of Things, while these methods and tools can continue to be used as the basis of the design of mechatronic modules and subsystems, an extension is needed to accommodate particular requirements for integration at the system level (characterized by the integration of components) that characterises CPS as well as the information exchange and sharing associated with the IoT.

Starting from the fundamentals of mechatronics design, the chapter has considered the extension of the associated methods, and particularly modelling approaches, to support the inclusion of the resulting mechatronics components within the larger integrated system concepts of both CPS and the IoT. Particularly emphasis is given to four design areas, namely data (the IoT), CPS, mechatronics and people (including not just users but others impacted both directly and indirectly by the system) chosen to characterise, and differentiate between, categories of system.

Building on this, four case studies have been presented, each of which focuses primarily on one but encompasses all domains. The nature of the case studies is such that they not only illustrate the design decisions associated with them, but the interaction with and between the various hierarchical levels. This then allows consideration to be given to design elements invisible to the designer of the individual systems components, as for instance the autonomous selection at the level of the IoT of system elements based on context and need.

Thus viewed from the perspective of both a CPS and the IoT, mechatronic design approaches can be considered as a key driver for the development of a wide range and variety of future systems. The concepts present designers with the challenge of implementing structures within information rich environments where information and communications are increasingly the drivers of the design process. This in turn requires designers to have access to new and novel means of simulation capable of representing such situations.

What is also clear is that the advent of both CPS and the IoT is introducing a new range of challenges for designers and the design process, many of which are centred around the need to ensure the privacy of individual users whilst supporting the integration of data across multiple users, as for instance in healthcare, to facilitate the early detection of situations and conditions and the appropriate responses to these. In doing so, designers are obliged to consider not only the hard aspects of privacy, namely ensuring the security of data during storage and transmission, but the soft or people oriented aspects of ensuring an individual's control over their own data. These are increasingly important consideration when considering the increasingly distributed nature of data, and the dynamic nature of systems, particularly at the level of the Internet of Things.

Acknowledgements The authors want to thank Thomas Natschläger for helpful comments. The research reported in this Sect. 7.4.4 has been supported by the Austrian Ministry for Transport, Innovation and Technology in the frame of the FFG project Smart Maintenance.

References

1. IoT Special Interest Group (2013) Internet of things and machine to machine communications—challenges: final paper May 2013 @ https://connect.innovateuk.org/web/internet-of-things. Accessed 15 Oct 2016
2. Lee J, Bagheri B, Kao Hung-An (2015) A cyber-physical systems architecture for Industry 4.0-based manufacturing systems. Manuf Lett 3:18–23
3. Bagheri B, Lee J (2015) Big future for cyber-physical manufacturing systems. Des World, September 2015 @ www.designworldonline.com/big-future-for-cyber-physical-manufacturing-systems/#_. Accessed 12 Aug 2016
4. Hehenberger P, Bradley D (2016) Mechatronic futures, challenges and solutions for mechatronic systems and their designers. Springer, London
5. Internet of Things Global Standards Initiative (2016). http://www.itu.int/en/ITU-T/gsi/iot/Pages/default.aspx. Accessed 12 Aug 2016
6. Harashima F, Tomizuka M, Fukuda T (1996) Mechatronics—what is it, why, and how? IEEE/ASME Trans Mechatron 1:1–4
7. Tomizuka M (2000) Mechatronics: from the 20th to 21st century. In: Proceedings IFAC conference on mechatronic systems, Darmstadt, Germany
8. VDI Guideline 2206 (2003) Design methodology for mechatronic systems
9. Hehenberger P, Poltschak F, Zeman K, Amrhein W (2010) Hierarchical design models in the mechatronic product development process of synchronous machines. Mechatronics 20:864–875
10. Isermann R (2005) Mechatronic systems. Fundamentals. Springer, London, p 2005
11. Hehenberger P, Bricogne M, Le Duigou J, Eynard B (2015) Meta-model of PLM for Design of Systems of Systems, In: Proceedings of the PLM international conference (PLM2015), Doha, Qatar, 19–21 Oct
12. Lee EA (2008) Cyber physical systems: design challenges. In: Proceedings 11th IEEE symposium on object oriented real-time distributed computing (ISORC)
13. Ordinez L, Alimenti O, Rinland E, Gómez M, Marchetti J (2013) Modeling and specifying requirements for cyber-physical systems. IEEE Latin America Trans 11(1)
14. Simko G, Lindecker D, Levendovszky T, Jackson EK, Neema S, Sztipanovits J (2013) A framework for unambiguous and extensible specification of DSMLs for cyber-physical systems.

In: 20th annual IEEE international conference and workshops on the engineering of computer based systems (ECBS)
15. EIDD Stockholm Declaration (2004) @ dfaeurope.eu/what-is-dfa/dfa-documents/the-eidd-stockholm-declaration-2004/. Accessed 16 May 2016
16. Goldsmith S (2000) Universal design. Taylor & Francis Ltd. (Architectural Press)
17. Marshall R, Case K, Porter M, Summerskill S, Gyi D, Davis P, Sims R (2010) HADRIAN: a virtual approach to design for all. J Eng Des 21(2–3):253–273
18. Whitney G, Keith S, Bühler C, Hewer S, Lhotska L, Miesenberger K, Sandnes FE, Stephanidis C, Velasco CA (2011) Twenty five years of training and education in ICT design for all and assistive technology. Technol Disabil 23(3):163–170
19. Clarkson PJ, Coleman R, Keates S, Lebbon C (2003) Inclusive design: design for the whole population. Springer
20. Communication from the Commission to the Council, the European parliament and the European Economic & Social Committee and the Committee of the Regions—eAccessibility @ eur-lex.europa.eu/legal-content/EN/TXT/PDF/?uri=CELEX:52005DC0425&from=EN. accessed 12 Aug 2016
21. Helsper E (2011) The emergence of a digital underclass: digital policies in the UK and evidence for inclusion @ eprints.lse.ac.uk/38615/1/LSEMPPBrief3.pdf. Accessed 16 May 2016
22. Helsper E (2008) Digital inclusion: an analysis of social disadvantage and the information society. Department for Communities and Local Government
23. Ordonez TN, Yassuda MS, Cachioni M (2011) Elderly online: effects of a digital inclusion program in cognitive performance. Arch Gerontol Geriatr 53(2):216–219
24. Aleixo C, Nunes M, Isaias P (2012) Usability and digital inclusion: standards and guidelines. Intl J Public Administration 35(3):221–239
25. Mervyn K, Simon A, Allen DK (2014) Digital inclusion and social inclusion: a tale of two cities. Inf, Commun Soc 17(9):1086–1104
26. Suh NP (2001) Axiomatic design, advances and applications. Oxford Series on Advanced Manufacturing, New York
27. Weber C (2005) CPM/PDD—an extended theoretical approach to modeling products and product development processes. In: Proceedings of the 2nd German-Israeli symposium on advances in methods and systems for development of products and processes, Stuttgart, Germany
28. Thramboulidis K (2005) Model-integrated mechatronics, toward a new paradigm in the development of manufacturing systems. Trans Ind Inf 1(1)
29. Hehenberger P (2012) Advances in model-based mechatronic design. Trauner Verlag, Linz ISBN 978-3-99033-041-8
30. Hehenberger P (2014) Perspectives on hierarchical modeling in mechatronic design. Adv Eng Inform 28(2014):188–197
31. Hehenberger P (2015) An approach to model-based parametric design of mechatronic systems. Comput-Aided Des Appl, Taylor & Francis, 12(3):282–289. https://doi.org/10.1080/16864360.2014.981456
32. Veltink PH, Koopman HFJM, van der Helm FCT, Nene AV (2001) Biomechatronics—assisting the impaired motor system. Archives Physiol Biochem 109(1):1–9
33. Miatliuk K, Siemieniako F (2011) Conceptual model for design of human-exoskeleton biomechatronic system. In: 2011, Summer simulation multiconference (SummerSim'11), 27–30 June 2011, World Forum, The Hague, Netherlands, GCMS 2011
34. Bly S, Schilit B, McDonald DW, Rosario B, Saint-Hilaire Y (2006) Broken expectations in the digital home. In: CHI'06 Extended Abstracts on Human Factors in Computing Systems, pp 568–573
35. Kirchhof M, Linz S (2005) Component-based development of web-enabled eHome services. Pers Ubiquit Comput 9(5):323–332
36. Norbisrath U, Armac I, Retkowitz D, Salumaa P (2006) Modelling eHome systems. In: Proceedings 4th international workshop middleware for pervasive & ad-hoc computing (MPAC2006), 4

37. Dobrev P, Famolari D, Kurzke C, Miller B (2002) Device and service discovery in home networks with OSGi. IEEE Commun Mag 40(8):86–92
38. Rose B (2001) Home networks: a standards perspective. IEEE Commun Mag 39(12):78–85
39. Kamilaris A, Pitsillides A, Trifa V (2011) The smart home meets the web of things. Int J Ad Hoc Ubiquitous Comput 7(3):145–154
40. Makonin S, Bartram L, Popowich F (2013) A smarter smart home: case studies of ambient intelligence. IEEE Pervasive Comput 1:58–66
41. Cook DJ (2012) How smart is your home? Science 335(6076):1579–1581
42. Pendyala VS, Shim SSY, Bussler C (2015) The web that extends beyond the world. Computer 48(5):18–25
43. Atzori L, Iera A, Morabito G (2010) The internet of things: a survey. Comput Netw 54(15):2787–2805
44. Cavoukian A (2012) Privacy by design and the emerging personal data ecosystem. In: Information and Privacy Commissioner Ontario @ www.ipc.on.ca/images/Resources/pbd-pde.pdf. Accessed 10 June 2016
45. Perera C, Zaslavsky A, Christen P, Georgakopoulos D (2014) Context aware computing for the internet of things: a survey. IEEE Commun Surv Tutor 16(1):414–454
46. CONNECT Advisory Forum (2014) @ ec.europa.eu/digital-agenda/en/connect-advisory-forum-working-groups. Accessed 18 Oct 2015
47. Tischer H, Verbic G (2011) Towards a smart home energy management system—a dynamic programming approach. In: IEEE conference innovative smart grid technologies Asia (ISGT), pp 1–7
48. Fadlullah ZM, Fouda MM, Kato N, Takeuchi A, Iwasaki N, Nozaki Y (2011) Toward intelligent machine-to-machine communications in smart grid. IEEE Commun Mag 49(4):60–65
49. Bhuiyan M, Picking R (2011) A gesture controlled user interface for inclusive design and evaluative study of its usability. J Softw Eng Appl 4(9):513–521
50. Portet F, Vacher MGolanski C, Roux C, Meillon B (2013) Design and evaluation of a smart home voice interface for the elderly: acceptability and objection aspects. Pers Ubiquit Comput 17(1):127–144
51. Kühnel C, Westermann T, Hemmert F, Kratz S, Müller A, Möller S (2011) I'm home: Defining and evaluating a gesture set for smart-home control. Intl. J Human-Computer Studies 69(11):693–704
52. Edlinger G, Holzner C, Guger C (2011) A hybrid brain-computer interface for smart home control. In: Human-computer interaction—interaction techniques & environments. Springer, Berlin, Heidelberg, pp 417–426
53. Alam MR, Reaz MBI, Ali MAM (2012) A review of smart homes—past, present, and future, IEEE Trans. Syst, Man Cybern, Part C: Appl Rev 42(6):1190–1203
54. De Silva LC, Morikawa C, Petra IM (2012) State of the art of smart homes. Eng Appl Artif Intell 25(7):1313–1321
55. Li X, Lu R, Liang X, Shen X, Chen J, Lin X (2011) Smart community: an internet of things application. IEEE Commun Mag 49(11):68–75
56. Wilson C, Hargreaves T, Hauxwell-Baldwin R (2015) Smart homes and their users: a systematic analysis and key challenges. Pers Ubiquit Comput 19(2):463–476
57. Albino V, Berardi U, Dangelico RM (2015) Smart cities: definitions, dimensions, performance, and initiatives. J Urban Technol 22(1):3–21
58. Zajac FE (2002) Biomechanics and muscle coordination of human walking, part I. Gait and Posture 16:215–232
59. Power Knee (2016) http://www.ossur.co.uk/Pages/16600. Accessed 16 May 2016
60. Sup F (2008) Design and control of an active electrical knee and ankle prosthesis. In: Biomedical robotics & biomechatronics. 2nd IEEE RAS & EMBS international conference, pp 523–528
61. Fairbanks DM (2012) Semi-actuated transfemoral prosthetic knee. US Patent
62. Awad MI, Abouhossein A, Dehghani-Sanij A, Richardson R, Moser D, Zahedi S, Bradley D (2016) Towards a smart semi-active prosthetic leg: preliminary assessment and testing, In: 7th IFAC symposium on mechatronics & 15th mechatronics forum international conference, Loughborough, UK, 5–8 Sept

63. https://en.m.wikipedia.org/wiki/Industry_4.0. Accessed 2 June 2016
64. Vogel-Heuser B, Hess D (2016) Guest editorial industry 4.0–prerequisites and visions. IEEE Trans Autom Sci Eng 13:411
65. Hehenberger P, Egyed A, Zeman K (2010) Consistency checking of mechatronic design models. In: Proc ASME 2010 international design engineering technical conferences & computers and information in engineering conference IDETC/CIE 2010, Montreal, Quebec, Canada
66. Ding SX (2008) Model-based fault diagnosis techniques. Springer, Berlin, Heidelberg
67. Isermann R (2006) Fault-diagnosis systems. Springer, Berlin, Heidelberg, p 2006
68. Reiter R (1987) A theory of diagnosis from first principles. Artif Intell 32(1):57–95
69. Kogler A, Traxler P (2016) Locating faults in photovoltaic systems data. SCCH Technical Report
70. Traxler P (2013) Fault detection of large amounts of photovoltaic systems, in Proceedings of the ECML/PKDD 2013 workshop on data analytics for renewable energy integration
71. Firth SK, Lomas KJ, Rees SJ (2010) A simple model of PV system performance and its use in fault detection. Sol Energy 84:624–635
72. Chouder A, Silvestre S (2010) Fault detection and automatic supervision methodology for PV systems. Energy Convers Manag 51:1929–1937
73. Rousseeuw PJ, Leroy AM (2005), Robust regression and outlier detection. Wiley
74. Pearl J (2000) Causality—models, reasoning, and inference. Cambridge University Press
75. Koller D, Friedman N (2009), Probabilistic graphical models—principles and techniques. MIT Press
76. Niggemann O, Windmann S, Volgmann S, Bunte A, Stein B (2015) Using learned models for the root cause analysis of cyber-physical production systems. In Proceedings of the 24th international workshop on principles of diagnosis (DX-2015)
77. Niggemann O, Stein B, Vodencarevic A, Maier A, Büning HK (2012) Learning behavior models for hybrid timed systems. AAAI 2:1083–1090
78. Traxler P, Gomez P, Grill T (2015) A robust alternative to correlation networks for identifying faulty systems, in Proceedings of the 26th international workshop on principles of diagnosis (DX-2015)
79. Kogler A, Traxler P (2016) Efficient and robust median-of-means algorithms with applications to fault detection. SCCH Technical Report

Chapter 8
Emergence of Product-Service Systems

Margherita Peruzzini and Stefan Wiesner

Abstract Product-Service Systems (PSSs) are a new emergent way to innovate traditional products and to extend the company portfolio, by reducing time and cost while offering high quality and meeting the expectations of both customers and stakeholders, which have to be considered during the design and development process (Complex systems concurrent engineering. Springer, London, pp. 321–328, 2007 [1]). A further challenge is to close loops between Product Lifecycle Management (PLM) and Service Lifecycle Management (SLM) by providing feedback from service delivery to the beginning-of-life phase of products, or defining a structured procedure to coordinate product and service development activities. The objective of this chapter is to provide a common understanding about PSSs, to deepen the Servitization process and its main features, and to understand how PLM and SLM can be integrated to define future organization of PSS-oriented companies. The final aim is to present PSS as a new business model, which companies can adopt to innovate their products and to enlarge their offer to the market, according to a consumer-oriented approach.

Keywords Product service system (PSS) · Servitization · Differentiation · Product-service lifecycle management (P-SLM) · Business model

8.1 PSS Concept and Definition

The concept of Product Service System (PSS) came from a publication published in 1999 in the Journal of Cleaner Production by Goedkoop et al. [2]: it was a report concerning sustainability, where he defined PSS as "a marketable set of products and services capable of jointly fulfilling a user's needs". This work can be considered a milestone in PSS literature, since it provided a clear evidence of a spreading trend

M. Peruzzini (✉)
University of Modena and Reggio Emilia, Modena, Italy
e-mail: margherita.peruzzini@unimore.it

S. Wiesner
BIBA University, Bremen, Germany
e-mail: wie@biba.uni-bremen.de

J. Stjepandić et al. (eds.), *Systems Engineering in Research and Industrial Practice*,
https://doi.org/10.1007/978-3-030-33312-6_8

Table 8.1 PSS concept in literature review

PSS concepts	PSS components	PSS typology	Authors
Definition of extended product	Product, services, system	–	[6–9]
Definition of PSS components	Product, related services, infrastructure, partners' network	–	[2, 10]
Integration of product and service to reach customer needs	Product, related services, infrastructure, partners' network, customers' requirements	–	[11–13]
Definition of the main PSS typology	–	PSS product-oriented, PSS use-oriented, PSS results oriented	[14–16]
PSS as a mean to reduce the environmental impacts	Product, related services, infrastructure, partners' network, environmental impacts	–	[12, 15]

in different industrial sectors. Furthermore, it also defined the PSS characterizing elements, which are:

- *Product*, representing the tangible commodity manufactured to be sold and capable of fulfilling the users' needs;
- *Service*, the "activity" delivered to generate an economic value by its exploitation and often done on a commercial basis;
- *System*, contributing to realise the collection of the two elements after definition, including their relations.

After this first characterization of the PSS concept, in literature, several authors have over time discussed and extended this innovative way to join product and service offers. Table 8.1 below gathers these contributions and tries to discriminate the authors' focus: if they are mostly interested in the definition of the PSS by its main components, or in the identification of its typology according to what it is able to deliver to customers.

According to Table 8.1, PSS concepts and relative definitions are mainly centred on the keywords of integrated bundles of products and services, and concerns directly the customer, aiming at the achievement of sustainability [3]. This first literature review on PSS proves that even if several terms used to identify the integration between product and service exist (e.g., extended products, technical services, product-service systems (PSSs)), they represent the same concept: a mix of tangible products and intangible services designed and combined to increase the value for customers [4]. The value creation is realized through the extension of the current business network, involving different stakeholders having the knowledge and skills required to design, develop and deliver the new PSS offer.

The shift of interest of both industries and academia towards an integrated offer of products and services starts from the idea of the Extended Product, where intangible services are integrated into a core product to add value for customers and improve company's profits and competitiveness. Moreover, such concept is illustrated by the Servitization process. Vandermerwe and Rada introduced a formal definition of Servitization. They referred not directly to the concept of PSS (which was not born yet, but it may be a consequence of this definition), but to several models specifically designed for those enterprises that would create a new value for their products, and having as a result the increase of their profitability and market shares [5].

Some years later, this concept was conceptualised in a paradigm of a transition represented along a linear axis comprising four different steps. In particular, the common idea is moving from the traditional customer experience (i.e., consumers buying products) to a new customer experience (i.e., consumers buying solutions and benefits in respect to their needs). Figure 8.1 shows the Servitization process as conceived by Thoben et al. [6] and it involves the following four steps: (1) tangible product, (2) product and supporting services, (3) product and differentiating services, (4) product as a service. Steps 2 and 3 are defined also as Product + Service, and they mean the selling of product plus several services; while the fourth step Product2Service refers to selling only the services [9]. According to this view, PSS is defined like a combination and integration of product and services into a system to deliver required functionalities in order to satisfy the customer needs [17, 18] and it is able to produce synergies among profit, competitiveness, and environmental benefit. The so defined PSS is composed by four main elements: the product, the related services, the ICT infrastructure required, and the partners' network to involve [10].

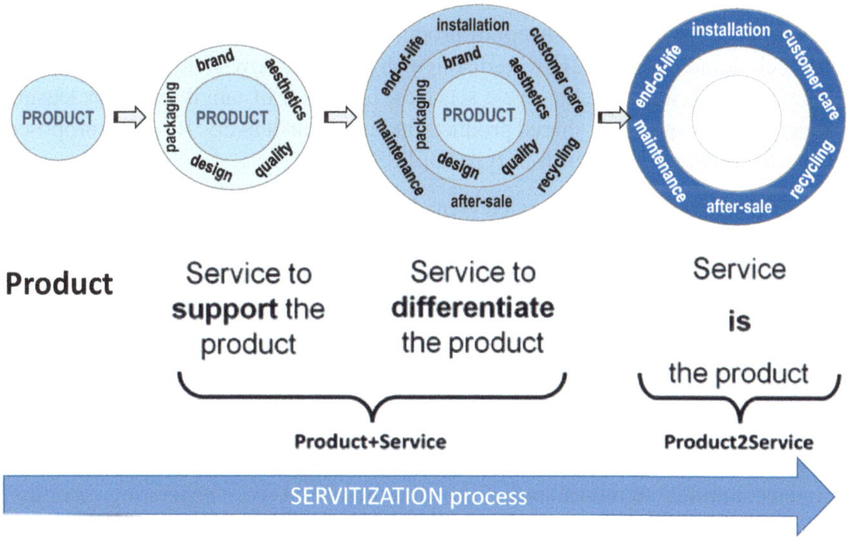

Fig. 8.1 Servitization process. Adapted from [6]

In recent years, clear evidence shows that service plays an increasingly important role in many manufacturing industries, especially in industries that produce complex products. The concept of servitization directs the strategy transformation of manufacturers in high-value-manufacturing. As an immediate consequence, most of them have moved from selling products to delivering product-service systems. Indeed, the Servitization process is a fundamental means for manufacturing companies that would find new business opportunities and involve new customer segments, increasing their market share [9, 19]. This process not only affects the company business model, but also the whole enterprise, in terms of those internal processes and standard procedures that support the design, development and delivery of the new value proposition.

As reflected in previous definitions, PSS is a business means that allows manufacturing companies to create a new value for those products which have become mature. In the market, several mature products exist (e.g. household appliances, phones, cars); they represent products that, after a period of introduction into the market and the consequent growth in term of sales and market shares, are becoming old for reasons of technology, aesthetics or so on. For such products, it is required to create a new business value, in order to extend and reinvent their lifecycle trend. The application of PSS concept to these products is a way to improve their business value and extend their lifecycle.

In literature, several typologies of PSS have been defined by different authors. They aim to describe different options of product-service offer within a certain company or for a particular manufacturer. For example, Wise and Baumgartner identified four types of PSSs [20]: those with embedded services, comprehensive services, integrated solutions or distribution control. This classification is very useful for the description of the service content but the authors do not consider the relative product ownership. Instead, the concept of product ownership is one of the topics faced by Michelini and Razzoli [21] who distinguish different provision forms: provision of tangibles with included life cycle services, provision of tangibles under leasing arrangements, provision of shared products and function delivery. Roy [22] proposed a categorization consisting of four types of PSS:

- *Result services*, where the service provider is responsible of all physical aspects of the system, providing a 'result' instead of a product;
- *Shared utilization services*, consist of sharing products among different users or a community of users in order to increase their utilization rate;
- *Product-life extension services*, where the service provider is responsible of the maintenance, repair, reuse and recycling activities related to products to increase their useful life;
- *Demand side management* (or integrated resource management), which was originated in the field of energy supply in US as an evolution of the idea that it was often more economical to reduce energy demand than build more generating capacity.

Mont [11] stated that a PSS comprises products, services or their combinations and classified the services forming a PSS from the product life cycle perspective as:

- Services at the point of sale;
- Services related to product use;
- Services prolonging product life cycle;
- Revalorization services, which refer to products end-of-life and consisting of reverse logistics, reuse or recycling of products or their parts.

Oliva and Kallenberg [23] proposed the service space where different types of services can be considered according to two drivers: whether the services are related to a product or to end user's process, and whether the service is based on transactions or on relationships.

Even if other authors proposed different classifications, Tukker's classification of PSS and the product-service continuum is the most widely accepted by researchers and academia [14]. This classification identifies the following three PSS models:

- *Product-oriented PSS*: The physical product is sold to the customer in a combination with services such as maintenance, recycling and customer trainings, which guarantee the functionality and a long use-cycle. Main aspects in the development of this PSS type are the creation of a durable product to minimize service costs and optimize the product end-of-life through recycling and reusable parts.
- *Use-oriented PSS*: In this case the product is not owned by the customer anymore but is made available (e.g. through leasing) for customer-usage through the producer. High rates of usage as well as a long lifecycle of their products are the main goals for companies offering these product-service-systems.
- *Result-oriented PSS*: This is the most complex type of a PSS, selling a desired result in place of a product (e.g. the offering of washed clothes instead of selling washing machines). The ownership as well as the decision of technology, maintenance, disposal etc. stays with the producer. Thus, the development of this PSS has to focus on the changed business model for which the consumer only pays per obtained output.

Firms can move from one type of PSS offering to another by changing the relative share of product and service components according to user requirements. Figure 8.2 shows such classification and how the author, according to the PRODUCT and SERVICE concepts, conceives it.

Fig. 8.2 PSS continuum. Adapted from [14]

Often, researchers used Tukker's classification, while refining it and adding further elements, in order to describe what should be the kind of business cooperation between customers and suppliers [24–27]. Nevertheless, Tukker's classification about PSS is not able to capture the complexity of PSS model itself; for this reason, these categories may be explored more in depth to facilitate the most appropriate categorization for manufacturing companies that should apply such theoretical models in practice. According to this aim, Adrodegari et al. [28] has proposed a new PSS classification according to Tukker's model, where five PSS configurations are identified in two different groups:

- *Ownership-oriented (Group A)*: the focus of this group is that the product sales are the main source of revenue and the services are sold as an add-on to the product, through a transactional (e.g. technical assistance without any contractual agreement) or relational approach (e.g. maintenance contracts). Inside such group, two main configurations are highlighted by the authors:
  - *Product-focused*: the provider sells the product and separately it guarantees payment services during the product use phase (e.g. break-fix repair, maintenance contract, etc.). Companies have traditional 'tangible' production costs and the revenue is mainly generated from the product sale;
  - *Product and processes focused*: the company offers services, both in the pre- and after-sale phases in order to optimize and increase efficiency and effectiveness of customer's operations. Anyway, the main revenue stream still consists of product sales: in the product price is often included a pre-sales service component.
- *Service-oriented (Group B)*: the focus of this group is the service, strictly linked to the usage of a product, which represent the main source of revenue. Indeed, in this category the ownership of the product is not transferred to the customer.
  - *Access-focused PSS*: customer does not buy the product but pays a fixed regular fee to gain access to it. The fee is not related only to the product usage but includes the guaranteed additional services. The company usually keeps the product property rights and has the responsibility for its utilization during a given period of time;
  - *Use-focused*: customer does not buy the product but pays a variable fee that depends on the usage of the product (pay- per-usage time, pay-per-usage unit). The manufacturing company is responsible for all life cycle costs, stimulating the company itself to optimize the product costs. Customers are focused on the value-in-use, rather than on the value- in-exchange. For this reason, the company should be able to predict the customer behaviour, since otherwise no clear cost calculation can be made. Such configuration allows defining a new revenue model, where the focus is the definition of new selling parameters driven by customer perceived value instead of internal cost. The payback period of the value delivered is often longer than the payback period of traditional product sales.

– *Outcome-focused business PSS*: customer does not buy the product but pays a fee that depends on the achievement of a contractually set result in terms of product performance or outcome of its usage. Here the value for the customer is generated by the reduction of initial investment, the minimization of operational costs and risks to achieve an expected outcome with the product usage. An outcome-based contract could be contracted on a fixed payment basis tied to performance measures.

At the same time, Benedetti et al. [29] proposed an alternative Energy Services' classification proposal based on the definition of three different dimensions:

- axis x, represents the "intangibility", which basically corresponds to Tukker's PSS classification;
- axis z, represents the "scope" as defined in Sorrell's classification (Sorrell 2007);
- axis y, represents the "risk" accepted by both the client and the service provider.

Table 8.2 contains the main PSS classifications defined in literature by several studies and shows correspondence of meaning between the different classes, when possible. With respect to the Tukker's classification, other authors added a fourth class of PSS as highlighted in Table 8.2, while other authors reduced the number of classes.

Table 8.2 PSS classifications in literature

Source	PSS classification in literature			
[14]	Product-oriented	Use-oriented	Result-oriented	
[28]	Ownership-oriented	Service-oriented		
[29]	"Intangibility" (Tukker's categories)	"Scope" (Sorrell's classes)	"Risk" (from client and service provider)	
[20]	PSS through embedded services	PSS through comprehensive services	PSS through integrated solutions	PSS through distribution control
[21]	Provision of tangibles, included life cycle services	Provision of tangibles by leasing arrangements	Provision of shared products and function delivery	
[22]	Result services	Shared utilization services	Product-life extension services	Demand side management
[11]	Services at the point of sale	Services related to product use	Services prolonging product life cycle	Re-valorisation services
[23]	Services related to a product or end user's process	Service is based on transactions or on relationships		

The results of this first literature analysis about PSS proved that it is a new emerging trend for manufacturing companies, where the focus is proposing a solution and no longer selling a product (based on its ownership), but rather selling its usage (e.g. renting, pay-per-use, etc.) and performance (e.g. pay-per-performance). This phenomenon concerns the evolution from a "traditional" product-centred business model to a new service-oriented business model. According to this trend, several authors have conceptualized the shift from products to PSS through various concepts: "servitization" [3], "transition from products to services" [23], "going downstream in the value chain" [20], "product-service systems" [14], "moving towards high-value solutions, integrated solutions and system integration" [30, 31], "manufacturing/service integration" [32] and "service infusion in manufacturing" [33–35]. All these authors converge into the concept of solutions defined as innovative combinations of products and services leading to high-value and unified responses to customers' needs. The common factors and the main novelties are synthetized in Table 8.3. It critically compares the proposed PSS classifications and highlights the common factors and the main differences of the different concepts proposed in literature. It considers in particular the models able to represent PSS for industrial manufacturing companies,

Table 8.3 Common factors and novelties in PSS from literary review

Source	Common factors	Main novelties
[14]	• Need to define PSS categories in manufacturing • Different strategic drivers (e.g., economic optimization, sustainability improvement, energy efficiency) to categorise the PSS • New strategies in product service system design • Business model analysis to identify the successful manufacturing solution	PSS categorization according to the customers expected outcomes (product-oriented, use-oriented, result-oriented)
[28]		Revision of the PSS typologies defined by Tukker, according to the business model adopted by industrial companies (ownership-oriented, service-oriented)
[29]		Proposal of a new PSS classification to map different types of energy services, based on the existing Tukker's PSS categories
[20]		Proposal of four different successful business models in manufacturing and highlighting of servitization trend
[21]		–
[22]		Definition of four type of PSS according to the sustainability driver
[11]		Discussion about the applicability and feasibility of PSSs in manufacturing. The main uncertainties are the companies' readiness to adopt them, the consumers companies to accept them, and their environmental implications
[23]		–

because they are able to define if a proposal is more product- or service-oriented. Along the years, this classification has been adopted by other authors and revised according to different strategic drivers (e.g., economic optimization, sustainability improvement, energy efficiency) for better understanding all the PSS solutions' potentialities. This brief literature review has demonstrated that the main barriers to develop and adopt PSSs in manufacturing are the company readiness to adopt them, the consumers companies to accept them, and the PSS environmental implications [11]. These represent the main challenges in PSS manufacturing application. Indeed, several discussions and approaches exist for PSS design, but only at a conceptual level. Instead, concrete examples of their application in manufacturing are still missing.

8.2 Product Lifecycle Management (PLM) and Service Lifecycle Management (SLM)

According to the innovative trend to develop a PSS, a change in the innovation thought of manufacturing companies was realized. Indeed, the innovation concept has moved from the manufacturers' needs, complying the production costs and constrains, to the users' satisfaction. For a long time, producers have been regarded as the principal source of innovation and their motivation to innovate is driven by monetary profit expectations from selling products and services. Within the last decades, users are an important complementary source of innovation, and their motivation to innovate is driven by their own needs and expected benefits from using the innovation themselves rather than monetary profit expectations. In this context, PSS has moved to the forefront.

Talking about PSS, it is noteworthy that Product Lifecycle Management (PLM) and Service Lifecycle Management (SLM) must be combined together. Currently service development might not be identified as similar to product development, due to the differences concerning product lifecycles and service lifecycles. Therefore, the link between product and service activities is possible through the application of a lifecycle management approach. Stark defines the lifecycle of an offer (tangible or intangible offer) as follows: Imagine, Define, Realize, Support/Service, Retire [36]. This model enables to define a first distinction between product activities and service activities, but it does not show up the interactions between PLM and SLM. For this reason, it is worth to analyse both PLM and SLM separately, in order to understand what main intersections should exist to develop a PSS.

Product Lifecycle Management is defined in literature as a holistic approach to manage the product information along its lifecycle [37] supported by Product Data Management (PDM) applications, which focus on designing and engineering data [38]. Moreover, PLM is able to exploit the interoperability with other IS of an industrial company to manage the product information. Indeed, the final aim of a PLM is managing information in an integrated manner in a digital chain [39, 40]. Usually,

PLM covers the whole lifecycle of a product, from the first idea and concept to its recycling and disposal. There are many different lifecycle models found in literature. However, the majority is based on three main life cycle phases:

- Beginning of Life (BoL);
- Middle of Life (MoL);
- End of Life (EoL).

Service Lifecycle Management is involved in the Service Science, Management and Engineering (SSME) [19], which is a young research field that addresses the open questions and challenges coming from the servitization process. Indeed, service lifecycle concern appears in the literature often correlated to PLM. Service Lifecycle Management aims to create a link between Management and Engineering. Despite this topic being quite new and innovative in the literature, some approaches are arising to also manage service information. According to Freitag et al. (2013), the Service Lifecycle Management framework consists of four parts:

- Phases of Service Life Cycle Management;
- Role Model for Service Life Cycle Management;
- Methods and Tools for Service Life Cycle Management;
- Interactions between product and service lifecycle management.

The main model to compare PLM and SLM involves three Service Lifecycle phases [41]:

- Service creation, which consists of two main pillars: provision of conditions and ideation;
- Service engineering, which consists of four phases: service requirements, service design, service implementation and service testing;
- Service operations management, where the first task is to acquire customers. After this, the service needs to be delivered to the customers.

In Fig. 8.3 both PLM and SLM lifecycle phases are identified and displayed according to the most widespread idea that considers services and their lifecycles as aligned to the product, in order to be assessed and designed together. Nevertheless, currently, with the increasing interest in PSS approaches and methodologies, there is the need to have a strong interaction between PLM and SLM, in a systematic way and in both directions.

In the literature, a method named Product Service Lifecycle Management (PSLM) exists, able to unify product and services under one common approach, allowing also an effective collaboration of product and service actors. PSLM enables effective collaboration between all the stakeholders of products and services lifecycle management. It provides both services information management through a service-centred approach (i.e. SLM), and product information management through a product-centred approach (i.e. PLM). At the same time, this approach allows having a strong interaction between the two main entities (product and service), in both directions. Indeed, its final scope is to provide to and share with product and service stakeholders all the information required. The management of products and

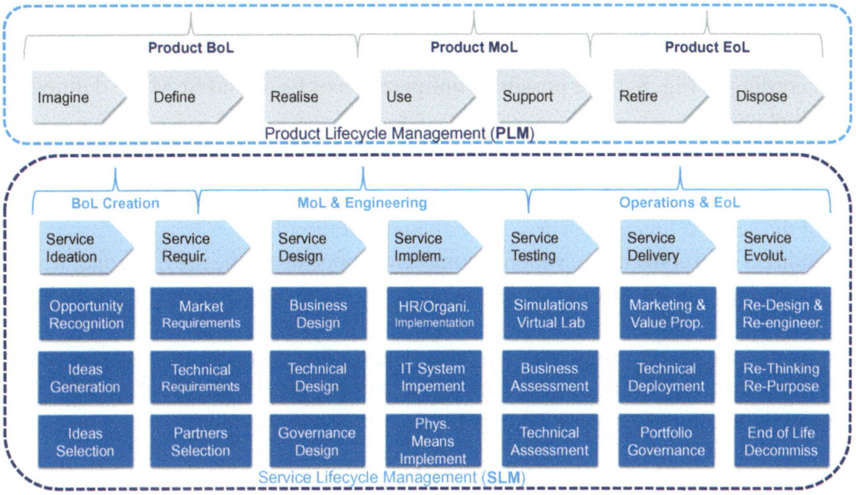

Fig. 8.3 PLM and SLM lifecycle phases

services activities requires a transversal collaboration among partners that should be supported by a collaborative framework.

In order to identify the interactions between PLM and SLM, different models have been developed by researchers. Mahut et al. [42] starts from the definition of PLM and SLM given by [36] and proposed two main categories to identify the possible interconnections between product and services:

- Major links, which represent the substantial link between product and service activities. It reveals the necessity to construct products and services in a very strong collaboration;
- Minor links, which are necessary but not predominant interactions (they can identify a purpose).

Such approach does not assume how PLM and SLM should be managed and what is the typology of their interconnection. Indeed, in literature four alternative typologies exist and have been formalised [43]. Figure 8.4 shows them, which represent how the interactions between Product and Service Lifecycle Management could be identified. This categorization has manifested itself inside the development of a European project, namely Manufacturing SErvice Ecosystem (MSEE), which

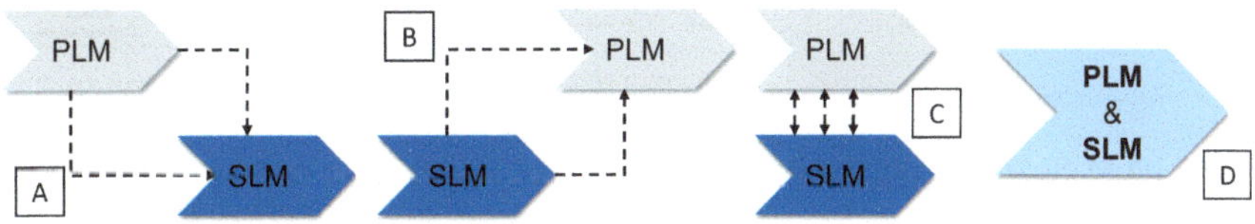

Fig. 8.4 PLM and SLM interactions proposed by the MSEE European project

has the scope to design and develop several tools able to support product-oriented manufacturing companies to design a new PSS.

According to Fig. 8.4, the possible configuration of PLM and SLM are defined in the following:

A. Direct interconnection, which is the most common situation in the manufacturing industry, where SLM is triggered by PLM and depends on it. The management of the service lifecycle is driven by changes of the PLM;
B. Indirect interconnection, which is completely opposite to the previous one, where PLM depends on SLM (the management of the product lifecycle is driven by SLM);
C. Parallel interconnection, where product and service lifecycle are managed at the same time. Mostly, the product and the according service lifecycle are the same length but the interactions take part only if they are necessary;
D. Coordination, where both lifecycles are managed in a highly integrative way and the managerial boundaries between PLM and SLM disappear. Decisions always have influence on both components of the integrated life cycle, until the highest degree of integration is reached, where products and services do not looked at separately anymore but treated as integrated PSS. This interconnection is the best desirable, thinking to a PSS offer.

Peruzzini et al. [44] proposed a first example of Product Service Lifecycle Management (P-SLM), which fits a different definition of PLM and SLM interactions and shows how Product and Service Lifecycle can be managed concurrently. Figure 8.5 shows the comparison between the different PLM, SLM and P-SLM models proposed in literature. Defining a new approach as P-SLM to manage the integration of PLM and SLM with the aim of proposing a PSS instead of a traditional product, a new challenge is to identify the methods and tools able to support each phase inside the P-SLM.

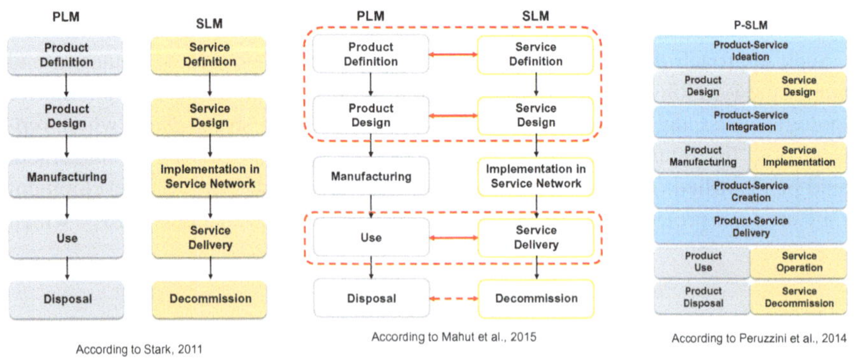

Fig. 8.5 PLM, SLM and P-SLM concepts

8.3 PSS: A New Business Model

A Business Model (BM) describes the rationale of how an organization creates, delivers, and captures value [45]. According to this definition, BMs in the manufacturing industry have focused on the fabrication or assembly of more or less customized products and have generated revenue from their sales. The required machines, materials and qualified personnel cause high fixed costs, so supply chain organization and efficiency have a high influence on competitiveness. The levels of standardization, automation and the technological advance have been important indicators for the success of a manufacturing company [46].

The shift from providing only physical products to integrated solutions, able to increase market share and customer satisfaction, expands the role in the value chain by seeking to innovate and design new products and services in order not to compete on the basis of cost alone [47]. Product-Service Systems lead to a new business model definition that aims to sell not only goods, but also value added service propositions like training, system integration and consulting. The PSS business model, instead of traditional products, changes the manufacturer's perspective about the costs and revenues arising during the PSS lifecycle. This issue represents a challenge for industrial companies and offers opportunities for investigation [48].

As manufacturers often lack the competencies needed for the provision of services, the development of a PSS necessarily requires the creation of a structured network of partners and stakeholders, able to exploit the necessary tangible and intangible assets and create valuable solutions to share among all partners [49]. This means moving from the traditional concept of manufacturing enterprise to a new idea of Global Production Network (GPN), which represents an aggregation of several partners with different knowledge and capabilities, focused on the realization of a specific PSS value proposition. Moreover, the GPN implies the definition of a proper Business Model in order to recognize the strategic factors for each partner as well as the key resources and activities and mechanisms for risk and profit sharing to involve in the new PSS scenario to develop [46].

According to the aim of designing, configuring and developing a new PSS, business-modelling techniques are the most appropriate to analyse the scenario to develop. They can be considered as conceptual tools able to support industrial companies to identify, understand, design, analyse, and change their current Business Models (BM) [50]. In literature, several research studies identify the same method to develop a new BM for a PSS; it involves four main research steps [27]:

- Identification of PSS characteristics and typology;
- Investigation of business model concepts;
- Development of the framework;
- Application of the developed framework by means of a case study.

In this context, the Business Model Canvas [51] is a well-defined concept that allows the company to easily describe and configure business models to create new strategic alternatives. The model consists of nine elements or business areas, shown in Fig. 8.6.

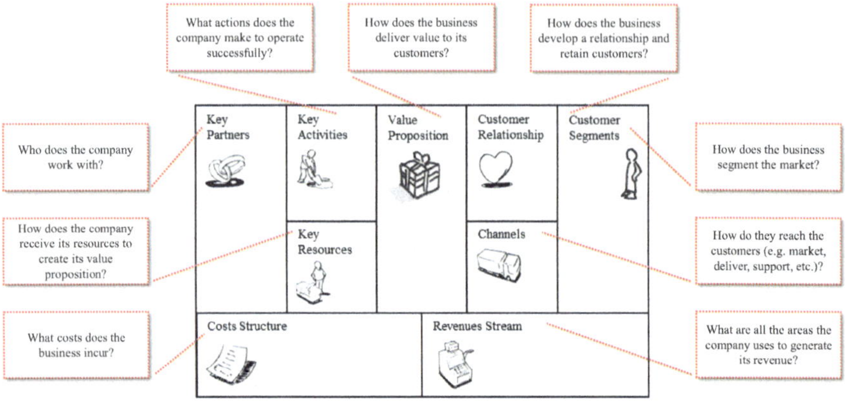

Fig. 8.6 Canvas business model. Adapted from [45]

A manufacturing enterprise that changes from the fabrication of products to offering PSS solutions and transforms its supplier base into an ecosystem of network partners will have to analyse and adapt the elements in all model parts to create a new and competitive BM.

- Value proposition, that is the definition of the offer proposed to customers (what the manufacturing company would offer to market). The product as output of the manufacturing process is replaced by a guarantee of the functionality, availability or outcome of the product usage [52];
- Customer segments, that represent the groups of expected people or organizations to reach through the defined value proposition (who the manufacturing company would reach). For a successful BM, it is important to identify and address potential customer segments outside the current boundaries of the manufacturing industry;
- Channels, which are the company's interfaces with its customers (how the manufacturing company reaches its customers). Pure physical delivery of the product has to be extended with new channels for service provision;
- Customer relationships represent the types of relationships the manufacturing company establishes and maintains with specific customer segments. The selling transaction has to be replaced by a permanent relationship to the customer to generate constant streams of value and information;
- Key resources, that are the assets required to offer and deliver the value proposed. Additional human, financial, physical and intellectual resources are required. This includes competencies in service development, product-service integration and collaboration;
- Key activities, those involved in offering and delivering the value proposed have to change from manufacturing to service provision and the creation and management of a suitable network of partners for each customer demand;
- Key partners (i.e. network of suppliers and partners that support the business model execution) must be complemented by service providers and other stakeholders of

the PSS. An ecosystem has to be created, in order to be able select the appropriate network partners for the realization of each value proposition;

- Revenue streams, that represent the revenue that the manufacturing company is able to generate from each customer segment. Revenue then will not be generated by a one-time sale of a product, but it should be concentrated on generating a constant revenue stream through service or usage fees;
- Cost structure, that represents the costs incurred when operating a business model. PSS are value driven. The focus should not primarily lie on reducing the costs for manufacturing the product, but to combine products and services in a way to deliver the largest possible value to the customer.

This business model has been applied in several organizations widespread around the world (e.g. IBM, Deloitte, Ericsson, etc.) and it is adopted both by industrial companies to identify, design, analyse, and change their current business models, and by researchers, as an empirical analysis. The main challenge for manufacturing enterprises is to integrate the new and unknown value proposition of a PSS and the associated collaborative arrangements into their BM without experience in this field. Building networks with unconventional business partners is difficult and can bring incalculable risks [53]. New information and communication technologies (ICT) have to be utilized for service provision and to develop closer relationships to the customer. New stakeholders in the ecosystem affect the cost structure and require new kinds of revenue models, which are currently not elaborated in manufacturing industries.

The Canvas model alone is not sufficient to understand the transition towards a more service-oriented business model. It is necessary to integrate this business approach together a technical approach able to design the value proposition. A new practical methodology, which helps manufacturers to adapt their strategy and BM according to a vision of servitization and collaboration, has been developed by [41]. The methodology has been instantiated in a workshop concept and evaluated in four manufacturing case studies.

During the first phase (Fig. 8.7), the current strategy and BM of the manufac-

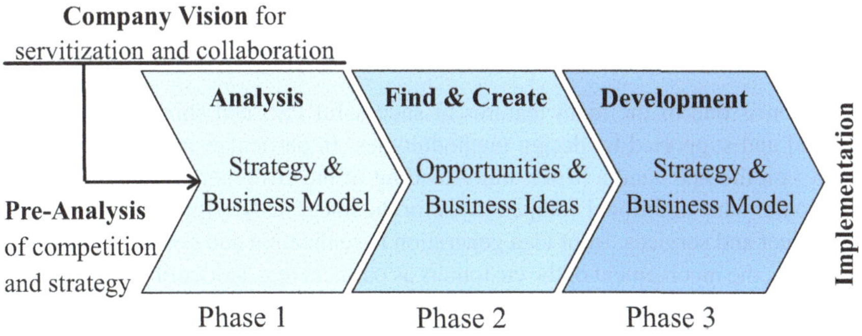

Fig. 8.7 Methodology for business model development

turer are analysed in detail, based on the generic strategy framework of Porter [54]. Additionally, a competitor's analysis makes potentials and market boundaries visible [55]. The analysed manufacturer needs to identify the relevant strategic factors of its industry, which can be mapped out directly on a Strategy Canvas as an analytical tool [56]. The current BM is analysed in the next step and is mapped out on the Business Model Canvas [51].

Phase 2 consists of a creative process to identify opportunities for new strategies and BMs based on trends and environmental changes. Therefore, a simplified Social, Technological, Economical, Environmental and Political (STEEP) analysis [56] is used to capture future trends. Based on the Six Paths Framework [57], company representatives discuss in a creative process how to change the business according to the STEEP factors. The next step of the methodology is to create a new Strategy Canvas, using the Four Actions Framework [57] as a tool to eliminate, reduce, raise and create strategic factors.

In Phase 3 a new strategy and BM were developed, based on the superior vision of servitization and collaboration. Out of the new Strategy Canvas, a new Business Model Canvas is created. The new BM can rely upon results of the STEEP-Analysis and the Six Paths Framework as well. The company vision of providing a PSS through collaboration with partners is now visualized and becomes comprehensible. The creation of the new strategy and the BM is interrelated and is understood as an iterative process. Finally, the practicability of the new BM is evaluated.

The implementation of PSS BMs requires disruptive changes in the existing organization of a manufacturer. This includes the company structure, business processes and IT environment, as well as changing the mindset from a product-centric to a more collaborative, service-centric perspective. However, as the organization consists of many different stakeholders, it is difficult to transform. Thus, to overcome internal resistance to the implementation of a new PSS BM, a suitable change management approach is critical. It is necessary to analyse the changes required for the implementation of the new BM, define actions for servitization and collaboration and execute them in a structured process.

8.4 Approaches to Support PSS Innovation

Innovation is one of the main features of successful PSS and should be properly analysed and supported by design methodologies. In particular, innovation of PSS requires on the one hand a better understanding of the customer requirements, and on the other hand additional competencies for the integrated design and provision of the product and services, from idea generation to realization and commercialization [58]. Thus, the involvement of the customers as partners (e.g. to identify maintenance needs) and collaborative arrangements with other enterprises (e.g., local maintenance service providers) become more and more important [59]. Manufacturing enterprises do not only have to support service innovation and provide the required physical resources, organizational structures, as well as IT tools. They have to furthermore ensure the interoperability of their products to existing and newly developed services.

There are not just services that are developed specifically for the physical product. Rather, the PSS can make use of services that already exist independently from the particular physical product. However, some of these services need adaptations to work with the physical product. There are for example existing service centres that monitor data from sensors and take defined actions in case of alerts or if the received sensor data show critical values. If the machining centre mentioned above should be monitored by such a service provider, it is necessary to establish the data connection and to define threshold and corresponding actions.

Another category of services is based on available standard services that are applied without further adaptations. However, it may be necessary to prepare the physical product with interfaces according to the correct standards to make the service work. In the machining centre example, such a standard service could be express spare part logistics that enable the exchange of components and modules within 24 h, 365 days per year. To apply the service, it could be necessary to align the size and weight of machine components to the standards of the specialized logistics provider that offers the service.

According to this view on PSS, there are different options for service innovation. Usually, it is assumed that service innovation leads to new services. For instance, Toivonen and Tuominen defined service innovation as "a new service or such a renewal of an existing service which is put into practice and which provides benefit to the organization that has developed it" [60]. However, in the context of PSS this does not cover all options of service innovation: additional possibilities are new combinations of existing services with a certain physical product. Innovation also includes new adaptations of universal services or of the physical product to enable the application of standard services. This leads to certain challenges for manufacturing enterprises, if they are aiming at innovative PSS:

- they need competencies in service development if they want to develop new services for their product or new products with related services;
- for new combinations of existing services with their products, they need to get to know potential service candidates. This means that they have to "look beyond their own backyard" into branches that are not yet related to their product;
- in both cases it could be necessary to have competencies in equipping the already existing product with standard interfaces to services. Therefore they need some "service thinking" that they can obtain from cooperation with services providers.

In the past, approaches towards innovation in manufacturing enterprises have been focusing on the physical product, as their outcome has been rare in services [13]. Manufacturing and distribution as well as maintenance, repairs and recycling (in case such services are offered) were organized in linear deterministic supply chains. A new approach to support service innovation in manufacturing enterprises will have to overcome these rigid structures and address the challenges identified in the previous section. Innovation can be defined as: "*the implementation of a new or significantly improved product (good or service), or process, a new marketing method, or a new organizational method in business practices, workplace organization or external relations*" [61]. The following attributes were mentioned in most definitions for innovation [62]:

- New: it defines something that did not exist before or has changed fundamentally. Even if the solution is already established in one branch or country it can be regarded as new in another branch or country [63];
- Improved: it usually defines an added value, in comparison to what was already available before;
- Beneficial: it refers to the creation of a benefit for the "creator" of the innovation, the innovator. In business market success is usually regarded as the decisive benefit.

This first attribute "new" has an essential impact. The higher the degree of novelty the higher is the uncertainty and therefore the risks. On the other hand the degree of novelty could contribute significantly to the added value perceived by the customer. So the enterprise that is aiming at innovations has to decide if they choose a high degree of novelty that could lead to strong competitive advantage but bears also a high risk or if they "play safe" and choose a low level of novelty. For PSS there are three general levels of novelty [62]:

- A new combination of existing services with an existing product without adaptation and virtualisation of physical value components, this means the substitution of physical components by IT-based services;
- Either product or service is new or adapted (the other part stays as it is). This could be based on virtualization of physical value components;
- Both product and services are new or adapted, including virtualization of physical value components.

The degree of novelty is not a "quality criteria" for an innovation. Even the combination of existing product and existing service can lead to a success on the market and disrupt other solutions. But the tussle is not only addressing the money that is needed for investment. A sometimes even more important aspect is the effect the introduction of new PSS could have on existing business. Generally the PSS could be competitive, complementary or neutral to the existing business. These effects can occur externally on the market, e.g. if the PSS and the existing products compete for the same customer budgets ("cannibalization" effects), or internally if they apply the same resources, a situation that could produce conflicts or synergies.

Traditionally innovation processes are regarded as critical processes taking place "in the 4 walls" of a company. The intention was to protect intellectual property and to build up internal knowledge that creates advantage over competitors [64]. There are a lot of examples for companies that were successful with this approach in the past. However, several changes require a different approach today from companies. Customers require solutions that are in many cases so complex that the required knowledge cannot be provided by one company alone in an efficient way. In addition, there is a shift from "one-time market launch of a fixed product" to a more continuous development and improvement of products and/or services that requires an almost permanent investigation of developments on the market regarding technologies and corresponding fast reactions. This is hardly accomplishable by a single company. As a consequence, many companies open their "4 walls" to obtain external input and to involve partners in the innovation processes. A well-known and established approach of inter-organisational innovation is the approach of Open Innovation brought up by Chesbrough and described as follows:

> Open Innovation is the use of purposive inflows and outflows of knowledge to accelerate internal innovation, and expand the markets for external use of innovation, respectively. [This paradigm] assumes that firms can and should use external ideas as well as internal ideas, and internal and external paths to market, as they look to advance their technology. [65]

This definition shows that Open Innovation is not a one-way. This means that a company is not only receiving input but is also willing to provide impulses and input to partners that can use this for their own innovation processes. The wall of the "classical innovation funnel" that describes the way from a big number of ideas to a new innovative product that is successfully introduced on the market becomes permeable. Gassmann and Enkel [66] described three ways of exchange with partners:

- *Outside-in* process is the process of internalizing external knowledge, e.g. when a company involves research institutes or when customers are involved to obtain their feedback and their ideas.
- *Inside-out* process is the process of exploiting internal knowledge through opening of organizational boundaries, e.g., the commercialization of IP.
- *Coupled process* is a cross-over of an outside-in and inside-out process. It describes collaborative research and commercialization with an external community. Innovation networks and joint ventures are typical examples for this approach.

The Internet provides important opportunities to follow the approach of Open Innovation and exchange knowledge and impulses with others. Keywords are for example Netnography [67], that is an online ethnography in particular to analyse customers and how they act by using the internet, and Crowdsourcing [68], where delegate tasks to a big group of users in the internet to receive their contributions as input for the innovation process.

To identify opportunities for new, innovative combinations of products and services it is necessary to clarify where to look for these opportunities. Due to limited resources it is necessary to focus on promising search areas. The Method to define Search Areas should help SMEs to look beyond well know domains that are already served ("looking out-of the box") and do this in an efficient way. There are different strategies an enterprise can choose to search for servitization opportunities. Figure 8.8 gives an overview of these strategies.

The differentiating aspect between the strategies is the potential "distance" from the own existing product and the "home market" that is served:

- *Same business area*: Same or comparable customers, overlapping of Value Chains;
- *Other business area*: Different customers, different/independent Value Chains;
- *Comparable physical products*: Products that are serving the same purpose, offering the same functionality or that are based on the same technology;

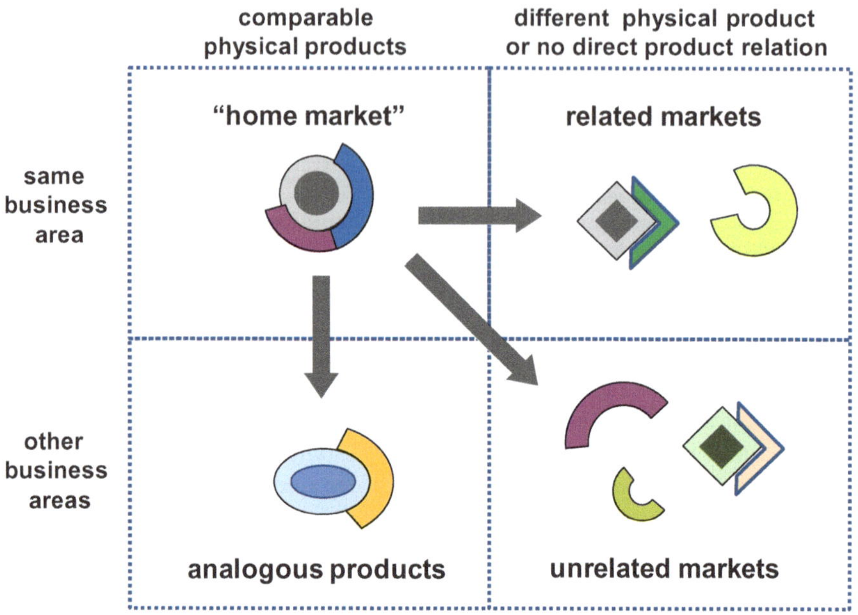

Fig. 8.8 Different search areas for servitization opportunities

- Different *physical products or no direct product relation*: There could be stand-alone services that are not related to a physical product yet. Or there are products that serving a clearly different purpose.

8.5 Summary and Outlook

This chapter introduces the concept of PSS and describes the different forms by which a product-service solution can be handled and managed by industry. A lot of factors need to converge to order to make a PPS appreciated by the market; among them, innovation is a key factor. At the same time, however, the PSS strategy appears also as one of the main drivers of innovation in modern companies. Furthermore, a PSS can be used to easily differentiate the company portfolio and to improve the market share [69].

The creation of PSS is based on a deep understanding of the customer requirements, and the capacity to collect additional competencies, usually outside the company, to design and provide both product and services and support the overall process, from idea generation to realization, commercialization and after sales [70].

The strategies resulting from the combination of product and service aspects can be different: from "home market", to analogous products, related markets und unrelated markets. The selected method offers, for each strategy, a set of questions that help to identify and clarify the business goal and the business model [71].

In this context, the first step is starting with the right questions to define robust requirements for the new PSS. This concept can be easily explained by an example in the field of analogous products: the requirements upon hygiene are the same for devices in food industry as for equipment in the healthcare sector, so it could be interesting to analyse corresponding services in healthcare to trigger ideas for food industry, and viceversa. Relevant questions to identify analogies are as follows: What are essential functions and characteristics of the own product? What products are comparable according to aspects that are essential for the considered product (comparable functionality, used under comparable conditions, etc.)? Can the PSS exploit existing channels? Which is the related business? On the basis of these questions, a successful PSS can be conceived and designed. Chapter 10 will provide a set of methods and tools for the systematic design and development of PSSs.

References

1. Elgh F (2007) Modelling and management of manufacturing requirements in design automation systems. In: Loureiro G, Curran R (eds) Complex systems concurrent engineering. Springer, London, pp 321–328
2. Goedkoop MJ, van Halen CJG, Te Riele HRM, Rommens PJM (1999) Product service systems—ecological and economic basics
3. Baines TS, Lightfoot O, Benedettini O (2009) The servitization of manufacturing: a review of literature and reflection on future challenges. J Manuf Technol Manag 20(5):547–567
4. Furrer O (2007) Le rôle stratégique des services autour des produits. Revue Française de Gestion 113:98–108
5. Vandermerwe S, Rada J (1988) Servitization of business: adding value by adding services. Eur Manag J 6(4):314–324
6. Thoben KD, Jagdev H, Eschenbaecher J (2001) Extended products: evolving traditional product concepts. In: Proceedings of 7th international conference on concurrent enterprising, Bremen
7. Manzini E, Vezzoli C (2003) A strategic design approach to develop sustainable product service systems: examples taken from the 'environmentally friendly innovation' Italian prize. J Clean Prod 11(8):851–857
8. Brady T, Davies A, Gann DM (2005) Creating value by delivering integrated solutions. Int J Project Manage 23(5):360–365
9. Wiesner S, Guglielmina C, Gusmeroli S, Dougmeingts G (2014) Manufacturing service ecosystem. Mainz Verlag, Aachen
10. Mont O (2004) Product–service system: Panacea or myth? (Doctoral thesis). Retrieved from the National Library of Sweden database 91-88902-33-1
11. Mont O (2002) Clarifying the concept of product-service system. J Clean Prod 10(3):237–245. https://doi.org/10.1016/S0959-6526(01)00039-7
12. Brandstötter M, Haberl M, Knoth R, Kopacek B, Kopacek P (2003) IT on demand-towards an environmental conscious service system for Vienna. In: Proceedings of EcoDesign'03: third international symposium on environmentally conscious design and inverse manufacturing, Japan, pp 799–802
13. Aurich JC, Mannweiler E, Schweitzer E (2010) How to design and offer services successfully. Proc CIRP J Manuf Sci Technol 2(3):136–143. https://doi.org/10.1016/j.cirpj.2010.03.002
14. Tukker A (2004) Eight types of product-service system: eight ways to sustainability? Experiences from SusProNet. Bus Strategy Environ 13(4):246–260

15. Baines TS, Lightfoot HW, Evans S, Neely A, Greenough R, Peppard J, Roy R, Shehab E, Braganza A, Tiwari A, Alcock JR, Angus JP, Bastl M, Cousens A, Irving P, Johnson M, Kingston J, Lockett H, Martinez V, Michele P, Tranfield D, Walton IM, Wilson H (2007) State-of-the-art in product-service systems. Proc Inst Mech Eng, Part B: J Eng Manuf 221(10):1543–1552. https://doi.org/10.1243/09544054JEM858
16. Alix T, Zacharewicz G (2012) Product-service systems scenarios simulation based on G-DEVS/HLA: generalized discrete event specification/high level architecture. Comput Ind 63(4):370–378
17. Aurich JC, Fuchs C, Wagenknecht C (2006) Life cycle oriented design of technical product-service systems. J Clean Prod 14(17):1480–1494. https://doi.org/10.1016/j.jclepro.2006.01.019
18. Aurich JC, Schweitzer E, Fuchs C (2007) Life cycle management of industrial product-service systems. In: Takata S, Umeda Y (eds) Advances in life cycle engineering for sustainable manufacturing businesses. Springer, London, pp 171–176
19. Spohrer J, Maglio P (2010) Toward a science of service systems. Handbook of Service Science. Springer, New York, pp 157–194
20. Wise R, Baumgartner P (1999) Go downstream: the new profit imperative in manufacturing. Harvard Bus Rev 77(5):133–141
21. Michelini RC, Razzoli RP (2004) Product-service eco-design: knowledge-based infrastructures. J Clean Prod 12(4):415–428
22. Roy R (2000) Sustainable product-service systems. Futures 32(3–4):289–299
23. Oliva R, Kallenberg R (2003) Managing the transition from products to services. Int J Serv Ind Manag 14(2):160–172
24. Azarenko A, Roy R, Shehab E, Tiwari A (2009) Technical product-service systems: some implications for the machine tool industry. J Manuf Technol Manag 20(5). http://dx.doi.org/10.1108/17410380910961064
25. Cook MB, Bhamra TA, Lemon M (2006) The transfer and application of product service systems: from academia to UK manufacturing firms. J Clean Prod 14(17):1455–1465
26. Copani G, Marvulli S, Lay G, Biege S, Buschak D (2010) Business model innovation paths and success in the machine tool industry. In: Proceedings of CIRP IPS2 conference, Linköping, Sweden, pp 437–444
27. Barquet APB, de Oliveira MG, Amigo CR, Cunha VP, Rozenfeld H (2013) Employing the business model concept to support the adoption of product–service systems (PSS). Ind Mark Manage 42(5):693–704
28. Adrodegari F, Alghisi A, Ardolino M, Saccani N (2015) From ownership to service-oriented business models: a survey in capital goods companies and a PSS typology. In: Proceedings of 7th IPSS conference, Saint Etienne, France
29. Benedetti M, Cesarotti V, Holgado M, Introna V, Macchi M (2015) A proposal for energy services' classification including a product service systems perspective. In: Proceeding of 7th IPSS conference, Saint Etienne, France
30. Davies A (2004) Moving base into high-value integrated solutions. A value stream approach. Industrial and Corporate Change 13(5):727–756
31. Windahl C, Lakemond E (2010) Integrated solutions from a service-centred perspective: applicability and limitations in the capital goods industry. Ind Mark Manage 39(8):1278–1290
32. Schmenner RW (2009) Manufacturing, service and their integration: some history and theory. Int J Oper Prod Manag 29(5):431–443
33. Kowalkowski C, Kindström D, Brashear Alejandro T, Brege S, Biggerman S (2011) Service infusion as agile incrementalism in action. J Bus Res 65(6):765–772
34. Gustafsson A, Brax S, Witell L (2010) Setting a research agenda for service business in manufacturing industries. J Serv Manag 21(5):557–563
35. Ostrom AL, Bitner MJ, Brown SW, Burkhard KA, Goul M, Smith-Daniels V, Demirkan H, Rabinovich E (2012) Moving forward and making a difference: research priorities for the science of service. J Serv Res 13(1):4–36
36. Stark J (2011) Product lifecycle management, vol 34. Springer, London, pp 1–16

37. Saaksvuori A, Immonen A (2008) Product lifecycle management. Springer, Berlin, Heidelberg
38. Eynard B, Gallet T, Roucoules L, Ducellier G (2006) PDM system implementation based on UML. Math Comput Simul 70(5–6):330–342
39. Le Duigou J, Bernard A, Perry N (2011) Framework for product lifecycle management integration in small and medium enterprises networks. Comput-Aided Des Appl 8(4):531–544. https://doi.org/10.3722/cadaps.2011.531-544
40. Bricogne M, Troussier N, Rivest L, Eynard B (2011) PLM perspectives in mechatronic systems design. In: Advances in production management systems, Cernobbio, Como, Italy, Italy, p 110
41. Wiesner S, Padrock P, Thoben KD (2014) Extended Product Business Model development in four manufacturing case studies. In: Proceedings of 6th CIRP international conference on industrial product-service systems, vol 16, pp 110–115, Windsor, Canada. https://doi.org/10.1016/j.procir.2014.01.014
42. Mahut F, Bricogne M, Daaboul J, Eynard B (2015) Servicization of product lifecycle management: towards service lifecycle management. In: Proceeding of PLM conference, Doha, Qatar
43. Wiesner S, Freitag M, Westphal I, Thoben KD (2015) Interactions between service and product lifecycle management. In: Proceedings of 7th CIRP IPSS conference, Saint Etienne (France)
44. Peruzzini M, Marilungo E, Germani M (2014) A QFD-based methodology to support product-service design in manufacturing industry. In: Proceedings of 2014 international conference on engineering, technology and innovation: engineering responsible innovation in products and services ICE 2014, Bergamo, Italy, 23–25 June pp 1–7. https://doi.org/10.1109/ice.2014.6871572
45. Osterwalder A, Pigneur Y (2013) Business model generation: a handbook for visionaries, game changers, and challengers. Wiley, Hoboken, NJ
46. Boyer R, Freyssenet M (1995) The emergence of new industrial models. Actes du GERPISA 15:75–144
47. Porter ME, Ketels CHM (2003) UK Competitiveness: moving to the next stage, DTI Economics Paper
48. Mont O (2004) What is behind meagre attempts to sustainable consumption? Institutional and product-service systems perspective. In: Proceedings of the international workshop, driving forces and barriers to sustainable consumption, Leeds, UK
49. Wiesner S, Westphal I, Hirsch M, Thoben KD (2013) Manufacturing service ecosystems. In: Emmanouilidis C, Taisch M, Kiritsis D (eds) Advances in production management systems. Competitive manufacturing for innovative products and services. IFIP advances in information and communication technology. Springer, Berlin, Heidelberg, pp 305–312
50. Ghaziani A, Ventresca M (2005) Keywords and cultural change: frame analyses of business model public talk, 1975–2000. Soc Forum 20(4):523–529
51. Osterwalder A, Pigneur Y, Tucci CL (2005) Clarifying business models: origins, present, and future of the concept. Commun Assoc Inf Syst 16:1–40
52. Meier H, Roy R, Seliger G (2010) Industrial product-service systems—IPS2 CIRP annals. Manuf Technol 59(2):607–627. https://doi.org/10.1016/j.cirp.2010.05.004
53. Gebauer H, Fleisch E, Friedli T (2005) Overcoming the service paradox in manufacturing companies. Eur Manag J 23(1):14–26. https://doi.org/10.1016/j.emj.2004.12.006
54. Porter ME (2008) On competition. The Harvard Business Review Book Series. Harvard Business School Pub, Boston, MA
55. Bergen M, Peteraf MA (2002) Competitor identification and competitor analysis: a broad-based managerial approach. Manag Decis Econ 23(4–5):157–169. https://doi.org/10.1002/mde.1059
56. Mauborgne R, Kim WC (2005) Blue ocean strategy: how to create uncontested market space and make the competition irrelevant. Harvard Business School Press
57. Fleisher CS, Bensoussan BE (2015) Business and competitive analysis: effective application of new and classic methods. Pearson Education, Upper Saddle River, NJ
58. Miller D, Hope Q, Eisenstat R, Foote N, Galbraith J (2002) The problem of solutions: balancing clients and capabilities. Bus Horiz 45(2):3–12. https://doi.org/10.1016/S0007-6813(02)00181-7

59. Windahl C, Andersson P, Berggren C, Nehler C (2004) Manufacturing firms and integrated solutions: characteristics and implications. Eur J Innov Manag 7(3):218–228. https://doi.org/10.1108/14601060410549900
60. Toivonen M, Tuominen T (2009) Emergence of innovations in services. Serv Ind J 29(7):887–902. https://doi.org/10.1080/02642060902749492
61. OECD (2005) Oslo manual: guidelines for collecting and interpreting innovation data, 3rd edn. SourceOECD. OECD, Paris
62. MSEE: D2.11—State of the Art report for requirements and innovation principles. MSEE Project (2013)
63. Fischer B (2006) Vertikale Innovationsnetzwerke, 1st edn. Forum Produkt- und Produktionsmanagement. DUV Deutscher Universitäts-Verlag, s.l
64. Gassmann O, Enkel E, Chesbrough H (2010) The future of open innovation. R&D Manag 40(3):213–221. https://doi.org/10.1111/j.1467-9310.2010.00605.x
65. Chesbrough H (2003) The logic of open innovation: managing intellectual property. Calif Manag Rev 45(3):33–58. https://doi.org/10.2307/41166175
66. Gassmann O, Enkel E (2006) Open innovation: Externe Hebeleffekte in der Innovation erzielen Zeitschrift Führung Organisation, vol 3, pp 132–138
67. Kozinets RV (2002) The field behind the screen: using netnography for marketing research in online communities. J Mark Res 39(1):61–72. https://doi.org/10.1509/jmkr.39.1.61.18935
68. Howe J (2006) The rise of crowdsourcing. Wired Mag 14(6):1–4
69. Wognum N, Bil C, Elgh F, Peruzzini M, Stjepandić J, Verhagen WJC (2019) Transdisciplinary systems engineering: implications, challenges and research agenda. Int J Agile Syst Manag 12(1):58–89
70. Elfirdoussi S, Jarir Z (2019) An integrated approach towards service composition life cycle: a transportation process case study. J Ind Inf Integr (in press)
71. Peruzzini M, Marilungo E, Germani M (2015) Structured requirements elicitation for product-service system. Int J Agile Syst Manag 8(3–4):189–218

Part III
Applications

Chapter 9
A Meta-Model for Intelligent Engineering Design of Complex City

Fabien Pfaender, Egon Ostrosi, Alain-Jérôme Fougères and Bin He

Abstract A city is a complex system, requiring the input of multiple disciplines for its (re)design. It shares some properties of two kinds of objects: empirical objects as well as theoretical objects. As city emerges as a complex object for multi-disciplinary studies, it is of the highest importance to adopt a systemic and global approach in order to bring new knowledge to this field. To master the growing complexity of cities and to consider in the same spot heterogeneous ways of thinking of city, we need intellectual tools and models. The goal of this paper is to propose a model for describing engineering modelling knowledge with relationships and transformations between four domains: (1) *citizen*, (2) *functional*, (3) *physical and* (4) *process*. The proposed model is structured on four levels of modelling: (1) *conceptual* (2) *mathematical* (3) *computational* and (4) *experimental*. These *network of models* should be necessary intelligent for managing the engineering design of a smart city. For overall city design, the paradigm should change from *planner-centric* to *citizen-centric*. However, while these models are potentially relevant, data that may feed these models is lacking most of the time. Moreover, filling and detailing each of the models, requires additional input from different experts and theories. In this chapter smart city

F. Pfaender
UTSEUS, Shanghai University, 333 Nanchen Rd, Shanghai 200444, People's Republic of China
e-mail: fabien.pfaender@utc.fr

Costech EA2223, Université de Technologie de Compiègne, CS 60319 - 60203, Centre Pierre Guillaumat, Compiègne, France

E. Ostrosi (✉)
Pôle ERgonomie et COnception des Systèmes ERCOS/ELLIADD EA4661, Pôle Industrie 4.0, Université de Bourgogne Franche-Comté, UTBM, 90010 Belfort Cedex, France
e-mail: egon.ostrosi@utbm.fr

A.-J. Fougères
ECAM Rennes, Campus de Ker Lann - Bruz, 35091 Rennes, France
e-mail: alain-jerome.fougeres@ecam-rennes.fr

B. He
Shanghai Key Laboratory of Intelligent Manufacturing and Robotics, School of Mechatronic Engineering and Automation, Shanghai University, 149 Yanchang Rd, Shanghai 200072, People's Republic of China
e-mail: mehebin@shu.edu.cn

J. Stjepandić et al. (eds.), *Systems Engineering in Research and Industrial Practice*,
https://doi.org/10.1007/978-3-030-33312-6_9

engineering design will focus on three interrelated approaches: (a) *data that should be gathered,* (b) *models that can be used by means of these data, and* (c) *interpretation methods and tools to elaborate knowledge and decision from the results these models can produce*. The paper presents some findings from the application of the proposed meta-model.

Keywords City modelling · Complex city · Smart city · Holon city

9.1 Introduction

The rapid urbanization of the world and the increasing complexity of urban systems make it necessary to study the cities by engineering sciences approaches. General inquiries about climate change, urban sprawl and the necessary densification of present cities, as demands for energy savings, better mobility, better information, or protection against disasters, have transformed the city into a physical problem of major concern [1, 2]. However, as a multi-physic, multiagent and multiscale system interacting with soil, atmosphere, biosphere and global networks, the city remains a rather unknown and poorly predictable object.

Structurally, the world's cities are more and more similar, day after day, through the convergence of construction techniques, streets adaptation to modern forms of mobility, comfort standards in housing, and an increasing amount of tertiary activities. They differ, however, in climate, and in the historical, institutional and cultural context of their long evolution.

The Brundtland report raised the issue of sustainable city development, which has been the focus of the planning community. Seeking a sustainable city form has become the goal of the planning community [3]. With the development of city informatization, the complexity of the city has increased dramatically. It has brought valuable opportunities and brought new challenges, such as "information islands" and design discontinuities. In addition, as the size of the city continues to expand, city systems are increasingly demanding information sharing, system interoperability, and software reuse. These relatively independent and standard different systems are no longer sufficient and have exposed more and more short-comings, and an integrated city system is urgently needed.

A city is a complex system. Engineering design and planning of the city is also a complex problem. This complexity results from the conjugation of a huge amount of heterogeneous data, from the area of natural, technical, human and social sciences, interacting with each other. Three key properties of the city as a complex system can then be drawn [4]:

Property 1 *City is a multi-physic, multi-agent, multi-stratified and multi-scale system.*

Indeed, city is a multi-physic object because it is characterized by multi-flux of energy, materials, information and human activities behavior [5]. City is a multi-agent object because it is formed by the populations (of citizens) and different actors of the urban scenery. City is a multi-stratified object from historical, institutional and cultural context of its long evolution. Cities is a multi-scale object because it is a whole system, that is part of a vaster whole, and which at the same time contains subsystems, of which it is composed and which provide its structural and functional meaning, interconnected by networks as well as characterized by social, cultural, political and economic aspects.

Property 2 *City can be considered an evolving living system in complex interaction with its citizens, its artificial physical environment, and its natural physical environment.*

Indeed, city is a living complex geometrical and topological object, limited by its artificial physical environment, and its natural physical environment. It is lived by its citizens and therefore is constrained by sociological, societal, political and economic parameters.

Property 3 *City is also an intersecting system. It shares some of properties of two kinds of objects: empirical objects as well as theoretical objects.*

Indeed, we face now objects that share properties of these two kinds of objects: empirical and theoretical. They are called the *intersecting objects*. These objects are empirical entity: they are not the result of a conceptual construction. These objects are also a meeting point for several scientific disciplines and so can be studied theoretically. City satisfies these characteristics. Therefore, it is an intersecting engineering design object.

The city as an intersecting object requires interdisciplinary research, where each discipline may potentially be reconfigured thanks to its consideration of inputs coming from the other disciplines. By reformulating problems raised by other disciplines in this interdisciplinary perspective, a research managed in a discipline will be led to think differently its basic concepts in order to take into account those problems. It is the reason why city as an intersecting object is a good spot to elaborate a paradigm shift by proposing additional dimensions that are able to account for its complexity and globality. Thus, a transdisciplinary research is needed which can lead in the elaboration of a new discipline, which considers the city, an intersecting object, as its own reality. Now, a new scientific community is emerging, strongly multidisciplinary [6], bringing together physicists and geographers, city planners, sociologists, economists and policymakers, congregated around computer sciences and various engineering disciplines (civil, electrical, materials, mechanics, systems…). As cities emerge as an intersecting complex object, it is of the highest importance to adopt a systemic and global approach in order to bring new knowledge in this field as well as to build the bridges between the disciplines. Coping with complexity in its conceptualization, in its modelling, in building up theoretical and practical arrangements to intervene and modify certain aspects, implies first to enlighten the representations generally hidden in the frameworks that gave birth to the methods and tools.

Thus, city offers exceptional scope for original engineering research and applications, with critical impact for the well-being of its citizens. Developing better strategies for the engineering design of the smart cities can be considered a global imperative [7]. The goal is to propose new models, tools and bridges between these models and tools either to lead to a better design and planning of cities or to give a better predictive approach for a better decision-making process.

Sustainable city modelling is a key issue discussed in the planning circle in the 21st century. In recent years, the development and application of sustainable city system design have become a hot spot. As far as the whole world is concerned, due to the lack of unified norms and standards, it is not uncommon to see system development fails in operation and fails to achieve the design effect [8].

With the development of computer technology as the core of the comprehensive technology, city planning management information system will be integrated, networking, multimedia, intelligent and directly oriented to decision-making development [9]. Sustainable urban modelling needs to be based on large-scale graphical information database, with a variety of city information, how to accurately obtain, comprehensive statistics, analysis and application, has become a top priority. Its sustainable development is mainly reflected in the following three following aspects:

Process sustainability. According to the current status of many urban construction and planning, it is necessary to define the entire urban design modelling system from multiple fields [10]. By designing some conceptual models and specific models to cross-analyze various fields, the material and function of the constituent elements are highly integrated to achieve sustainable development in the application field.

Data sustainability. The advent of the era of big data has brought opportunities for data sustainability research, but also brought new challenges such as problem complexity and computational efficiency. City modelling is a comprehensive technology that relies on computer technology for information acquisition, storage, display, analysis, and conversion. It provides decision support for various user coverage, including system integration, advanced networks, and large databases [11]. However, the system's core content data determines that the construction of the data information system will be a long-term and arduous process with sustainable development characteristics.

Technology sustainability. Technology has penetrated into every aspect of people's production and life, and they are at the core of sustainable development [12]. How to use technology to play a sustainable role in the current design to improve the efficiency of resource allocation becomes the most important issue. Throughout the history of technological development, we can see that the driving force for the continuous development of technology lies in demand, and the demand for spatial information in cities is driving the continuous development of city modelling.

This chapter presents a model for describing the engineering modelling knowledge in the design of smart cities in light of a detailed understanding of "*what we want to achieve*" and "*how we want to achieve*" relationships and transformations. It is based on the principles of axiomatic design theory structured here on different levels of modelling. Axiomatic design provides a scientific basis for the design of engineered complex systems [13]. The complexity of the systems can be reduced based on axiomatic design based complexity theory [14–16]. Axiomatic design approach considers design as the interplay between what we want to achieve and how we want to achieve it [13]. Engineering design progress through interplay and iterations between "*what we want to achieve*", representing functional modelling, and "*how we want to achieve*", representing structural modelling. This model can be applied to great advantage in the design of the complex city, considered as an intersecting object. The proposed model can better capture the engineering modelling knowledge, scientific and empirical, because it provides: (a) *a disciplined way of thinking,* (b) *a disciplined way of modelling and* (c) *modelling tools and techniques.*

However, when models have been identified with the help of axiomatic design theory, each of them need to be detailed and specified for the purpose for which it is meant. For this detailing and specification additional expertise, disciplines and accompanying methods and tools are needed. This process also requires a transdisciplinary approach [17]. In this paper, the initial model will be identified, while some of them are further detailed.

The remainder of the paper is organized as follows: In the second section, the model for engineering modelling analysis is proposed. The third section describes the engineering models. The fourth section proposes an application of the proposed model. In the last section, the conclusions show the interest of the proposed approach.

9.2 Meta-Model for Engineering Modelling Analysis

Science usually considers empirical objects as well as theoretical objects. While the first is real, given through scientific experience, the second ones are ideal and given through formal reasoning or conceptual argumentation. Empirical objects are those that drive our interest for practical purposes. While any knowledge on them is potentially useful, they lack rigour and precise formalisms to understand them and to validate our knowledge. On the contrary, theoretical objects may be known thanks to rigorous demonstrations and precise concepts. Examples of theoretical objects are the individual in the economy, the concept in psychology, an electron in physics. Empirical objects are any object of our real environment, for example, people, animals etc. This tension between these two kinds of object relies on the traditional legacy of an ideal epistemology (what we can know does not exist) and empirical ontology (what exists cannot be known). However, we face now objects that share some of the properties of these two kinds of objects: *the intersecting objects*. These objects are an empirical entity: they are not the result of a conceptual construction. They are intersecting insofar as they are a meeting point for several scientific disciplines and so can be studied by theoretical objects proposed by those disciplines. However, intersecting objects have a double transcendence: an empirical transcendence and an epistemic one. According to the first transcendence, intersecting objects exceed any experience we may have of them. While they exist, it is not possible to circumscribe them through empirical or scientific experience. According to the epistemic transcendence, intersecting objects exceed any conceptual characterization we may propose of them which even cannot be used as a reasonable approximation to study them. Intersecting objects are objects of special interests since they require relying on many disciplines while they exceed the sum of them.

The city is as well an empirical and theoretical object [4]. By reformulating problems raised by other disciplines, a research managed in one discipline in an interdisciplinary perspective will be led to think differently about its basic concepts to take into account those problems. However, this reconfigured discipline will be still incomplete regarding the city as an intersecting object. It is the reason why a city as an intersecting object is a good spot to elaborate a paradigm shift by proposing a new concept, an additional dimension that is able to account for the city in its complexity and globality. An engineering meta-model for describing the engineering modelling knowledge can help the multidisciplinary teams to adopt a systemic and global approach in order to bring new knowledge in this field as well as to build the bridges between the disciplines [18].

Engineering design of the city of the future for its citizens of all ages with its fast and dynamic evolving structure is a complex problem. Our first claim is that engineering design theories and practices can successfully be adopted in the engineering design of the complex city [4]. The second claim is that modelling knowledge plays an important role in enhancing the rationality and usability of engineering models in the engineering design of smart cities like Shanghai. Engineering design can be analysed, synthesized and validated through dynamic engineering models. Developing models for describing the engineering modelling knowledge can improve the quality of complexity management in the practice of the engineering design of smart cities.

In the past, there were many attempts to draw up models to handle the complexity management of design process in systematic steps [13, 19–23]. Both functional modelling and structural modelling have been investigated [24–27]. The goal of engineering design is the conversion of a perceived need or a technical problem into information from which a product can be built with sufficient quality and reasonable cost to meet the needs or to overcome the problem [28]. A design process usually starts with the identification of a need, proceeds through a sequence of activities to seek for a solution to the problem, and ends with a detailed description of the product or the technical system.

From the axiomatic design point of view, engineering design of city can be considered as a process of transformation of abstract models into concrete models from citizen domain to process behaviour domain. The analysis of the design from the coevolution function/structure point of view is a central point for reflexive management of design activity [29]. In this research, the city, an intersecting object, is considered and conceived as an evolving living system in complex interaction with its citizens, its artificial physical environment, and its natural physical environment in a time. Indeed, the size of the evolution of a city is not only measured in terms of space but also in terms of time. Time is a determining marker. In addition, to systematize the engineering design knowledge for the design of a city, the concept of domains is used [13]. The world of design modelling of a city as a living system can be defined in four domains: citizen domain, functional domain, physical domain and process domain.

The citizen domain is characterized by attributes a citizen desires. With this viewpoint, a city is seen for what it gives to the citizen, without or with very few knowledge on the life of the city or how the city, as a living system, works. For instance, a citizen does not necessarily know which type of energy is more comfortable and efficient. But he or she can describe the characteristics he or she wants considering its effects on the pollution of the city. The citizen domain includes attributes related to acoustics and noise, heat transfer and ventilation, daily and artificial light.

The functional domain specifies functional requirements. In the functional domain, desired performances are specified for the city as a living system. This domain refers to the actions and interactions that the city as a living system has with its environment during its life. These interactions are multiple, complex and dynamic. They define the behaviour and functions of the city. Functions are abstractions of behaviour which is coherent with Yoshikawa's concern on the notion of abstraction [30].

To satisfy functional requirements, designers and engineers imagine and define design parameters in the physical domain. Design parameters are physical variables in a physical domain. Physical components of the city, physical fields in the city and the structure of the city as a living system (metabolism) represent these design parameters. The structure of the city is a strategic element in the development of renewable energies.

Finally, the process domain is characterized by process variables. Process variables describe the process developed to realize the city as a living system specified in terms of design parameters.

Two cross-referencing modelling levels with four domains are proposed: (1) *Conceptual models* and (2) *Concrete models*. The latter is decomposed into *Mathematical model, Computational model and Experimental model* (Table 9.1). Thus, *engineering design of a complex city is seen as a process of building engineering models from citizen domain to process behaviour domain*. The matrix of engineering modelling analysis (Table 9.1) classifies 4 × 4 types of engineering models corresponding to the intersection of the four levels of modelling with the four engineering domains.

A *conceptual model* refers to designer's models which are represented by concepts or related concepts for city modelling. A conceptual model for city modelling is a model of designer's qualitative understanding and predictions of some knowledge of the city as a living system. A conceptual model can exist prior to building a mathematical model [31]. It is formed after a conceptualization process and can be used for representing interplay and iterations between domains. Conceptual models offer flexible and adaptive tools for discussing and representing qualitatively designer intentions and semantics by using more formal languages. Conceptual models can also use natural language. They obey semantic rules and are codified.

Concrete models give a fine and accurate description of the knowledge for city modelling. Quite systematically, they used a formal description with a codified and specific language quite different from the natural one. We can distinguish three subclasses of such models: (a) *mathematical models;* (b) *computational models* and (c) *experimental models*.

Table 9.1 Matrix of engineering modelling analysis

	Citizen domain	Functional domain	Physical domain	Process domain
Conceptual model	Conceptual model of citizen domain	Conceptual model of functional domain	Conceptual model of physical domain	Conceptual model of process domain
Mathematical model	Mathematical model of citizen domain	Mathematical model of functional domain	Mathematical model of physical domain	Mathematical model of process domain
Computational model	Computational model of citizen domain	Computational model of functional domain	Computational model of physical domain	Computational model of process domain
Experimental model	Experimental model of citizen domain	Experimental model of functional domain	Experimental model of physical domain	Experimental model of process domain

A mathematical model represents empirical objects, phenomena, and physical processes encountered during city modelling in a mathematical or logical way. Mathematical models can take many forms such as dynamical systems, statistical models, differential equations, game theoretic models, category theory or mathematical structures.

A computational model is the representation of a mathematical model in a computational way. Like in engineering design, analytical solutions are not readily available for the design of the complex city. Therefore, design engineers use computational models by computer simulation. When a computational model is available, experimentation with the model is possible by adjusting design parameters in the computer, and by studying the differences in the outcome of the simulation. Often, computational models are used as "*black box*" models.

An experimental model represents empirical objects, phenomena, and physical processes encountered during city modelling in an experimental way. It is concerned with the observation of phenomena, social and physical, in order to gather data about the design of the city. Experimentation can be done on the city itself or on a physical model representing part of its characteristics for accessing ad hoc interactions with the designer, with external actions or with the citizen. Even if the experimental model tends to disappear, experimental models in city design offer the possibility to represent interactions that have not been modelled, and even not been identified. It has a wealth of discovering new problems and can build on considering the city as a living laboratory.

9.3 City System Engineering Design Models

This section describes the 4 × 4 types of engineering models corresponding to the four levels of engineering in the intersection of four domains (Table 9.1) in relationship with data. Indeed, one of the main issues in today's city sciences relies on the ability to get enough high-quality data to enable scientific and engineering modelling.

The data methodological focus is central for urban studies that aim to give an enlightened response to city complex issues. Data necessary for city sciences are typically heterogeneous with a spatial and temporal component and in a very large scale. This is especially true nowadays as the scale of cities is increasing together with the availability of city digital data and the growing need of it. Hence, data is a key aspect for the laboratory aiming to study the complexity of cities and tackle its various issues. The data problem can be divided into three complementary fields of interest: collecting data, exploiting them, mining and exploring them.

- First, data need to be collected, as broadly as possible, and methods should be put forward to reduce uncertainty these data are subject to. Already identified sources of data are existing databases (open data), data that can be explicitly obtained from sensors, simulation or serious gaming, and finally, data built or captured from online resources from the website and smartphones enabled services to the social network.
- Second, data need to be exploited in built models. In order to allow researchers to make use of the data in optimal conditions, it is necessary to investigate the two basic functions that are high performance and non-destructive data storage along with high performance and distributed computing. The effort in this field is essential, given the complexity and the size of datasets that inhibit classical techniques from working accordingly. Moreover, mining should enable a rich collaborative approach in an international decentralized context.
- Third, data need to be selected, synthesized mined and explored with different methods and models. The specificity of these data makes it essential to design new ad hoc indicators with their associated estimators and to propose innovative data interfaces. The ability to represent and visualize the on-going investigation through custom open multi-purpose platform or specialized software constitutes a strong research interest.

9.3.1 Conceptual Models

Smart cities are based on smart communities whose citizens can play an active part in their design. *Citizen science* is thought to provide a powerful participatory context to the future development of the smart city [32]. The conceptual model of citizen domain is thus a qualitative representation of citizen attributes or desired city characteristics. The conceptual model of citizen domain is crucial to the kind of citizen science that should provide a powerful participatory context to the future development of the smart city. Today, city characterized by multi-flux of energy, materials, information and human activities, should be design for emotional satisfaction of citizens [33, 34]. Thus, from emotional engineering design point of view [35], the main focus for overall city design should change from planner-centric to citizen-centric. When using oral language, citizen requirements should be expressed with the citizen's own words and other abstract terms [36]. The language used shows that there is an uncertain determination of citizen attributes or desired city characteristics, the citizens relying on qualitative linguistic information. Nevertheless, a more formal language could be useful here. The notion of affordance [37] offers such a language if we restrict it to the citizen domain. An affordance, in the context of the design of a city, is the abstract expression of what a city, as a living object, offers to its citizens. As for functions, a codified language for expressing affordances could be developed, in space and time, in the form "the city offers X to Y", where X is an ability/capability and Y the citizen.

The conceptual model of functional domain captures qualitatively the intended functionality of the city, as a living object. Sky, atmosphere and orography generate variable meteorological conditions over the city (solar, wind, rain…). Urban structures affect the induced fields (masks, reflections, absorption and re-emission, rainwater runoff…). Human activities produce new issues (heat, noise, pollution…). People receive these flows through the filters of their perceptions (hearing, visual, thermal…). They implement controls and protections, at the individual level (clothing, umbrella, parasol…), at the architectural one (heating, cooling, sunshades…) and at the urban one (orientation and length of streets, shading, sewers…). The conceptual model of the functional model represents the desired functions of the city and theirs decompositions. There are many different interpretations of the notion of the function in engineering design [38, 39] which can be extended in the design of a city. Functions can be defined as the abstracted behaviour [40, 41] of a city, considered as an artefact, that is intended by its citizens. Functions are described in terms of the logical flows of energy, material and signals. Functions and subfunctions can correspond to well-defined basic operations on well-defined basic flows of materials, energies, and signals leading to a taxonomy of functions [42–44]. Functional architecture of a city is a form of a conceptual model of the functional domain. A conceptual model of the functional domain is a qualitative representation of the physical behaviour of the physical structure of a city. The physical structure in interaction with a physical environment gives rise to the city's behaviour. For instance, "the amount of heat to be dissipated per square meter in summer is on a par with the amount of heat required for heating in winter when high levels of thermal insulation are realized" is a qualitative representation of the behaviour of physical structure at a particular instant. Behaviours are related to structural-physical descriptions of the city. Behaviours come out in some way the city's functions. For instance, how to improve crisis management (before, during and after the crisis) in an urban context [45], especially to preview and anticipate risks associated with middles-size or major natural risk phenomenon? Since catastrophic events cannot be completely avoided by prevention and mitigation policy, authorities also have to prepare and manage crisis situations, both for the immediate security of the population but also to make the city resilient to such events, which means the capacity to recover an acceptable state and normal activity in a limited time [46]. Thus, such crisis situations can be viewed as collaborative multidisciplinary ones, where highly stressed actors are managing and sharing a huge amount of knowledge in conceptual modelling.

To study all flows of energy, material and signals at the urban scale, it is necessary to have a geometric model with adaptive levels of detail and a semantics describing the main physical properties of urban surfaces. The conceptual model of the physical domain is a qualitative representation of design principles and physical principles in the design of a city. It represents structural-physical descriptions of the city. Rough drawings, drafts, schematic representations, etc., can also be considered as conceptual representations of a physical structure providing they do not contribute to the fixation of the city description, leaving a place for some interpretation. As for affordances and functions, conceptual models of physical domain obey to some semantic, for instance, conventions used in drafts and diagrams. Conceptual physical models are often used for discussing possible options for solutions. Especially, the first goal in the conceptual phase is to find as many concepts as possible that can provide each function identified in the conceptual model of the functional domain. Since many models are currently necessary, they have to be "cheap" and easy to produce and transform. Often, simplifications are necessary for study different functions of the city and its city topology (Fig. 9.1).

The second goal is to configure these individual concepts into an overall concept. The laws of TRIZ [29, 47, 48] can be used to determine the city structure satisfying its functions and evolutions. Big cities around the world are more and more sensitive to numerous natural and man-induced risks, especially with the impact of climate change that will increase the number and the intensity of extreme events (floods, drought, storms ….). Making the city sustainable implies integrating risk prevention policy in long-term urban design and planning.

The conceptual model of the process domain is a qualitative representation of process variables that can control design parameters. As for the other conceptual

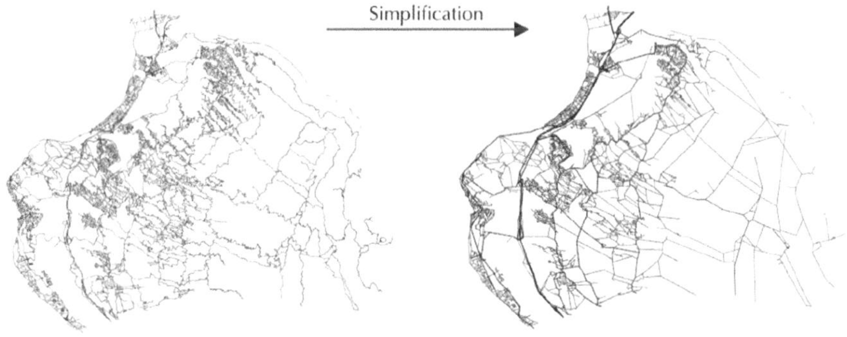

Fig. 9.1 Simplification of a very sinuous city: Saint-Paul of Réunion Island (France) with open library policosm

models, the use of flexible languages such as verbal language and drawing is useful. However, one can also use classes of processes such as the classification of processes.

9.3.2 Mathematical and Computational Models

In recent years, considerable progress has been made in data acquisition, in handy 3D models development, in the dynamic modelling of physical phenomena and human behaviour, and in the simulation of complex large-scale coupled problems. Giving way to strong interactions with the issue of data, computational modelling involves different scientific domains and various potential applications. Therefore, it is highly relevant to the broad field of smart and sustainable cities. As a matter of fact, computational models are developed and used in many fields such as economy, social sciences, geometry and geography, urban and building physics, environment, transportation, construction life cycle and management, energy consumption, greenhouse gas emission, the impact of climate changes, evacuation of population in case of emergency. Computational models are requiring pertinent data and are used to produce indicators for decision support by architects, urban planners, economists, city authorities and managers through simulation software. In addition, computational models and their associated software are necessary for optimization, improved design and decisions of various kinds.

Computational modelling is strongly related to mathematical modelling. A mathematical model of citizen domain is a formal representation of citizen attributes or desired performances of design. For instance, the definition of linguistic values and their transformation into a degree of membership or a degree of belief is still an open question in the real world of design. It tolerates imprecision, which can be exploited to achieve tractability, robustness, low solution cost, and better rapport with reality [49]. The quantification and categorization of semantic similarities between linguistic items based on their distributional properties in this large sample of language data allow finding semantic similar towns and cities from the description with language dependent bag of words. A similarity graph useful for understanding the similarities of towns of France in the web of photographers Flickr using the computational model Word2vec is shown in Fig. 9.2.

A mathematical model of functional domain represents functions formally. A formal representation of functions is a prerequisite for representing functions in computers. A mathematical model of the behaviour represents formally the behaviour of the physical model. In engineering science, many models are available. The mathematical model of the physical domain is a formal representation of physical variables, design principles and physical principles in the design of a city. Figure 9.3 shows the eleven kinds of graphlets [50] for studying the topology of the city. For instance, for physical objects, the topology of a city can be described using formal representations. Related to the output of the previous model, the mathematical model of the process domain represents formally process variables and process physical principles.

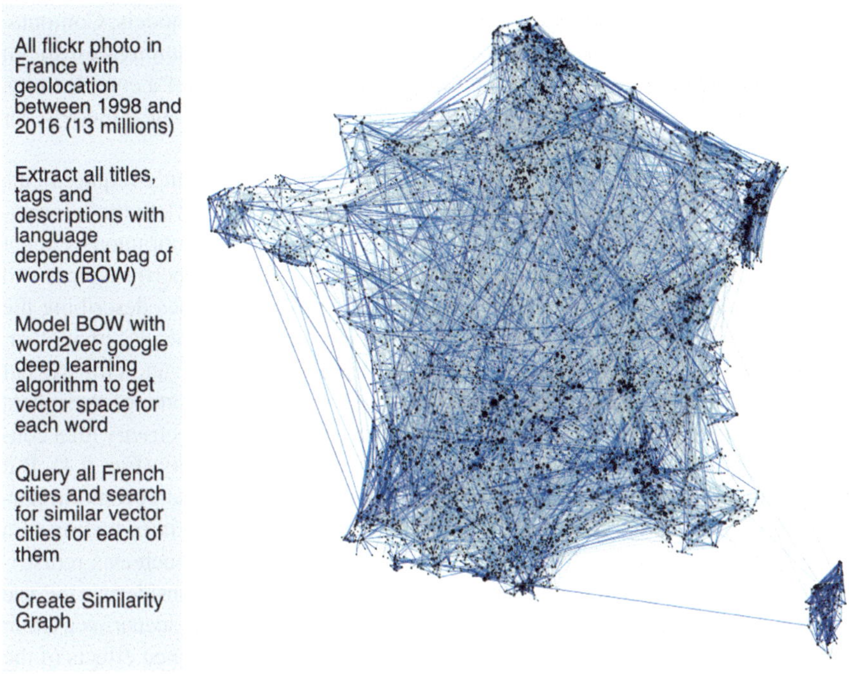

Fig. 9.2 Building the similarity graph of the towns of France

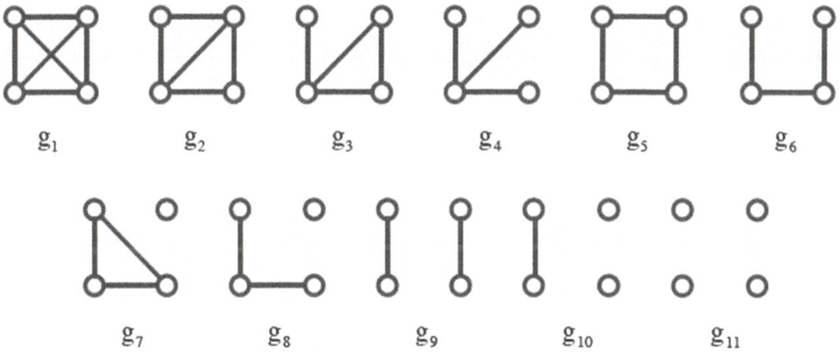

Fig. 9.3 The set of graphlets of the degree 4

The computational model is based on grounded mathematical models. Computation is no more than a way to achieve results when mathematical solutions are difficult to obtain, either impossible or too long in the case of the design of a complex city. They can use some simplifications and approximations of mathematical models, for instance using a discrete model approaching a continuous one.

The computational model of citizen domain represents citizen's requirements computationally. Computing with linguistic values is a necessity when the information is subjective, hence imprecise [49]. Users of large cities must continuously adapt to a range of urban systems of which they perceive only partially performances and issues. To model these systems, it is necessary to produce statistics describing the uses (transport, energy, water,…), which can vary greatly from one city to another, depending on the installed systems, the urban density, geography and climate, but also habits, education, land ownership and activities of people. Some of these data can be collected automatically, as for example, smart meters for electricity final consumption, or dynamic property prices in the different area of a city (Fig. 9.4). But other data require very careful field surveys to prevent bias. These become important when it is intended to predict responses to changes (e.g. the introduction of a new mode of transport, with potentially systemic consequences, such as a redistribution of the habitat). To gather such information, one way is considering the use of online serious games techniques. This should help to refine the scenarios used in functional and physical modelling (for example, predict the combined effects of the installation costs, regulatory incentives and education for sustainable development on the options chosen by users in terms of energy efficiency and the introduction of renewable energy).

These techniques can be used to resolve two main problems in innovating in smart solutions: (a) gathering collective intelligence of a great number of people

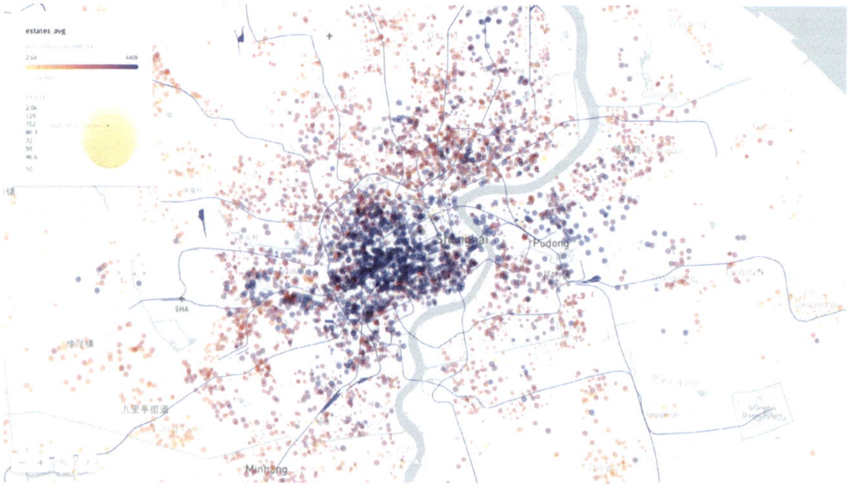

Fig. 9.4 Automatic extraction of property prices (RMB/m^2) in Shanghai city on 8 December 2017

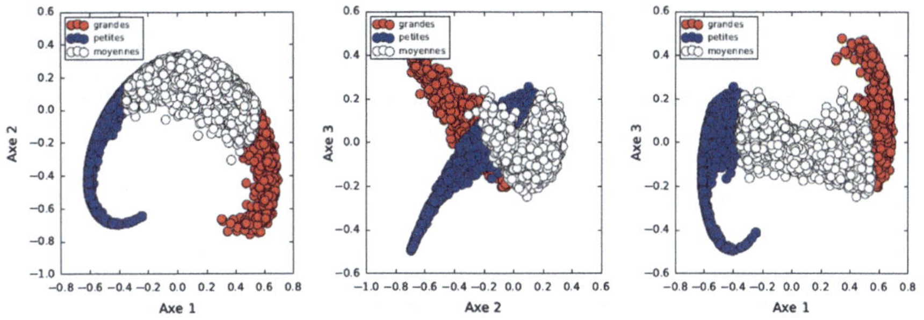

Fig. 9.5 Comparison of a set of cities in France. Blue colour represents small cities; white colour represent middle cities and red colour represent big cities

collaborating; (b) producing trustable data from bottom-up information and inductive approach instead of top-down categories checked by very restricted empirical polls.

The computational model of the functional domain represents functions computationally [51]. It offers a solution of mathematical representation of functions. A computational model of the physical domain is a computational representation of the corresponding mathematical model. It is a computational representation of design principles and physical principles. Figure 9.5 shows the results of a Principal Component Analysis of the shortest path kernel applied to the road network of a range of French cities of different sizes. The clusters are colour highlighted and reveal a strong relationship between the city size in term of population and its road network considering the shortest path property [52]. The first three axes explain 95% of the inertia. Axis 1, axis 2 and axis 3 represent, respectively, 63%, 25% and 7% of inertia. Axis 1 represents the size of cities. The bigger a city is, the greater is its coordinate on this axis. Axis 2 represents the frequencies of the shortest paths and finally, axis 3 represents the diversity of the nature of the shortest paths.

The computational model of physical domain represents structural-physical descriptions of the city in computers. Computational modelling at the city level is three dimensional and also time dependent. It requires to visualize space items with advanced GIS (2D and 3D), providing several scales for editing and representing dynamically objects, events, logistical data and interaction data, according to several actors' viewpoints. These computational models are needed to study the built environment to better understand and quantify the complex phenomena in interaction and, in the end, to help architects and town planners to satisfy the goals of conceptual modelling as for instance sustainable development and better quality of life for cities inhabitants in the context of worldwide increasing urbanization. However, cities present very difficult multi-physics phenomena and multi-scale aspects, (from individual buildings to the urban territory), with strong interaction between soil and atmosphere and involving fluid mechanics, thermal sciences, or solar radiation effects. The applications of advanced models and simulation tools are urban acoustics, diffusion of pollutants, fire propagation, flood simulation, urban temperature (heat island), urban comfort, among others. Therefore, simulation tools involve, in general, large computer resources, dedicated algorithms and solvers. Not only physical models are difficult to build and to represent a city or even a small part of it, but the geometrical representation is also requiring different levels of details (LOD) depending on the precision and type of the problem to be addressed.

A computational model of the process domain is a computational representation of process physical principles. These computational models are needed for the evaluation of the building and infrastructure life cycle and management. The elaboration of domain-oriented models based on a set of concepts, their properties and their relationships can help describe the complex city as a systemic model on which it becomes possible to implement diagnosis and decision support process including experts' heuristic knowledge (expert systems) and agent-based reasoning [53, 54]. This kind of approach can be applied for example for crowd behaviour in case of emergency (flood, cyclone ...) and mass evacuation, and also to a lot of varied domains in engineering design and diagnosis such as holon city [55], infrastructure design, maintenance design.

9.3.3 Experimental Models

The experimental model of citizen domain is an experimental representation of attributes that citizens desire. The involvement of citizens in a design process is a way to obtain information, and experimental models can offset the lack of mathematical and computational models of citizen domain. Tests with effective citizens or representatives of them are necessary as soon as the acceptation or the differentiation of components of a body of a city as an evolving living system is a key feature for innovation. Feedbacks from effectively used components are also a means for capturing the information from the citizens.

An analysis carried out on the usage of microblogging Weibo by women and man showed that different patterns of behaviour were detected [56]. It was found that women tend to use more frequently the microblogging Weibo than men and their distribution of the usage in the same area is different (Fig. 9.6).

The usage of the microblogging Weibo by men is more concentrated in some area. Otherwise, the usage of Weibo by women is more distributed. Some remarkable overlapping of usage by two genders is also found. The switching from a pattern of behaviour to another by women and men is also interesting. This remarkable finding bears the gender issues in the design of microblogging service.

A method for collecting reliable and pertinent data for complex issues is researched through engaged scholarship fostering shop floor participation. In such tools, one should notice that subjective evaluation and real behaviour in real time are taken into account. However, this is seldom the case in traditional quantitative studies which aims at measures of objective and purely physical data. Intangibles such as care, trust, taste, friendship, cooperation, tolerance, appropriation of techniques and apprenticeship in learning are important variables that have been largely underweighted in urban equilibrium and achievements.

Fig. 9.6 Distribution of usage of microblogging: women versus men

The experimental model of functional domain represents the behavior of functions of a body of a city experimentally. The test in real or simulated conditions or tests of parts of a body of a city can be used. Feedbacks from effectively used can also give technical information in real conditions. In order to improve the information extracted from citizens, the experimental modelling can also host more fundamental studies on the representations of the city formed by the populations and different actors of the urban scenery. Indeed, the quality of the information is part of the engineering design of urban systems, as it leads new habits, new uses and new ways to experience the functional city together. The experimental model of physical domain is an experimental representation of design principles and physical principles encountered or applied in the design of the city. A physical mock-up is such a model. It can be manipulated by designers in order to have a concrete representation of the structure of the city. This experimental domain is related to the experimental model of the process domain. The experimental model of the process domain is an experimental representation of process physical principles applied in the building of the city or in the interaction of the city with different products like cars, bicycles…etc.

For instance, the maintenance and managing of the *city–car system* should deliver the right the services right all the time. This is typically a *City–Car System Service* engineering scenario [57] (Fig. 9.7). The interaction between the user and *mixed space* of the city (for instance: circulating space, intelligent road), the interaction between the artefact (for instance: non motor vehicles, motor vehicles, etc.) and user, and the interaction between the mixed space and the artefact can been identified (Fig. 9.8). The purpose is then to co-design the *city–car system*. The intelligent decision of what is acceptable by the *city–car system* involves striking a balance between traffic capacity, the environment, speed, safety and *city–car* user comfort. The *city–car system* should resolve conflicts and accommodate the competing demands made upon it.

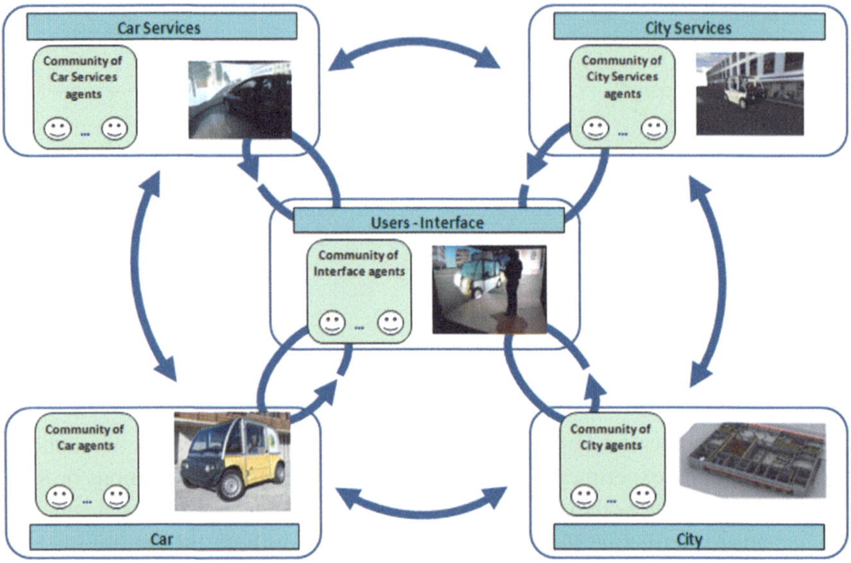

Fig. 9.7 Architecture city–car service system

Fig. 9.8 A view of a virtual simulation of the city–car service system

9.4 Application—Complex City of Shanghai

This section describes the usability of proposed engineering models in the engineering design of complex city like Shanghai. Understanding of "what citizens want", its progress and advancement can be achieved by observing the dynamics of interactions between different citizens in real time. Within the engineering design of a complex city like Shanghai, large quantities of information and knowledge are widely distributed across citizens. Therefore, in this application, it is assumed that Shanghai is a living laboratory. This living source of information on citizen interactions can allow design researchers to develop richer models of designing which in turn will provide the basis for a better understanding of engineering design problems of a complex city and developing intelligent tools to support this process.

9.4.1 Citizen Problem Space and Citizen Models

The definition of the design problem in terms of what citizens like is an important part of the design process. The rapidity of densification and growth of city, changing and the evolution of different actors of the urban scenery makes the problem definition as never final. Therefore, the movement of the problem in time depends on the movement of populations (citizens) and different actors of the urban scenery. The space of problem defined from the citizens is called Citizen Problem Space.

The experimental model of citizen domain is the first model developed in this application. The goal of our research presented in this study, consists in discerning, from the real interactions, the different citizens' problems on one side, and the dynamics of citizen organizations to these problems on the other. In the frame of our research, the experiences with citizens are used as a situation of observation.

The interaction between the citizens and different objects considering the task of citizens and the roles of these objects is also carried out (Fig. 9.9). During interactions, citizens communicate their thoughts verbally or in writing. Experiences show that the majority of real problems appear through verbalizations and writings. Therefore, the verbal and written communication offers us a direct path to the citizen requirements. For that reason, we consider a message as being a form of the representation of a problem. It can be characterized by a set of syntactic elements with a specific semantics to a domain of knowledge. The category of these elements is called analysis entities [58].

Computational model of citizen domain and *Mathematical model of citizen domain* are used to study both citizen and automated organization as computational entities. Interactions have been viewed as inherently computational. Every interaction can be filtered by means of analysis entities. Clustering the entities of analysis can be considered a principle for *citizen problem* discovering. Clustering permits to identify families of analysis entities (Fig. 9.9).

Fig. 9.9 Real interactions between citizens on popular social network Sina Weibo

Mathematically, the search for interaction families and analysis entities families is a problem of search for simultaneous partitions of the two sets, the filtered interactions set and analysis entities set in correspondences or in the quasi-correspondences of a class of partition to a class of partition. Hence, this correspondence allows characterizing an interaction family by the corresponding analysis entities family that is by the corresponding *citizen problem*. If the families of state-problems are mutually exclusive, it is clear that the state-problems are completely independent. In practice, depending on the particular nature of the citizen problems, some or all of the state-problems result in either being mutually independent or not being as such. This means that interactions create "*state-problems within a state-problem*".

The conceptual model of citizen domain is developed from the interpretation of the results of the computational model of citizen domain. Computational analysis permits a better understanding of the interactions between citizens, the nature of problems, the emergent patterns and structures of the organization during interactions. The distribution of services and shops for a district allows to recuse or validate the hypothesis of availability of daily life services in the 500 m radius (health, food, schools, entertainment, etc.) (Fig. 9.10). The design for configuration of the city should consider the optimal distribution of services inside the city of inside a district like in Fig. 9.10 for Shanghai downtown Jing'An district.

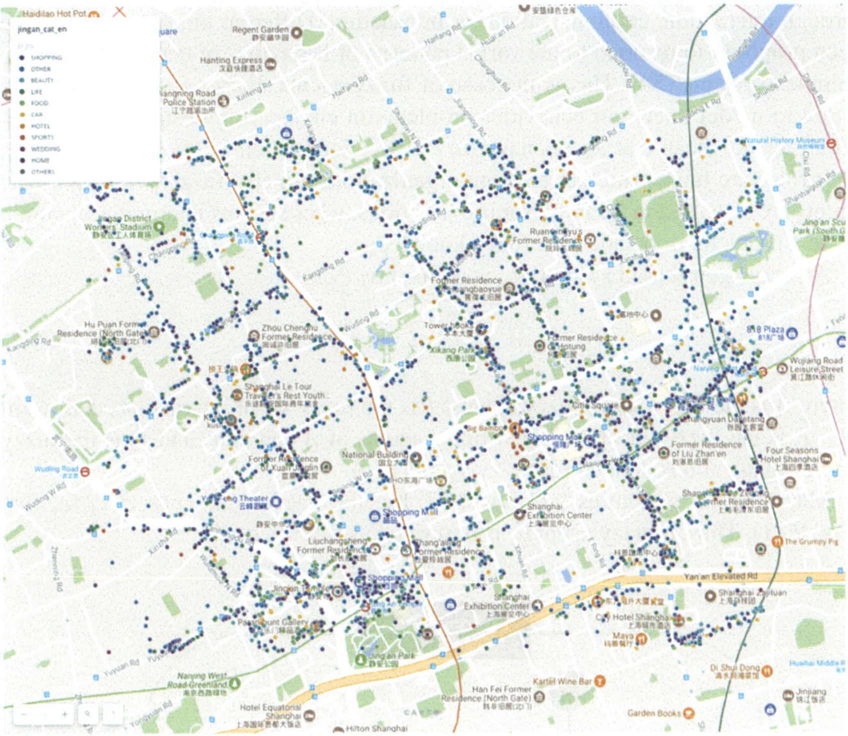

Fig. 9.10 Distribution of services according to crowd-sourced platform DianPing.com website in Jing'An district in Shanghai (2016)

9.4.2 *Agent-Based Computational Models*

The Citizen Problem Space is bridged to Functional Problem Space. The functional problem is formulated in response to the citizen problem. The functional problem is reformulated also in response to intermediate solutions and co-evolves with the design solution. Design solution belongs to the Solution Space. Process Space also interacts with Solution Space. The solution problem is formulated in response to the process problem (for instance, the maintenance of city). Thus, the design solution can only be consensual: satisfying both functional problem and process problem. This model of design depicts an evolutionary system composed of four evolutionary spaces. The evolution of each space is guided by the most recent population in the other space. It is a co-evolution. It provides the basis for a multi-agent computational model of engineering design of the city.

From the field of Distributed Artificial Intelligence, agent-based systems are characterized by the distribution of knowledge and information needed to solve a problem on a set of interacting agents, able to continue and reach a global goal. An agent-based system is a society of autonomous agents cooperating to achieve a global objective through interaction, communication, or transaction. Different kinds of agents have been proposed to respond to the varied patterns of behaviour of entities making up complex systems [59]. This is the case of fuzzy agents. Fuzzy agents emerged as a tool to model uncertain behaviour problems in engineering design [53, 54, 60]. Fuzzy agents mean that agents have fuzzy behaviours, their interactions are fuzzy, their roles are fuzzy, and the resulting organizations are also fuzzy [61, 62]. Fuzzy agents are also used in fuzzy reasoning situations, where agents interpret a situation, solve a problem, or decide with fuzzy knowledge.

A fuzzy agent-based system $\widetilde{M}_\alpha$ is defined by (9.1):

$$\widetilde{M}_\alpha = \langle \widetilde{A}, \widetilde{I}, \widetilde{P}, \widetilde{O} \rangle \tag{9.1}$$

where $\widetilde{A}$ is the fuzzy set of fuzzy agents, $\widetilde{I}$ is the fuzzy set of interactions defined in $\widetilde{M}_\alpha$, $\widetilde{P}$ is the fuzzy set of roles that fuzzy agents of $\widetilde{A}$ can play, and $\widetilde{O}$ is the fuzzy set of organizations defined for fuzzy agents of $\widetilde{A}$.

Many agent structures are inspired by the cycle $\langle observe, decide, act \rangle$ (Fig. 9.11). Thus, a fuzzy agent $\widetilde{\alpha}_i$ is described as follows (9.2):

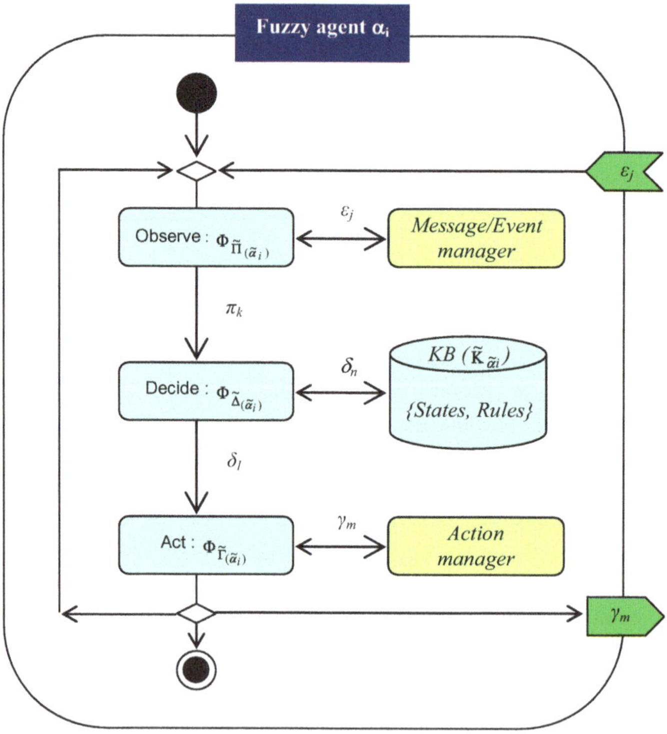

Fig. 9.11 The behavioral architecture of fuzzy agents

$$\widetilde{\alpha}_i = \left\langle \Phi_{\widetilde{\Pi}(\widetilde{\alpha}_i)}, \Phi_{\widetilde{\Delta}(\widetilde{\alpha}_i)}, \Phi_{\widetilde{\Gamma}(\widetilde{\alpha}_i)}, \widetilde{K}_{\widetilde{\alpha}_i} \right\rangle \tag{9.2}$$

where:

$\Phi_{\widetilde{\Pi}(\widetilde{\alpha}_i)} : \widetilde{\Sigma} \times \widetilde{\Sigma}_{\widetilde{\alpha}_i} \rightarrow \widetilde{\Pi}_{\widetilde{\alpha}_i}$ is the function of perceptions of $\widetilde{\alpha}_i$: $\widetilde{\Sigma}$ is the fuzzy set of states of $\widetilde{M}_\alpha$; $\widetilde{\Sigma}_{\widetilde{\alpha}_i} \subseteq \widetilde{\Sigma}$ is the fuzzy set of states of $\widetilde{M}_\alpha$ that $\widetilde{\alpha}_i$ knows, $\widetilde{\Pi}$ is the fuzzy set of perceptions in $\widetilde{M}_\alpha$, and $\widetilde{\Pi}_{\widetilde{\alpha}_i} \subseteq \widetilde{\Pi}$ is the fuzzy set of perceptions of $\widetilde{\alpha}_i$;
$\Phi_{\widetilde{\Delta}(\widetilde{\alpha}_i)} : \widetilde{\Pi}_{\widetilde{\alpha}_i} \times \widetilde{\Sigma}_{\widetilde{\alpha}_i} \rightarrow \widetilde{\Delta}_{\widetilde{\alpha}_i}$ is the function of decisions of $\widetilde{\alpha}_i$: $\widetilde{\Delta}$ is the fuzzy set of fuzzy decisions defined in $\widetilde{M}_\alpha$, and $\widetilde{\Delta}_{\widetilde{\alpha}_i} \subseteq \widetilde{\Delta}$ is the fuzzy set of decisions of $\widetilde{\alpha}_i$;
$\Phi_{\widetilde{\Gamma}(\widetilde{\alpha}_i)} : \widetilde{\Delta}_{\widetilde{\alpha}_i} \times \widetilde{\Sigma} \rightarrow \widetilde{\Gamma}_{\widetilde{\alpha}_i}$ is the function of actions of $\widetilde{\alpha}_i$: $\widetilde{\Gamma}$ is the fuzzy set of actions which can be performed in $\widetilde{M}_\alpha$, and $\widetilde{\Gamma}_{\widetilde{\alpha}_i} \subseteq \widetilde{\Gamma}$ is the fuzzy set of actions that $\widetilde{\alpha}_i$ can process;
$\widetilde{K}_{\widetilde{\alpha}_i} \subseteq \widetilde{K}$, with $\widetilde{K}_{\widetilde{\alpha}_i} = \widetilde{\Delta}_{\widetilde{\alpha}_i} \cup \widetilde{\Sigma}_{\widetilde{\alpha}_i}$, is the fuzzy set of fuzzy knowledge of $\widetilde{\alpha}_i$: $\widetilde{K}$ is the fuzzy set of fuzzy knowledge defined in $\widetilde{M}_\alpha$. Knowledge of $\widetilde{\alpha}_i$ is composed of decision rules, values on the domain, acquaintances and dynamic knowledge, as observed events or internal states.

The proposed platform to assist the problem of complex city intelligent design is called F-ACCID (Fuzzy Agents for Complex City Intelligent Design). This platform (Fig. 9.12) is composed of three levels:

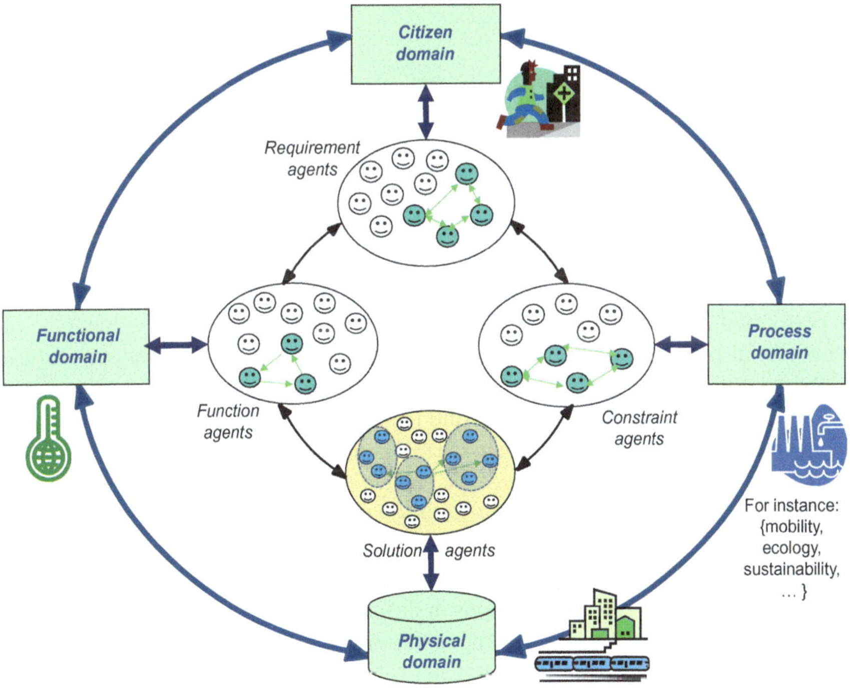

Fig. 9.12 Agent-based architecture of F-ACCID platform

(1) *Communication and cooperation level.* It implements services of communication and cooperation for fuzzy agents of F-ACCID (interface agents and design agents).
(2) *Design fuzzy agents' level.* It is divided into four fuzzy communities of agents:
    (a) fuzzy community of citizen requirements agents that interact with the fuzzy community of function agents, in response to requests from the citizen requirements agents,
    (b) fuzzy community of function agents that interact with each other and with fuzzy communities of citizen requirements agents and solution agents,
    (c) fuzzy community of solution agents that may interact with each other and with fuzzy communities of function agents and city constraints agents, and
    (d) fuzzy community of city constraints agents that interact with the fuzzy community of solution agents, in response to requests from the city domains agents.
(3) *Interface level.* It supports the connection of different human actors involved in complex city design (experts and customers) through the use of collaborative software like micro-tools [63]. These micro-tools communicate the orders' actors to associated city domains agents, who might transmit them to the fuzzy communities of citizen's requirement agents and fuzzy city constraints agents.

9.4.3 Multi-scale and Holonic City

From the second property, the city is a multi-levelled hierarchy of semi-autonomous sub-wholes, branching into sub-wholes of a lower order, and so on to form a holon. Each sub-whole within the hierarchic tree has two properties: it is a whole relative to its own constituent parts, and at the same time a part of the larger whole above it in the hierarchy. A *city cell* is defined as a holon entity [55]. All city functions must be performed and completed in their entirety as independently as possible. One of the essential requirements of the city cell is the capacity for independent actions. Each city cell must itself be a city cell. This means creating "*city cells within a city cell*" [55]. The city cells largely structure themselves and together serve the whole system of the city. Reference can be made to the principle of regulating city functions and/or city solutions that can control the behaviour of independent city cells. Thus, the internal relationships within a city are closer and more intensive than the relations with the outside. City cells are self-similar also. Here, the city functions are grouped so they are performed and completed in their entirety as independently as possible. The relationship between city solutions in different levels of the "*city cells within a city cell*" (Fig. 9.13) allows finding regulating function or solutions that can control the behavior of autonomous city cells.

Consensus, first mentioned in social and sociological sciences [64] can be used for working out an agreement and solving conflicts in complex city engineering

Fig. 9.13 City cell-within-city cell

design. Consensus in functional requirements is assumed and consensus in configurations fulfilling consensual requirements is sought [65]. Figure 9.14 shows the fuzzy solutions agents of F-ACCID platform during the seeking the consensual configurations [4]. A context of exchanges between Citizens agents and the "Ecology" and "Mobility" domains is shown in Fig. 9.15. Then, the city cells can largely organize themselves in consensual configurations.

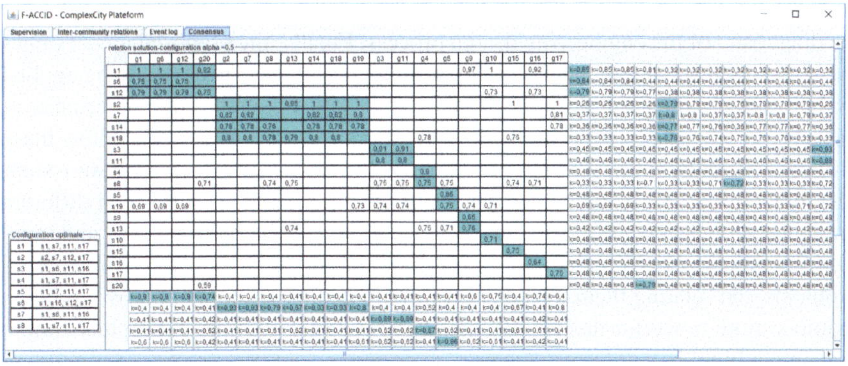

Fig. 9.14 Fuzzy solutions agents of F-ACCID platform during consensual configurations seeking

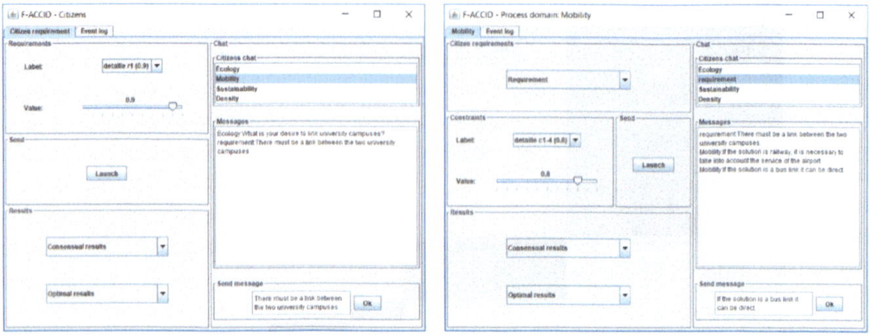

Fig. 9.15 Exchanges of messages in F-ACCID

9.5 Conclusions

The city is an intersecting engineering design object. A model for describing the engineering modelling knowledge in the design of a complex city is presented in this paper. A complete engineering modelling analysis is necessary for understanding and discovering the problems. The framework captures the knowledge of design engineers through the building models. Furthermore, honouring a transdisciplinary approach, the collaboration between different disciplines from various scientific and practical fields facilitates the final result. The application shows that considering the city as a living body, the experimental model is key to discovering the problems in the design of the city. The application has also demonstrated in practice, that iterative developing citizen model, from concrete to abstract, permits to establish and to evaluate the conceptual model. The proposed framework can be used also for the diagnostic of the engineering design process. Finally, the proposed model draws the way how engineering design theories and practices can successfully be applied in the future engineering design of the complex city. Assessing the importance of the roles of citizens, how the citizens' differences can be considered and how these differences can be integrated into the proposed approach are some relevant issues. The dynamics of the fuzzy behaviour of citizens for consensus seeking in different problems is the subject of further research. It will comprise knowledge of and input from social-science disciplines. The task of modelling, in particular dynamic, these problems and relating them to each other would be a challenging endeavour which requires more powerful theory. Managing the complexity in its conceptualization, in its modelling, in building up theories, experiences and best practices for building strategy, management and operations in complex city design implies also to give priority in projects that could increase and innovate in the following domains:

(a) Screening of convenient information among bid datasets through new techniques (data mining, mapping of digitalized and geolocalized information) in order to build up pertinent data.

(b) Construction of permanent banks of data available for better governance in some points of the city.
(c) Construction of data suitable for complex issues and sustainable growth (for example product life management; building life management through bottom-up techniques and collaborative platforms like online serious gaming.

References

1. Campbell S (1996) Green cities, growing cities, just cities? Urban planning and the contradictions of sustainable development. J Am Plan Assoc 62(3):296–312
2. Hillier B (2012) The city as a socio-technical system: a spatial reformulation in the light of the levels problem and the parallel problem. In: Digital urban modeling and simulation, Springer, pp 24–48
3. Keeble BR (1988) The Brundtland report: 'our common future'. Med War 4(1):17–25
4. He B, Ostrosi E, Pfaender F, Fougeres A-J, Choulier D, Bachimont B, Tzen MZ (2014) Intelligent engineering design of complex city: a co-evolution model. In: Cha J et al. (eds) Advances in transdisciplinary engineering, vol 1. IOS Press, Amsterdam, pp 434–443
5. He B, Luo T, Huang S (2019) Product sustainability assessment for product life cycle. J Clean Prod 206:238–250
6. Cha J, Chou S-Y, Stjepandić J, Curran R, Xu W (2014) Moving integrated product development to service clouds in the global economy: proceedings of the 21st ISPE Inc. International conference on concurrent engineering, 8–11 Sept 2014. Advances in Transdisciplinary Engineering, vol 1. IOS Press, Amsterdam
7. Derix C, Gamlesæter Å, Miranda P, Helme L Kropf K (2012) Simulation Heuristics for Urban Design. In: Arisona SM, Aschwanden G, Halatsch J, Wonka P (eds) Digital urban modeling and simulation. Springer, Berlin, Heidelberg, pp 159–180
8. Fouda YE, Elkhazendar DM (2019) A criterion for modelling the 'live-and-work' city index using sustainable development indicators. Int J Urban Sustain Dev 11(1):24–47
9. Hijikata M, Takagi A (2010) A communication-based planning system for a modern information society: an innovative approach to city management. Int J Urban Sci 14(2):176–190
10. Lim TK, Ignatius M, Miguel M, Wong NH, Juang H-MH (2017) Multi-scale urban system modeling for sustainable planning and design. Energy Build 157:78–91
11. He B, Pan Q, Deng Z (2018) Product carbon footprint for product life cycle under uncertainty. J Clean Prod 187:459–472
12. He B, Niu Y, Hou S, Li F (2018) Sustainable design from functional domain to physical domain. J Clean Prod, 197:1296–1306
13. Suh NP (2001) Axiomatic design: advances and applications, 1st edn. Oxford University Press, New York
14. Suh NP (1999) A theory of complexity, periodicity and the design axioms. Res Eng Design 11(2):116–132
15. Suh NP (2005) Complexity in engineering. CIRP Ann 54(2):46–63
16. Suh NP (2005) Complexity: theory and applications. Oxford University Press on Demand
17. Wognum N, Bil C, Elgh F, Peruzzini M, Stjepandić J, Verhagen WJC (2019) Transdisciplinary systems engineering: implications, challenges and research agenda. Int J Agile Syst Manag 12(1):58–89
18. Stjepandić J, Verhagen WJC, Liese H, Bermell-Garcia P (2015) Knowledge-based engineering. In: Stjepandić J et al (eds) Concurrent engineering in the 21st century: foundations, developments and challenges. Springer, London, pp 255–286
19. French MJ (1971) Engineering design: the conceptual stage. Heinemann Educational Publishers, London

20. Pahl G, Beitz W, Feldhusen J, Grote K-H (2007) Engineering design: a systematic approach, 3rd edn. Springer, London
21. Hubka V, Eder WE (1988) Theory of technical systems: a total concept theory for engineering design. Springer, Berlin and New York
22. Suh NP (1998) Axiomatic design theory for systems. Res Eng Design 10(4):189–209
23. Albano LD, Suh NP (1992) Axiomatic approach to structural design. Res Eng Des 4(3):171–183
24. Erens F, Verhulst K (1997) Architectures for product families. Comput Ind 33(2–3):165–178
25. Ulrich KT (2003) Product design and development. Tata McGraw-Hill Education
26. Jiao J, Tseng MM (2000) Fundamentals of product family architecture. Integr Manuf Syst 11(7):469–483
27. Welch RV, Dixon JR Representing function, behavior and structure during con-ceptual design. In: 4th international conference on design theory and methodology, 1992, pp 11–18
28. Hales C (1993) Managing engineering design. Longman Scientific & Technical, Harlow, UK
29. Choulier D, Weite, P-A (2011) Découvrir et appliquer les outils de TRIZ. Université de technologie de Belfort-Montbéliard
30. Yoshikawa H (1981) General design theory and a CAD system. In: Proceedings IFIP WG5. 2-5.3 working conference on man-machine communication in CAD/CAM, pp 35–58
31. Nomaguchi Y, Taguchi T, Fujita K (2010) A framework of describing and managing engineering analysis modeling knowledge for design validation. Concurr Eng, 18(3):185–197
32. Batty M et al. (2012) Smart cities of the future. The Eur Phys J Spec Top 214(1):481–518
33. S. Fukuda (2011) Emotion: a gateway to wisdom engineering. In: Emotional engineering. Springer, London, pp 1–20
34. Fukuda S (2013) Emotion and innovation. In: Emotional engineering, vol 2. Springer, pp 11–21
35. Fukuda S (2013) Emotion and satisficing engineering. In: Emotional engineering, vol 2. Springer, pp 1–10
36. Ullman DG (1992) The mechanical design process, vol 2. McGraw-Hill, New York
37. Maier JR, Fadel GM (2009) Affordance-based design methods for innovative design, redesign and reverse engineering. Res Eng Des 20(4):225
38. Chakrabarti A, Blessing L (1996) Special issue: representing functionality in design. Artif Intell Eng, Des, Anal Manuf 10(04):251
39. Erden MS, Komoto H, van Beek TJ, D'Amelio V, Echavarria E, Tomiyama T (2008) A review of function modeling: approaches and applications. AI EDAM, 22(2):147–169
40. Chandrasekaran B, Josephson JR (2000) Function in device representation. Eng Comput 16(3–4):162–177
41. Chandrasekaran B (2005) Representing function: relating functional representation and functional modeling research streams. AI EDAM 19(2):65–74
42. Hirtz J, Stone RB, McAdams DA, Szykman S, Wood KL (2002) A functional basis for engineering design: reconciling and evolving previous efforts. Res Eng Des 13(2):65–82
43. Stone RB, Wood KL (2000) Development of a functional basis for design. J Mech Des 122(4):359–370
44. McAdams DA, Stone RB, Wood KL (1999) Functional interdependence and product similarity based on customer needs. Res Eng Des 11(1):1–19
45. Shagieva AK, Makarov AS, Karpova NV, Vagazova GI, Madyshev IS (2018) Information infrastructure components of anti-crisis management in the city economy. Int J Civ Eng Technol 9(11):1709–1719
46. Wang H, Tanaka K (2016) Management of marine logistics in the case of emergency or disaster. Int J Agile Syst Manag 9(3):251–268
47. Altshuller G (2002) 40 principles: TRIZ keys to innovation, vol 1. Technical Innovation Center, Inc.
48. Orloff MA (2006) Inventive thinking through TRIZ: a practical guide, 2nd edn. Springer, Berlin, New York
49. Zadeh LA (1996) Fuzzy logic = computing with words. IEEE Trans fuzzy syst 4(2):103–111
50. Shervashidze N, Vishwanathan SVN, Petri T, Mehlhorn K, Borgwardt K Efficient graphlet kernels for large graph comparison. In: Artificial Intelligence and Statistics, 2009, pp 488–495

51. Wood WH A computational representation of functions in engineering design. In: Meijers A (ed) Philosophy of technology and engineering sciences, Elsevier, North Holland, 2009, pp 543–564
52. Borgwardt KM, Kriegel HP Shortest-path kernels on graphs. In: Fifth IEEE international conference on data mining (ICDM'05), Houston, TX, USA, 2005, p 8
53. Ostrosi E, Fougères A-J, Ferney M, Klein D (2012) A fuzzy configuration multi-agent approach for product family modelling in conceptual design. J Intell Manuf 23(6):2565–2586
54. Fougères A-J, Ostrosi E (2013) Fuzzy agent-based approach for consensual design synthesis in product configuration. Integr Comput-Aided Eng 20(3):259–274
55. Issa H, Ostrosi E, Pfaender F, Lenczner M, Habib R, Tzen MZ Multi-scale design using a holonic approach. In: IET international conference on smart and sustainable city 2013 (ICSSC 2013), Shanghai, China, 2013, pp 1–6
56. Ostrosi E, Pfaender F, Choulier D, Fougères A-J, Tzen M Describing the engineering modeling knowledge for complexity management in the design of complex city. In: Proceedings of 19th international conference on engineering design: ICED13, Seoul, Korea, 2013
57. Zhang Z et al. City-product service system: a multi-scale intelligent engineering design approach. In: Moving integrated product development to service clouds in the global economy: proceedings of the 21st ISPE Inc. International conference on concurrent engineering, Beijing, China, 2014, pp 405–413
58. Movahed-Khah R, Ostrosi E, Garro O (2010) Analysis of interaction dynamics in collaborative and distributed design process. Comput Ind 61(1):2–14
59. Fougères A-J (2012) Modelling and simulation of complex systems: an approach based on multi-level agents. Int J Comput Sci Issues 8(6):8–17
60. Ostrosi E, Fougères A-J (2011) Optimization of product configuration assisted by fuzzy agents. Int J Interact Des Manuf (IJIDeM), 5(1):29–44
61. Ostrosi E, Fougères A-J, Ferney M (2012) Fuzzy agents for product configuration in collaborative and distributed design process. Appl Soft Comput 12(8):2091–2105
62. Fougères A-J A modelling approach based on fuzzy agents. Int J Comput Sci Issues 9(6):19–28
63. Fougeres A-J Agent-based micro-tools development for a co-operative design plat-form. In: ITI 3rd international conference on information and communication technology, (ICICT'05), Cairo, Egypt, 2005, pp 715–730
64. Cruikshank J, Susskind L (1989) Breaking the impasse: consensual approaches to re-solving public disputes. Basic Books, New York
65. Ostrosi E, Haxhiaj L, Fukuda S (2012) Fuzzy modelling of consensus during design conflict resolution. Res Eng Des 23(1):53–70

Chapter 10
Systematic Development of Product-Service Systems

Margherita Peruzzini and Stefan Wiesner

Abstract Main problems occurring in Product-Service Systems (PSSs), are due to an inadequate requirements analysis and lack of a strong PSS conceptual design. Problems vary from exceeding budgets, to missing functionalities, unsuccessful market launch, or even project abortion. Furthermore, the special characteristics of a PSS have to be considered already at an early stage of the development process. Requirements Engineering (RE) and design methodology as well as supporting Information and Communication Technologies (ICT) need to establish a common perception of the targeted PSS. At the same time, the inner complexity of PSS leaves requirements analysis, design activities and development tasks fragmented among many disciplines and sometimes conflicting, unstable, unknowable or not fully defined. In this context, a concurrent, transdisciplinary and collaborative design of PSS is required to create feasible and successful solutions. The objective of this chapter is to present a structured approach to face the specific challenges of PSS development in detail, to elaborate a general framework that features a systematic approach for PSS development, and to consider the effects of changes in specific product and service design on a systematic PSS development process.

Keywords Product-service systems (PSSs) · Requirements engineering (RE) · Design methods · Servitization

M. Peruzzini (✉)
University of Modena and Reggio Emilia, Modena, Italy
e-mail: margherita.peruzzini@unimore.it

S. Wiesner
BIBA University, Bremen, Germany
e-mail: wie@biba.uni-bremen.de

J. Stjepandić et al. (eds.), *Systems Engineering in Research and Industrial Practice*,
https://doi.org/10.1007/978-3-030-33312-6_10

10.1 Introduction

Design and development of Product-Service Systems (PSS) is a complex process, mainly due to the transdisciplinarity of the activities to be carried out and the cross-disciplinarity of input and output data. As a consequence, a concurrent, transdisciplinary and collaborative approach is required to promote the exchange of engineering knowledge about user and system requirements, design specifications and processing instructions between different stakeholders. In particular, this chapter provides and discusses a general overview of the design and requirements engineering (RE) methodologies, and proposes a high-level integrated model for P-S lifecycle management. Furthermore, a systematic PSS development framework is offered to manufacturing enterprises to determine their own position on the PSS maturity scenario and to plan future developments. It is based on a set of models determining the different aspects of servitization [1]. An Innovation Potential model describes the level of novelty of potential servitization approaches and gives the enterprise and its partners an impression of the general options for product service combinations, the required development work and the expected uncertainty. An Opportunity Potential model identifies the opportunities for innovative combinations of products and services and formulates a challenge to ask for ideas and trigger idea generation. Finally, a Search Areas model helps manufacturing enterprises to look beyond well-known domains that are already served ("looking out-of the box") and do this in an efficient way.

10.2 Design Methods for PSS

Usually manufacturing companies have well-defined and structured product development processes, but they lack a sufficiently definition of the service development processes as found in traditional service companies. Therefore, they are poorly equipped with appropriate approaches, methodologies and tools for supporting the development of PSSs in an efficient way. With manufacturer business models increasingly being extended to include the service phase of the product lifecycle, this poses a significant issue.

In literature several methodologies have been proposed to design a Product-Service System along its entire lifecycle [2]. Some of them are very theoretical and hard to implement in practice, while others are very specific and have a limited applicability range. The most significant existing PSS design methods to be applied to manufacturing contexts usually come from service design and engineering. Some of them are characterized by a transdisciplinary approach. The most significant are listed below:

- UML 2.0 model: this approach allows to concurrently conduct a systematic technical-services design and the corresponding product design process, as proposed by Aurich et al. [3];

- Model-based approach to allow Industrial PSS (namely IPSS) design modelling: it fosters functional behavior and modelling of PSS artefacts (Welp et al. [4]);
- Service Computer-Aided Design (CAD): this is able to support decision-making evaluation through the concepts design, prompting different alternatives scenarios, but it needs a structured Integrating Service CAD Lifecycle simulation (namely ISCL) to also allow a quantitative and probabilistic PSS design as suggested by Komoto and Tomiyama [5]. It is transdisciplinary because it merges technical issues and economical sciences;
- Software tools for designing service activity and products concurrently: these usually have to be adopted in a collaborative way from the early phase of PSS design. Different simulation tools have been recently proposed [6, 7], including service availability prediction [8]. They all consider a variety of aspects with a transdisciplinary perspective;
- Service Engineering based on Structured Analysis and Design Technique (SADT): SADT represents the PSS by its technical specifications by fully describing the object-service system, considering the different combinations of the two main aspects of total core products, from system architecture (i.e., hardware and service support system) to business issues (i.e., markets, risks, partnerships, business chains, agreements, sales and distribution) [8–12]. All mentioned studies report valuable examples of transdisciplinarity, thanks to the combination of technical and business aspects;
- Knowledge-sharing network: the traditional knowledge management systems adopted in the majority of manufacturing industries are product-oriented and can hardly be applied to PSSs, due to the wide set of competences required. In this context, the role of Web 2.0 technologies is a key factor in managing knowledge along the PSS life cycle in a transdisciplinary way [13];
- Lifecycle-oriented PSS approach: it implies a new definition of the product lifecycle phases, from ideation to delivery, use, and disposal in order to be in line with the service proposal. Indeed, the PSS lifecycle approach involves both the product and the service development in order to align designers and providers in delivering of the same solution that is more sustainable along all the lifetime [14–17];
- Layer-based Development Methodology for PSS: it is a new development methodology for Industrial Product-Service Systems (IPS2) that aims to integrate both products and services into a unique solution for creating innovative business models able to generate an added value for the customer. Such a method involves several models to support early PSS development phases, i.e. the PSS planning and requirements engineering [18].

On the basis of industrial case studies, a list of design guidelines have been defined in the current state of the art:

- Requirement Elicitation (RE) is a crucial method to identify the main requirements according to the target market. Indeed, offering PSS instead of a traditional product requires additional competencies to identify the service functionalities to enhance the product, and a better understanding of the customer requirements which must be reached [19]. RE has to be properly addressed for PSS by adopting user-centered approaches;

- Design Structure Matrix (DSM) can be used to define the main PSS functions, combined with Business Use Case (BUC) analysis, which defines the use-case model and a goal-oriented set of interactions between external actors and the involved system [20]. It is a transdisciplinary engineering tool since it merges technical, economical and social aspects;
- Serious Games are useful to investigate the PSS lifecycle and human-system interactions during the design stages [21];
- Quality Functional Deployment (QFD) technique allows mapping the customer needs with the PSS functions in order to elicit the final PSS requirements for the solution to be developed [22];
- UML 2.0 can be used to easily carry out PSS process modelling to define the main activities to achieve the process tasks, identify the enterprise's ability in capturing, sharing and transferring the involved knowledge. The main common techniques for process modelling come from static models, focusing on the information flow (i.e., UML, Petri-Nets, flowcharting, IDEF0, etc.) and dynamic models for process evaluation (i.e., Event-Process Chain), to provide a high-level technical view.

The combination of RE, DSM, QFD and Serious Games can assure a robust user-centred design process and the definition of reliable design specifications. Furthermore, the combination of process modelling techniques assures a deep process analysis to achieve a comprehensive mapping of PSS tangible and intangible assets.

However, existing research focuses on the description of PSS technical solutions, embedded technologies to enable product-related services, methods of data acquisition and elaboration, and software interfaces. However, the focus on technology often neglects the final customer needs. Applying a rigorous User-Centered Design (UCD) approach to PSS has the potential to create customer-oriented and adaptable services and, finally, more effective and efficient models to identify usability problems at the different stages of PSS design. A first example of an integrated transdisciplinary method to support technical and business issues in PSS design process is presented by Peruzzini et al. [23], which tries to provide a QFD approach that drives a designer along both axes of evaluation. Such a method is based on collaboration and knowledge exchange between practice and science, in particular between technical and social sciences (attention to social sustainability, human wellbeing, ergonomics, work organization, etc.).

In order to achieve a transdisciplinary view, UCD techniques (such as interview, questionnaires and role-playing) allow for directly involving users into the design process. In particular, role-playing highlights the PSS users' needs and tasks to be created to satisfy them by directly considering a set of "personas" representing sample users [24]. Role-playing is performed by experts in the specific PSS domain, who play as characters into the real context of use simulating the actions and moods of the consumers. Personas are widely used in the investigation of user experience as fictional characters representing different user types and experiences. Indeed, the combination of user-centered issues has been proven to benefit the final PSS by including the user needs and demands from the early design stages. It is an important issue to develop successful PSS, but it is still not deepened in existing methodologies. In particular, the analysis of human-system interaction is fundamental to predict the relationship that will be created between the user and the PSS and to optimize both product and service functions to have a higher business impact. In order to do that, functional prototypes enhanced with interaction features should be created. In this direction, Mengoni and Peruzzini provide a methodological design framework to support the design of PSS by adopting a UCD approach to involve end-users during the different stages of PSS development, using interactive virtual prototypes [25].

By adopting these methods to industrial cases, some problems still occur. They are mainly related to the definition and evaluation of the user experience generated by the PSS, the complexity to predict the PSS behaviour during the design stage, and finally the assessment of the human-system interaction. When interactive prototypes are used, by exploiting different software tools (i.e., CAD modelling, virtual prototyping developing platform, system simulation) to involve sample end-users to test PSS usability and performance, time and cost for PSS prototyping and optimization can be reduced. However, the creation of proper PSS simulation is still a hard task mainly due to system integration, proper interfacing with service provider platform(s), and reliable data analysis and service behaviour prediction on small test data samples.

10.3 Requirements Engineering for PSS

Understanding the customer and other affected stakeholders expectations, i.e. their underlying needs, and linking information from all phases of the product-service lifecycle to the development process is a prerequisite for successful solution engineering [26–28]. Inadequate RE is a main source for failure of development projects and leads to exceeding budgets, missing functionalities or even the abortion of the project [1]. Often the relevance of appropriate requirements is underestimated, which in turn leads to errors in the requirements specification, not to mention foregoing completeness, consistency, verifiability etc. of requirements. Such errors are mostly discovered late in the development process, thus substantially contributing to higher costs in order to compensate for and correct the errors [29].

Requirements are used to define the needs of stakeholders, such as organizations or individuals along with their environment and specify what a solution must provide to satisfy those needs. Their record, documentation and management are the main objectives of RE. It is "a process, in which the needs of one or many stakeholders and their environment are determined to find the solution for a specific problem" [30].

In traditional development approaches, mainly from the manufacturing domain, RE is seen as a discrete development phase. This has substantial disadvantages when dealing with the increasing complexity, dynamics and time constraints of PSS. Change requests in later phases will not be included in the requirements specification, so it is often insufficiently documented which parts of the original specification are really implemented at the end and which not. Thus, it is difficult to use the requirements specification for change management and testing. Furthermore, each development project will have its own RE phase, with no focus on requirements re-usability [31], increasing the overall development time. An example illustrating this challenge is that in traditional RE scenarios for simple products, the stakeholders are generally aware of their needs. Often, a specific functionality is requested and the product development is based on formalized requirements through a single enterprise. In contrary, a Product-Service System is expected to solve a particular customer problem without prescribing a specific functionality, allowing alternative usage. In addition, cross-linking with other systems and integration into the system environment increases the complexity of the system development even more. System integration leads to a fuzzy problem description which again influences the RE process [32].

PSS require temporary collaboration of different stakeholders in Systems Engineering, which increases the complexity of the RE process. Besides the customer and user of the system, actors like the project manager, product designers, software developers, service engineers, marketing experts, suppliers, quality assurance and many more have to be involved, often being spatially distributed. This induces a change in RE from a quasi-stable and simple environment to a more complex and dynamic variation, making the RE process more challenging, due to both different cultural issues, but also organizational issues like organization of meetings [33] and conflicts as well as interdependencies have to be assessed for a larger number of requirements.

10.3.1 RE Process for Products, Services and IT

In order to select or develop a suitable RE approach for PSS, it is necessary to identify the specific characteristics of such systems. The analysis of some widely used definitions of PSS found in the literature [34–36] reveals some characteristics that seem to be specific for PSS in general. These characteristics are listed below:

- Integration of product and service shares, including software components
- Mutual planning, development, provision and use of product and service shares
- Fulfilling an end user need by delivering value in use
- Provided by either a single company or by an alliance of companies
- Dynamic adoption of changing customer demands and provider abilities
- Enabling innovative function-, availability- or result-oriented business models.

As far as product development, RE approaches have already been implemented with a high degree of formalization. Structured fundamental models exist that provide a general development procedure including RE. However, they focus almost exclusively on requirements development as the main process, which is only conducted at the beginning of the development approach, e.g. by specifying the product requirements document [37]. Sometimes, also aspects of requirements management are adopted, but without explicit instructions for implementation [38].

The following fundamental contributions regarding product engineering approaches have been taken into account:

- RE for PSS—A State of the Art Analysis [39];
- Vorgehenszyklus für die Lösungssuche (i.e., Procedure Cycle for the Solution Search) [40];
- Engineering Design [37];
- Product Design and Development [41];
- Erfassen und Handhaben von Produktanforderungen (i.e., Capture and Manage Product Requirements) [42];
- Entwicklung technischer Produkte (i.e., Development of Technical Products) [43];
- Collaborative Product Design [44];
- New Product Development [45].

The analysis of the future development environment is commonly discussed among the analysed approaches. Possible influences on the product development are identified and the "overall objective of the development" is established. The stakeholders are identified in order to elicit the requirements. The elicitation of requirements is addressed in product engineering approaches; however, procedures for the elicitation of requirements for product-related services are not described. Moreover, the authors state that there are weaknesses in the derivation of requirements from the customer's value chain processes, and cross-domain knowledge is not considered [46]. The necessity of requirements translation of initial stakeholder's requirements to design requirements—concretized and in the language of the developers—is mentioned in the analysed approaches. However, procedures for the concretization of requirements are not mentioned explicitly. Quality Function Deployment is applied in product engineering, but the authors state that "it cannot be employed for new development or for the derivation of design requirements from customer requirements" [39]. Furthermore, the procedures described are not applicable to services due to focus on quantification of requirements. In source [42], the author analysed the approach of [37] and comes to the conclusion that the translation of initial requirements to design requirements is lacking the provision of concrete methods and procedures. However, the concretization is supported by a guideline of characteristic product features helping the developer to elicit the requirements systematically. The guideline is adaptable to varying problems. To derive the priority, the requirements are differentiated between wish and demand [42].

The product engineering approaches provide procedures for identification and resolution of conflicts (e.g. influence matrices). However, the proposed procedures are domain-specific and do not discover conflicts between requirements of different domains. Negotiation with stakeholders and developers is commonly mentioned to resolve conflicts. In source [43], the author proposes two methods to solve a conflict—either to find a compromise between the conflicting goals or to avoid the conflict by changing the concept. Procedures for the documentation of requirements are provided by product engineering approaches. However, cross-domain documentation is not considered. Influence or link matrices are used to trace the requirements to its origin. Interdependencies between requirements of different domains are not captured. Change management is not described in detail; only the necessity of it is observed [39]. Ahrens [42] argues that procedures are provided to structure requirements lists either by thematic affliction or alternatively by the product structure in the methodology of Pahl und Beitz [37]. Requirements validation is mentioned in the analysed approaches but not discussed in detail. Generally, validation is done through evaluation of design drawings by the customer. The authors stated that "the customer plays a central role during the entire development process" [46]. However, the integration of the customer and stakeholders is restricted to the early stages of the development—the requirements elicitation and agreement. The dimension of collaboration between domains is not covered by the state-of-the-art research. Modularization, specification of interfaces and the re-use of modules for different products are mentioned in the product engineering approaches [39]. The authors argued that manufacturers are attempting to reduce costs by "increasing the use of the same parts, or modules, across different products" [45]. Procedures are not explicitly mentioned. Liu et al. described the state-of-the-art on collaborative engineering design systems utilized in the collaborative design process—web-based CAD systems, and propose a new system including tools to resolve conflicts [44]. The web-based collaborative engineering design systems enable the developers from different locations and businesses to share and integrate design models, e.g. via collaborative modelling or multimedia tools such as online chat and meetings. Collaboration during the first phases of the product development is not mentioned explicitly. The utilization of collaborative networks is not mentioned in the analysed methodologies.

In the service area, numerous models for the systematic development of services have been proposed [47, 48]. However, no systematic procedures for the effective implementation of RE in industry have been established yet, because the characteristics of a service (e.g. its complexity, pose greater challenges). Thus, Service Engineering procedures do not integrate a holistic RE until now, but focus more on methods like "trial and error" [49].

The following literature discussing the service engineering state-of-the-art and approaches has been analysed:

- RE for PSS—A State of the Art Analysis [39];
- Design and Management Service Processes [50];
- Key Concepts for New Service Development [51];
- Ein Rahmenkonzept für die Entwicklung von Dienstleistungen (i.e., A Framework for the Development of Services) [47];
- Dienstleistungsproduktion (i.e., Service Production) [52];
- Anforderungsanalyse für produktbegleitende Dienstleistungen (i.e., Requirements Analysis for Product-related Services) [38];
- Service Engineering [47];
- Collaborative Service Engineering [53].

The elicitation process in service engineering comprises the tasks of identifying essential information—e.g. service ideas, possible customers and their expectations, and the sources of the requirements—and determining the goals, chances and risks.

The procedures are service-domain specific; cross-domain knowledge is not considered. Furthermore, no precise methods for the elicitation are provided. To this end, van Husen analysed the elicitation process of conventional service engineering methodologies in detail and comes to the conclusion that procedures for the requirements elicitation are described on a relatively general level [38]. The initial requirements are concretized "by assigning them to quantifiable attributes related to the implementation" [39] and classified into three dimensions—potential, process and result. Consequently, the activities and resources needed for the development as well as the result of the service provision can be derived. According to van Husen [38], only Ramasway provided a detailed design process procedure including the activities of prioritization of requirements according to their importance, specification of attributes which are required to fulfil the needs, and creating a link between the attributes and requirements [50]. The identification and resolution of conflicts is not described explicitly in the service engineering approaches. In the analysed approaches it is suggested to use the procedures known from software and product engineering. The approaches analysed by Berkovich et al. [39] provided a set of procedures to document requirements in natural language without giving detailed information about creating a requirements specification. Traceability of requirements and change management are only mentioned briefly according to the authors. It is argued that the validation of the requirements is described as a comparison of the service concept with the initial stakeholder requirements. The validation is not discussed in detail.

Furthermore, the customer requirements are captured by the elicitation procedures. However, the procedures are only vaguely mentioned. From the analysis it can be derived that the customer and stakeholders are actively involved in the RE processes. Explicit procedures for the collaboration are not described. The modularization of services and re-use is recognized by the approaches analysed by Berkovich et al. [39].

A third aspect to consider in PSS is the IT system. For the development of software systems, standard procedures have been increasingly established. Besides generic process models, specific methods for RE exist. According to the scope and risk of the project, a suitable development model can be selected. In direct comparison with product and service development, RE is integrated deeper and more comprehensive into software development [38]. The following literature discussing the software engineering state-of-the-art and approaches has been investigated:

- RE for PSS—A State of the Art Analysis [39];
- Requirements Engineering Framework [54];
- Requirements Engineering Process [55];
- Requirements Engineering Process [56];
- A Generic Process for Requirements Engineering [57];
- Engineering and Managing Software Requirements [58];
- Requirements Engineering [57];
- Collaboration in Distributed Software Development [59];
- Collaboration in Software Engineering [60].

Requirements and their sources are identified in the elicitation phase and customer-integration is emphasized in the software engineering approaches. Consequently, the focus is laid upon the software domain—interdisciplinary requirements are not considered. Similar to the considered product engineering approaches, the necessity of requirements concretization is recognized in the software engineering approaches. The procedures are not described explicitly and are not suitable for the development of new products or services. The procedures provided for the identification of conflicts focuses solely on the software domain; interdisciplinary conflicts are not discovered. Negotiation with stakeholders is suggested to resolve conflicts and find a compromise [39]. The description of requirements, changes and responsibilities are specified and documented. Model-based requirements documentation is commonly used in software engineering; however, Berkovich et al. [39] state that "there are no procedures and models for the representation of requirements on services, nor for the relationship between the requirements of different domains". Traceability procedures are provided, specifying the affiliation of the requirement towards the different layers of concretization (e.g. a requirement assigned to a component), and linking the design requirements to the initial requirements. The authors describe the utilization of traceability in detail [57]. Interdisciplinary traceability is not considered. It is recognized that requirements can change during any lifecycle phase and changes have to be captured and analysed to "check them for their feasibility by determining their costs and impacts on other requirements, to prepare them for further stages of development, as well as to ensure appropriate documentation" [39]. Change management is described as important during the whole lifecycle including the use phase of the product [57].

Finally, requirements validation is an important part of the RE process to check the requirements for ambiguity and falsity. The design requirements are validated against the initial (stakeholder) requirements to determine the fulfilment of the stakeholder needs. Validation procedures are discussed in detail in the software engineering approaches. The validation focuses solely on software engineering [39]. Customer integration is restricted to the requirements definition stage; integration in other phases such as the utilization of the software is not explicitly mentioned. Modularization is recognized in software engineering approaches. Li et al. state that "requirements encapsulation means organizing requirements into a set of clusters along with external interfaces such that each cluster can be ultimately implemented by a functional module" [61]. Moreover, Lanubile described the state-of-the-art of collaboration in the software domain, focusing on the collaboration between software engineers [59]. Collaboration in software engineering is taking place in various ways during the whole lifecycle of the development, e.g. collaboration with stakeholders to elicit requirements, identification of errors and collaborative working on the software design. The author mentioned knowledge centres as web-enabled tools to share knowledge and argued that "the quality of programmers is the most important factor in software work" and thus, developers are hired regardless of their location. Competencies from non-software domains are not explicitly mentioned. Furthermore, software companies outsource development work to programmers in low-cost countries to reduce development costs. The collaborative environment presented by Lanubile and Whitehead referred solely to the software domain [59, 60]. Nevertheless, collaborative networks are created.

10.3.2 RE Approaches for PSS

RE for PSS has to be conducted for a growing number of tangible and intangible components from a variety of distributed, multi-disciplinary stakeholders. Consequently, only robust and transdisciplinary RE approaches that can deal with the complexity of PSS, its openness and dynamics are suitable. Due the inherent complexity, the direct involvement of the end user and information exchange between the different stakeholders has to be enabled during RE. Thus, the domain specific formalisms and tools have to be made interoperable or substitutable. In this context, the following literature discussing the state-of-the-art and approaches of integrated product and service development has been analysed:

- RE for PSS—A State of the Art Analysis [39];
- Integrated Product and Service Engineering versus Design for Environment [62];
- Solution Approach in the Hybrid Product Development [49];
- Life Cycle Management of Product-Service Systems [63];
- Framework for Development of Product-Service Systems [64];
- Systematic Translation of Customer-specific IT Solutions in Integrated Product-Service Building Blocks with the SCORE Method [65];
- Review of PSS Design Methodologies [66];
- State-of-the-art in Product-Service Systems [35];
- PSS Design [67];
- Developing new product service systems [68].

According to Berkovich et al. [39], the literature about PSS development and design discusses the process of development only abstractly without going into detail. Firstly, the "organizational conditions are created in order to enable an integrated development of services and hardware/software". The stakeholder needs—in regard to products and services—are identified. Concrete techniques or methods are not mentioned. During the concretization the initial requirements are analysed and assigned to the respective domains; a requirements model representing the product structure is created and updated during the entire development. The development process focuses on the single components of the PSS. However, concrete procedures for the translation of initial requirements to domain-specific requirements are not provided. They argue that identification and resolution of conflicts is only mentioned briefly in the selected approaches.

Procedures for the documentation of requirements are provided; model based requirements documentation is not applicable to PSS due to missing procedures and models for the representation of service requirements. Furthermore, there are no procedures to capture the interdisciplinary relationship between requirements. Change management is not deepened in detail in any of the approaches [39]. The validation of requirements "is not discussed in detail" in the analysed approaches. The importance of customer integration in all lifecycle phases is recognized. However, specific methods or procedures are not specified. In addition, modularization is widely mentioned in literature to create standardized solutions [65]. Integration of the involved domains is neither supported in the first phases of the development (e.g. elicitation) nor in later phases such as requirements validation. The necessity of integration is recognized; concrete procedures are not examined [66]. Due to the nature of integrated products and services the necessity of additional competencies is recognized. Aurich et al. noted the "importance of extended value creation networks" [63]. Additionally, the authors argue that the importance of collaboration is only mentioned and "not detailed enough to understand the uniqueness of this process [the collaboration between stakeholders] and how to implement it in real-time" [66].

Engineering of PSS, in contrast to a centralized development process for simple products, requires the orchestration of distributed products, services and business processes for a common purpose. Therefore, organizational, technical and managerial interoperability is a prerequisite for the realization of the system. The RE methodologies of the product, service and software disciplines focus on the respective domain, neither consider the methodologies interdisciplinary requirements nor are interfaces for the handling of interdisciplinary requirements specified. In addition, procedures and methods are solely applicable to the respective domains, making it impossible to apply them to other domains, let alone PSS as a whole. The elicitation procedures in the product domain focus on technical requirements. The methods used to elicit requirements such as checklist are not suited for the elicitation of service requirements. As the concretization of requirements is mainly done by assigning quantifiable attributes, this is not applicable to the intangible part of the EP. Collaboration and integration of development processes with other business partners are not explicitly mentioned. In general, the lack of an interdisciplinary view and thus missing interfaces towards other domains, as well as the insufficient requirements documentation complicate the adoption of product engineering methodologies.

The service engineering methodologies display weaknesses—the procedures provided focus solely on the service domain; interdisciplinary requirements are not considered. The service engineering methodologies are not detailed enough to be used as a basis for PSS development. For example, for the identification of conflicts only a reference to the already existing methods and procedures of the software and product engineering is made. Indeed, the software engineering methodologies do not consider other domains and interdisciplinary collaboration. The procedures described for the prioritization of requirements are not suitable for the development of new products or services. Furthermore, the representation of service requirements is not possible with the provided procedures and modelling techniques. Collaboration is strictly within the software domain, e.g. through networks of companies spread worldwide.

The integrated approaches state the necessity of cross-domain knowledge, interfaces and interdisciplinary requirements. However, the RE methodologies of the integrated products and services are too vague and do not provide the procedures necessary in order to realize a PSS. The procedures are not explained in detail or similarly to service engineering, procedures of other domains are referenced.

To summarize, the adoption of existing requirements engineering methodologies of the product, software and service domain to the development of PSS seems to not be possible as they do not fulfil the requirements for a successful realization of such a complex solution. Especially the lack of a holistic, interdisciplinary view and the corresponding interfaces must be highlighted. A holistic view of the development process of the PSS is necessary. Thus, integration of the development processes of the individual components is mandatory. Missing interfaces to other domains make it difficult to apply domain-specific requirements engineering methodologies to the respective component of the PSS. Moreover, the methodologies of the product domain do not cover all the lifecycle phases required to realize a PSS. For example, change management is not intended after the product has been realized.

The requirements engineering methodologies of the individual domains do not cover the collaboration across domains and the integration of development processes. The requirements engineering methodologies of integrated products and services cover all phases of the PSS, however they do not provide the required integration interfaces. The selection of collaborative business partners depending on the configuration of the PSS and formation of business networks is not described in any of the methodologies.

Thus, two main aspects will have to be supported by a suitable transdisciplinary RE framework:

1. Collaboration and interoperability between stakeholders and PSS components from different domains, especially products, services and software;
2. Management of unstable and unknowable requirements, taking into account information from all PSS life cycle phases.

Integrating PSS components from different domains, like manufacturing and service, requires collaboration between previously separated stakeholders. These stakeholders needed for the realization of PSS typically have their own specific development methodology, standards and even "language". Thus, the "translation" of requirements between domains needs to be supported by the RE framework to enable a common understanding of the PSS. In order to be adaptable to changing requirements throughout the lifecycle, the value chains have to be flexible even in the PSS usage phase. To support interoperability between PSS components from different domains, new methodologies are needed, also to describe emergent system behaviour. Conflicting, unstable and unknowable requirements have to be identified across the different PSS domains. Methods and tools need to anticipate dynamic changes over the PSS lifecycle and environment. The interdependencies between the tangible product and intangible service components as well as between the stakeholders have to be described. In such a way, the PSS specification and stakeholder information needs could be comprehensively identified.

10.4 Roles in PSS Engineering

Pahl and Beitz defined a number of roles along the product lifecycle, from product origination to disposal or recycling [37]. The following roles are relevant for the product engineering process:

- The *Market/Customer* delivers information about the requirements and constraints in order to generate and select product ideas and create a requirements specification. Furthermore, he/she is the user of the product and gives feedback about product quality.
- The *Product Planner* defines the product portfolio of the manufacturer according to the information from the market (market pull) and the available technology (technology push). The aim of the strategic product planning is the development contract, specified by requirements and justified by a promising business plan.
- The *Product Designer* is responsible to specify the to-be product according to the customer requirements within the necessary documents for prototyping and production. He/she may also be responsible to create and review prototypes.
- The *Production Planner* allocates the necessary employees, materials and production capacity in order to realize the product portfolio created by the Product Planner and Designer. Thus, he/she plans the production and manufacturing processes for the OEM.
- The *Suppliers* deliver the necessary materials, components and missing competencies to realize the product portfolio together with the OEM.
- The *Product Development Team* is comprised of representatives from several of the roles defined above and deals with the coordination of the product development process. Therefore it is responsible for the project management for specific product lines and information exchange between the actors.

Moving to Service Engineering, Freitag et al. described a schematic role model for Service Engineering, with the different role owners in seven phases, from the idea contributor in the ideation phase to the service facilitator during market launch [69]. The role model aligns the specific, function-oriented roles from Scheithauer et al. to the service lifecycle [70]. As an example the owner of the role "Service Manager" is responsible for a set of tasks in the role model. He/she has to decide quickly on proposed service ideas on a strategic level, then allocate resources for the service development project and control the execution of the decision. This role can be fulfilled by an individual person as well as a team or department. Furthermore, the PSS Engineering process is characterized by the inclusion of competences in the form or various actors during the development phases [71]. During a PSS project, the involved actors are determined, and development teams are established and assigned to several PSS specific roles that can be found in literature [3, 71–74].

The application of agile development methods like scrum leads to the definition of additional roles for the model:

- *SE Project Manager*: Comprehensive and frequent communication with customers about Service Engineering results; monitoring of the project's economy regarding development efforts and added value for customer/revenues;
- *SE Project Team*: Shared aim of highly flexible reaction to short-term changes of customer demands, even in late development phases;
- *SE Project Moderator*: Control of the group members meeting working standards; taking care of personal relations in an interdisciplinary team;
- The *PSS Provider* is the focal point of all involved stakeholders and is responsible for the whole PSS lifecycle. The tasks of the PSS provider include the coordination and execution of design, development and production of the product, as well as planning and development of complementary services [75];
- The *Production Network* comprises various PSS suppliers who are responsible for provision of materials, parts and components or system modules for the PSS Provider;
- The *Service Network* contains distributors, subsidiaries and service partners, which are mainly material and service specialists. The main task of the Service Network represents is the PSS distribution, which includes the market-specific adaptation of the integrated service shares and the handling of client orders including the individual PSS configuration;
- The *Customer* plays another key role next to the PSS Provider. Especially in the early stages of development, he/she is considered as the initiating part, because demands towards the PSS will be drawn up and implemented based on the determined customer needs;
- The *PSS Project Manager* acts in various phases of the PSS development process and performs management activities. The main tasks of the PSS Project Manager include the establishment of the connection between the PSS project management and the PSS development process. In addition, it is a task to coordinate the PSS actors and their communication and networking over the phases along the development process;
- The *PSS Architect* can be defined as another PSS specific role. The role is characterized by its PSS specific knowledge and the overarching effectiveness in the PSS development process. The duties of a PSS Architect include, among others, the PSS idea generation, documentation and management of PSS concepts and making the link to the PSS project management. Thus, the activities of the PSS Architect also span over several phases of the development process.

All actors involved in the PSS development process need to communicate with each other in different phases for different reasons. According to the respective phases, thus there is a different distribution of tasks, competencies and responsibilities as well as changing communication needs [75].

10.5 Knowledge Management in PSS

The design phase in the lifecycle of products and services is characterized by an intense exchange of engineering knowledge [76]. This even increases if an integrated PSS shall be designed in a collaborative way, where the tangible and intangible components are entangled and dependent on each other [77–79]. Explicit and tacit knowledge for the engineering process, like user and system requirements, sentiments, competences, design specifications or processing instructions has to be exchanged between the involved stakeholders from different domains [80]. To this end, both knowledge from the product side as well as the service side must be shared in an appropriate way, combined and utilized, in order to create an attractive product-service bundle for the customer. On the one hand, it has to be elaborated which process steps are typically conducted in PSS design [79, 81, 82]. On the other hand, the involved stakeholders have to be identified and described as the relevant knowledge sources and targets [79, 80, 83]. Based on the results, the relevant types of knowledge and appropriate exchange mechanisms and standards have to be defined [79, 84, 85].

In the scientific discipline of Knowledge Management (KM), several approaches to capture, develop and apply knowledge effectively during product design have been developed. Knowledge-Based Engineering (KBE) for example is aiming at establishing engineering knowledge models, for application in product design and along the whole product life cycle [86]. First attempts have also been made to include service knowledge into a KM framework for PSS as well. These attempts are however focusing on using service knowledge for product design and service operations only. Furthermore, most approaches have been focusing on explicit formalized knowledge inside an individual organization [87].

In order to identify the knowledge exchanged in PSS design and develop an appropriate Knowledge Management approach, it is necessary to identify the stakeholders involved in the underlying processes, as they are the relevant sources and targets of knowledge. As the kind of stakeholders depends strongly on the products, services or industrial sector, a more generic classification is necessary. Thus role models are applied, which have shown to be useful for orchestrating the contributions of various stakeholders in innovation processes [88].

Aligned with the process models, role models describe a set of roles, which can be assigned to role owners, which can be internal or external stakeholders. Assignment of roles and tasks should thus be related to the underlying processes, organizational structure, and competences of stakeholders, respectively. The roles define the division of work between the stakeholders. They contain one or more tasks and can be assigned to one or more individuals or organizational units. The following sections will describe existing role models for product, service and PSS engineering.

In order to raise awareness, where relevant design knowledge can be found ("knowing who knows") and who might be interested in a specific knowledge asset ("knowing who should know"), it is required to define distinctive roles for a PSS design project. Such roles can be derived from the PSS lifecycle. Possible roles are

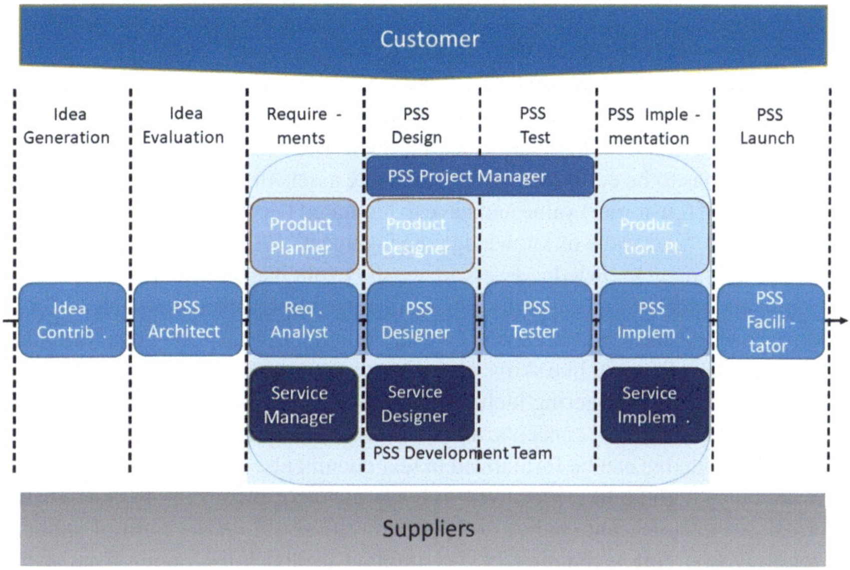

Fig. 10.1 Outline of roles in PSS lifecycle

illustrated in Fig. 10.1. The roles of the PSS Provider (in blue) act as coordinators and are responsible for the execution of design, development and realization of the PSS. The PSS Architect generates documents and manages PSS concepts. The PSS Project Manager coordinates the development team and their communication over the phases along the development process. The PSS Development Team is comprised of representatives from the different domains and deals with the coordination of the product and service development process.

The roles of the Production Network (in orange) are responsible for provision of materials, parts and components or system modules to the PSS Provider. The Product Planner defines the tangible portfolio for the PSS according to the information from PSS Architect. The Product Designer is responsible to specify the product components according to the PSS requirements. The Production Planner plans the production and manufacturing processes for the products. The roles of the Service Network (in green) include the market-specific adaptation of the integrated service shares and the handling of client orders including the individual PSS configuration. The Service Manager conducts comprehensive and frequent communication with customers and the PSS Provider about Service Engineering results; monitoring of the project's economy regarding development efforts and benefit for customer/revenues. The Service Designer reacts flexibly to short-term changes of customer and PSS Provider demands, even in late development phases. The Service Implementer plans the implementation of the services. The Customer plays another key role because demands towards the PSS will be drawn up and implemented based on the determined customer needs. Furthermore, he/she is the user of the PSS and gives feedback about quality. Finally, the Suppliers deliver the necessary materials, components and missing competencies to realize the product and service portfolio together with the PSS Provider.

About knowledge exchanges in PSS, it is worth to consider that PSS engineering is a dynamic process with fluctuating actors [80, 89], where the knowledge residing in individuals has to be combined with knowledge assets that are essential for creating the intended (customer) value and have to be shared between the roles, as centred in the so-called "2nd wave of knowledge management" [90]. While in "traditional" product development knowledge assets are mostly explicit and formalized in the form of documents, specifications and design etc. managed by applications such as CAD, PDM or PLM, during PSS engineering when the intangible aspects come into play knowledge is usually tacit, like skills, know-how, emotions and the like [84]. Explicit knowledge for PSS engineering includes market needs and customer requirements, product specifications and concepts, as well as the detailed product design or model [81]. This knowledge can be formalized in text documents, spread sheets, diagrams, CAD drawings and the like. However, only about 4% of organisational knowledge is formalized [91]. Recent studies on open innovation, e.g. in the form of application of crowdsourcing techniques [92] or implicit feedback leveraging from social media [93], have established the important role of open, crowd-oriented opinion and sentiment in enhancing products and services. This knowledge is mostly informal and unstructured, consisting of individual posts and discussions, ideas, comments and other interactions. Thus, it is difficult to codify and share, as it requires individual interaction to transfer. It is however equally important as knowledge for PSS engineering.

With respect to the ability to merge explicit knowledge from different domains, ontologies are capable in terms of multi-domain knowledge (e.g. Web Ontology Language—OWL [94]). Tacit knowledge, in the form of personal opinions and sentiments regarding PSS, poses extra challenges for the design and implementation of knowledge sharing. The informal nature of the relevant data and the inherent lack of formalization create additional issues [95]. The aspect of sharing explicit, formalized knowledge during PSS design is well covered with concrete approaches and frameworks in literature. Nemoto et al. described a framework to manage PSS design knowledge represented by five elements (core product, need, function, entity and actor) [79]. Zhu et al. and Zhang et al. formalized knowledge from previous PSS cases in a physical and a service model [81, 82]. Furthermore, Baxter et al. defined a KM framework for PSS design process knowledge, manufacturing knowledge, service design and service operations knowledge [31].

Concerning tacit or unstructured knowledge, some approaches can be found in literature, mainly on a conceptual basis. For instance, Bertoni emphasizes the importance of "bottom-up" knowledge sharing in PSS design and suggests Web 2.0 tools such as blogs, wikis or social networks to capture tacit and unstructured knowledge and tap into the "wisdom of crowds" [84]. This idea has been extended by Larsson et al. [77] into the concept of "Engineering 2.0", applying easy to use technologies for knowledge sharing, while Chirumalla explored the use of Web 2.0 tools for knowledge sharing [78]. Also Natural Language Processing (NLP) has been validly adopted to enable formalization of tacit knowledge, according to a transdisciplinary approach [95]. A balance has to be found that supports a "bottom-up" knowledge sharing without sacrificing an efficient way to search and identify relevant knowledge. Bertoni and Larsson have identified seven barriers for knowledge sharing in PSS design, which have to be overcome: acceptability and self-censorship, commitment and reward, resignation, time loss, awareness, language and models, and trust [96].

Furthermore, it is important that all members of the development team have access to the same knowledge in the right form [97]. For explicit knowledge sharing, SysML seems to be appropriate to be extended for the purpose of modelling and exchanging PSS design knowledge, as it is an established standard for systems engineering. A key advantage lies in its extendibility. The needed meta-model layer is provided by default in the specification of UML itself. Hence, an adaption to domain specific needs can be performed. As it is not feasible for all stakeholders to use a common standard for knowledge representation or work with models from other domains, ontologies can be used to share knowledge across domains. However, modelling a proper ontology can become very complex, in particular if a generic ontological representation of a PSS is envisaged. The ontology needs to be filled with product and service related knowledge from different domains. To define service features and software elements demands for specific expertise not only from the field of product design but informatics, service etc. The interface to the knowledge base has to become user-friendly to ensure an acceptance by the end-user.

10.6 PSS Evaluation in Industrial Contexts

For manufacturing companies, the measurement of PSS performance is a crucial aspect to identify the best solution to provide on the market and able to satisfy the customer needs, improving the product value proposition. Moreover, identifying the best PSS offer allows improving the company business model, its business performance, and thus its revenues. In order to promote PSS performance measurement, two main principles must be taken into account: "what cannot be measured cannot be improved" [98] and the Plan-Do-Check-Act (PDCA) approach [99], since a continuous monitoring is required during the entire P-S lifecycle.

According to the authors' perspective, PSS performance evaluation involves both the preliminary evaluation and the validation of the scenarios that will be implemented by the company. Indeed, the evaluation is determining whether the process in its entirety can yield an output that meets the desired requirements, while the validation is determining whether the process as implemented can yield an output that meets the specifications with acceptable capability. For validation, the process must be challenged using verified measurement systems. The areas investigated refer to the definition of a set of key performance indicators (KPIs), and the PSS assessment according to the performance achieved referring to the global system sustainability, according to its three main dimensions: economy, environment, and social well-being [100].

In literature, four different kinds of performance measures can be identified: Result indicators (RIs), Performance indicators (PIs), Key performance indicators (KPIs), and Key result indicators (KRIs). (K)RIs quantities the degree the company achieved its defined goals (they are measured over a long time period), while (K)PIs recognises the actions to do or that should be done in the future to increase the current performance and achieve the defined objectives (they usually are evaluated daily or at least weekly). KPIs in particular measure a current or future situation able to encourage the stakeholders to adopt any strategy in order to face up the scenario that arises. According to the aim of this research, KPIs are able to provide the guidelines to drive the company in the right business direction. Indeed, KPIs measuring the company performance regarding a certain business give information to company stakeholders during the decision-making process. Moreover, they are involved to discover what are the non-adding value activities (that approximately represent the 60% of a company's activities) inside a specific business [101]. Therefore, in order to identify the right KPIs to adopt for evaluating a certain business, literature proposes the adoption of the SMART principles, which are Specific, Measurable, Attainable, Realistic, and Time sensitive [102]. KPIs that comply with these five criteria allow companies assessing their real time performance, defining measures early enough before problems occur, and collecting the appropriate KPIs for PSS evaluation during the Design phase. This last one is a crucial aspect in PSS assessment, because the evaluation and validation of a new PSS offer during the design phase allows both reducing the time to market and successfully addressing the customers' needs. It is important to notice again that PSS assessment is different from product assessment, as in PSS, product characteristics and service functionalities influence one another. Currently in literature, few works about performance assessment in PSS exist; thus, it is an open issue. An interesting research was conducted by Mourtzis et al. [103], which classified KPIs with respect to the main PSS Design methods. Those classes involved in this paper are the following: Customers (C), Business (B), and Sustainability (S). Figure 10.2 shows how the KPIs classes refer to the related PSS design methods explained in the previous chapter. They are groups into three classes: B if referring to Business aspects, S if relating to Sustainability, and C if relating to the Customers. The main KPIs involved and the relative classes are listed in Table 10.1. Beyond the advantages that KPIs measurement offers to assess a PSS offer during the Design phase, some weaknesses remain. KPIs measurement demands lot of effort due to a frequent evaluation. For this reason, a critical aspect in the performance measurement system is to compare the value of an indicator with the effort required for its evaluation [104]. Furthermore, the number of indicators should be limited to ensure a meaningful overview of the current situation.

Generally a PSS will provide not only a higher customer satisfaction, but also a great advantage on the sustainability in respect to traditional products [105], according to its three main dimensions: economy, environment, and social well-being [100]. From the environmental viewpoint, PSS provides a more conscious product usage thanks to the service functionalities delivered, increasing resource productivity and a close loop-chain manufacturing. Moreover, because the PSS requires the involvement of different partners and stakeholders, they will deliver a solution able to create

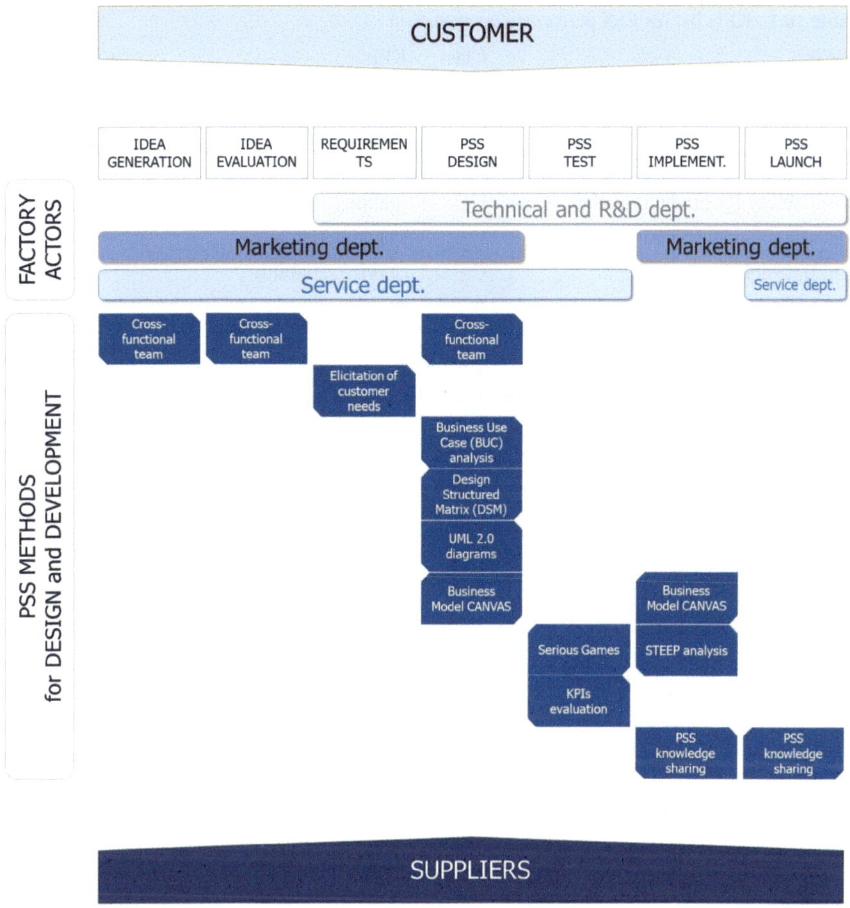

Fig. 10.2 Design and development framework for PSS

a sustainable supply chain, according to the service provided. From the economic viewpoint, PSS is able to create new market potentials and higher profit margins, and can contribute to higher productivity by means of reducing investment costs along the lifetime as well as reducing operating costs for the final users [35]. Finally, from the social viewpoint, PSS can build secure knowledge intensive jobs and contribute to a more geographically balanced distribution of wellbeing.

The PSS and the relative Servitization process extend the responsibility of the PSS provider to the whole lifetime of the product [106]. For this reason, it is required to perform assessment from a lifecycle perspective. In the manufacturing industry, product sustainability is calculated by adopting a lifecycle design approach: it allows quantifying product impacts and providing tangible commercial values in terms of efficiency and costs [107]. They are based on the definition of several indicators to assess the lifecycle performance and support comparative analysis. The technique to

Table 10.1 KPIs list for PSS performance evaluation

KPIs	Class	KPIs	Class
Customer satisfaction	C	Overall equipment effectiveness	B
Acceptability	C	Technical availability	B
Acceptance rate	C	Flexibility	B
Availability for production plan	C	Stability	B
Number of customer needs	C	Machine reliability	B
On-time delivery	C	Service reliability	B
Efficiency	C	Service assurance	B
Quality	B, C	Team qualification	B, C
Customer needs rate	C	Knowledge management	B
Requirement inconsistency	C	PS maintenance efficiency	B
Efficiency of collaboration	C	Development cost	B
Privacy	C	Service delivery costs	B
Product flexibility	C, B	Environmental quality cost function	S, B
Expansion flexibility	C, B	Energy efficiency	S
Sustainable product-service efficiency	S	Lease/reuse	S

support this described lifecycle design approach is defined as Life Cycle Assessment (LCA) according to ISO 14040:2006 (2006) [108] and it allows evaluating the environmental impacts on the product. Moreover, another lifecycle approach to assess the economic impact is Life Cycle Costing (LCC) [109], which has the scope to recognize all the economic impact during the product lifecycle. More recently, also the social impacts have been included in the lifecycle design approach by the so-called Social Life Cycle Assessment (SLCA) [110]. Such methods defined for product assessment could be "extended" and applied also to PSSs. However, the common indicators that assess economic, environmental or social domains separately will not approach and assess PSS sustainability in its complexity and wholeness. Indeed, the sustainability of a system cannot be assessed by the use of a single criterion mainly because of the intrinsic multidimensionality characteristic of sustainability. It is required to generate and assess a unique value that is the combination of all relevant criteria. In literature, some research calculates the sustainability of a PSS without adopting the lifecycle approaches, while other research has proposed to translate the lifecycle design approaches to assess the PSS sustainability demonstrating how to calculate the sustainability impacts of an integrated PSS by considering not only the impacts related to the product realization, usage and dismissing, but involving also the intangible assets and the ecosystem actors, as reported in Table 10.2. It shows several methods and relative indicators developed and used by different researchers in literature, which adopts a transdisciplinary approach. In the columns, the three main lifecycle indicators are identified (i.e. Environmental Impact, Economic Impact, and Social Impact), and also the integrated indicator to calculate the entire Sustainable impact.

Table 10.2 Lifecycle indicators in literature according to a transdisciplinary approach

References	Environmental impact indicator	Economic impact indicator	Social impact indicator	Sustainable impact indicator
[22, 111]	ENI (ENvironmental indicator) measured by Eco-indicator 99 point (EI-99)	ECI (EConomic Indicator) refers to all the lifecycle costs through the equivalent annual cash flow technique (EA)	SOI (SOcial Impact) considers separately Human Health contributions according to EI-99 methodology	SI = ENI + ECI + SOI (each indicator is normalized to obtained a monetary value (€), that is SI)
[112]	Total environmental impact along lifetime	Total lifecycle cost	None	None
[110]	Global warming potential	Lifecycle costs	QALY (measure of well-being)	QALY as a single-score alternative to direct monetisation
[72]	Life span, efficiency of resource consumption, closed cycle efficiency, and potentials for improvement	None	None	None

10.7 An Extensive Design and Development Framework for Industrial PSSs

On the basis of a critical analysis of the existing literature about the different aspects of PSS development, we can conclude that the design and development of successful PSS has to focus on the identification and interpretation of interactions between products and services to fully reflect stakeholder requirements. Design decisions cannot be merely technology-driven or manufacturing-related, but the user needs have to be the main focus. In this context, UCD techniques become a driving force. Thus, in the case of PSS, innovation relies on sharing knowledge between partners from different domains, maintain a common understanding of the design concept derived from customer needs and re-use experiences from other PSS projects; the usage of "downstream" knowledge from later phases of the life cycle and the inclusion of the customer into the design process is important as well. While in a conventional static OEM-supplier relationship contractual obligations set by the leading company define what and how knowledge is shared, such a model is not feasible for the dynamic collaboration required for PSS. Besides the missing lead-time required setting up such an arrangement, there might simply not be a partner powerful enough to impose its standards.

Based on this analysis, it can be stated that:

- Elicitation of customer needs is the key point and the first issues to solve in PSS design and development. It can be done by UCD techniques (ethnography, personas, role-playing);
- Technical analysis of PSS function can be executed by Business Use Case (BUC) analysis, which provides a user-centred investigation of the conceptual PSS model and a mapping of goal-oriented interactions between external actors and the PSS items;
- Design of the PSS functions by Design Structured Matrix (DSM);
- Definition of the PSS process model and system infrastructure by UML 2.0 diagrams and extended SysML models for systems engineering. The PSS meta-model layer is provided by UML, and extension of SysML allows exchanging PSS design knowledge, as it is an established standard. Tacit knowledge sharing can be supported using Web 2.0 tools for the PSS stakeholders on a dedicated social platform;
- Definition of the Business Model and involved stakeholders by CANVAS modelling;
- Identification of the new business strategies and trends by a simplified STEEP analysis;
- Management of the PSS knowledge sharing by formalization of explicit engineering knowledge and flexible exchange of unstructured and tacit knowledge between the stakeholders involved;
- Creation of cross-functional teams to foster knowledge sharing during the design phase, including people coming from the different functions, domains and organisations involved (i.e., stakeholders from product, service, or system integration). For this purpose;
- PSS model validation by Serious Games and hybrid PSS digital mock-ups in order to simulate the human-system interaction;
- Evaluation of PSS performances by proper key performance indicators (KPIs), investigating different areas such as Business, Sustainability and Customers. The latter include also the evaluation of the user experience by tests with samples users. Interactive prototypes can be adopted for this scope: high-fidelity mock-ups that combines realistic visualization (e.g., high quality aesthetic rendering, realistic environments, truthful use cases) with high level of interaction and simulation of the PSS behaviours according to the PSS conceptual model (e.g., movements of product parts according to some interaction with its interface or the service functions, real-time feedback connected to the service delivery).

The overall framework defined to support PSS design and development is represented in Fig. 10.2. It integrates the above-mentioned methods along the P-S lifecycle management (P-SLM) and considers the actors involved, as presented in Fig. 10.1.

In addition, Table 10.3 shows the selected KPIs for an overall evaluation of the PSS performances, according to the three identified areas: Business, Sustainability

Table 10.3 Selected KPIs for PSS evaluation

Area	KPIs	PSS methods for design and development
KPIs for business (B)	Overall equipment effectiveness	
	Technical availability	
	Flexibility	
	Stability	
	Machine reliability	
	Service reliability	IPS^2 models
	Service assurance	IPS^2 models
	Team qualification	
	Knowledge management	Layer-based methodology
	Knowledge sharing	Layer-based methodology
	PS maintenance efficiency	Lifecycle oriented approaches
	Development cost	Lifecycle oriented approaches
	Service delivery costs	Lifecycle oriented approaches
	Environmental quality cost function	Lifecycle oriented approaches
	Quality	Software tools
	Product flexibility	Software tools
	Expansion flexibility	Software tools
KPIs for sustainability (S)	Environmental quality cost function	Lifecycle oriented approaches
	Energy efficiency	Web-based models
	Lease/reuse	Web-based models
	Sustainable product-service efficiency	Lifecycle oriented approaches
	Sustainability assessment (SA)	Life cycle assessment (LCA) and costing (LCC)
KPIs for customer (C)	Customer satisfaction	Usability assessment
	Acceptability	Service engineering
	Acceptance rate	Service engineering
	Availability for production plan	IPS^2 models
	Number of customer needs	Requirements elicitation
	On-time delivery	Service engineering
	Efficiency	Usability assessment
	Quality	Software tools
	Customer needs rate	Requirements elicitation

(continued)

Table 10.3 (continued)

Area	KPIs	PSS methods for design and development
	Requirement inconsistency	Requirements elicitation
	Efficiency of collaboration	IPS^2 models
	Privacy	IPS^2 models
	Product flexibility	Software tools
	Expansion flexibility	Software tools
	Team qualification	
	Effectiveness	Usability assessment

and Customers. In respect to the state of the art, Business indicators have been extended to consider also knowledge management and knowledge sharing among all the stakeholders, while Sustainability indicators have been expanded according to the last researches in this field (as cited in Table 10.2) and Customers indicators have been integrated with usability, value and interaction indicators.

10.8 Summary and Outlook

This chapter discussed the most useful methods for PSS Requirement Elicitation (RE), design and development and proposes a structured way to manage such a complex and transdisciplinary process, with a special attention to transdisciplinary methods. The most significant RE approaches for PSS and the most successful design methods are presented, and numerous examples of transdisciplinary methods from recent literature are discussed. After that, the chapter focuses on knowledge management, which is a key aspect in PSS due to the increased complexity and the higher quantity of data and knowledge exchanged during PSS design and develop. The discussion highlights how to identify the stakeholders involved in the underlying processes, as relevant sources and targets of knowledge, and how to choose the best role models for orchestrating the contributions of various stakeholders in the innovation process. Finally, the measurement of PSS performance is presented as a crucial point to identify the greatest solution to provide on the market and able to satisfy the customer needs, improving the product value proposition. A set of Key Performance Indicators (KPIs) suitable for PSS are defined according to the reference area: business, sustainability, and customer. This overview about design methodologies and evaluation strategies allows having a high-level view and useful tools to develop a systematic PSS development framework, in order to help manufacturing enterprises to determine their own position on the PSS maturity scenario and to plan future actions toward servitization, according to knowledge exchange and transdisciplinary view [113].

References

1. Hauksdóttir D, Mortensen NH, Nielsen PE (2013) Identification of a reusable requirements structure for embedded products in a dynamic market environment. Comput Ind 64(4):351–362. https://doi.org/10.1016/j.compind.2012.10.008
2. Garetti M, Rosa P, Terzi S (2012) Life cycle simulation for the design of product-service systems. Comput Ind 63(4):361–369
3. Aurich JC, Fuchs C, Wagenknecht C (2006) Life cycle oriented design of technical product-service systems. J Clean Prod 14(17):1480–1494. https://doi.org/10.1016/j.jclepro.2006.01.019
4. Welp EG, Meier H, Sadek T, Sadek K (2008) Modelling approach for the integrated development of industrial product- service systems. In: Proceeding of 41st CIRP conference on manufacturing systems
5. Komoto H, Tomiyama T (2008) Integration of a service CAD and a life cycle simulator. CIRP Ann—Manuf Technol 57(1):9–12
6. Shimomura Y, Hara T, Arai T (2009) A unified representation scheme for effective PSS development. In: CIRP Ann—Manuf Technol, pp 379–382
7. Marilungo E, Coscia E, Quaglia A, Peruzzini M, Germani M (2016) Open innovation for ideating and designing new product service systems. In: Proceeding of 8th CIRP conference on industrial product service systems, Bergamo, Italy
8. Sakao T, Shimomura Y (2006) Service engineering: a novel engineering discipline for producers to increase value combining service and product. In: Huisingh D et al. (eds) J Clean Prod 15(6):590–604
9. Tomiyama A (2005) A design methodology of services. In: Samuel A, Lewis W (eds) DS 35: proceedings ICED 05, the 15th international conference on engineering design. Melbourn, Barton, pp 1970–2014
10. Komoto H, Tomiyama T (2009) Systematic generation of PSS concepts using a service CAD tool. In: Sakao T, Lindahl M (eds) Introduction to product/service-system design. Springer, London, pp 71–92
11. Sakao T, Shimomura Y, Sundin E, Comstock M (2009) Modeling design objects in CAD system for service/product engineering. Comput Aided Des 41(41):197–213
12. Shimomura Y, Arai T (2009) Service engineering—methods and tools for effective PSS development. In: Sakao T, Lindahl M (eds) Introduction to product/service-system design. Springer, London, pp 113–136
13. Chirumalla K, Bertoni A, Ericson A, Isaksson O (2013) Knowledge-sharing network for product-service system development: is it atypical? The philosopher's stone for sustainability. Springer, Berlin, pp 109–114. https://doi.org/10.1007/978-3-642-32847-3_18
14. Matzen D, Tan A, Andreasen MM (2005) Product/service-systems: proposal for models and terminology. In: Meerkamm H (eds) Design for X: Beiträge zum 16. Symposium. Erlangen: Lehrstuhl für Konstruktionstechnik, pp 27–38
15. Matzen D (2009) A systematic approach to service oriented product development. Stokkemarke: Scandinavian Digital Printing A/S, Ph.D. thesis
16. Tan A (2010) Service-oriented product development strategies. Scandinavian Digital Printing A/S, Stokkemarke
17. Peruzzini M, Marilungo E, Germani (2014b) Functional and ecosystem requirements to design sustainable P-S. In: Advances in transdisciplinary engineering, volume 1: moving integrated product development to service clouds in the global economy, proceedings 21st ISPE Inc. international conference on concurrent engineering (CE2014), pp 768–777. https://doi.org/10.3233/978-1-61499-440-4-768
18. Müller P (2013) Integrated engineering of products and services—layer-based development methodology for product-service systems. Fraunhofer IRB, Stuttgart
19. Miller D, Hope Q, Eisenstat R, Foote N, Galbraith J (2002) The problem of solutions: balancing clients and capabilities. Bus Horiz 45(2):3–12. https://doi.org/10.1016/S0007-6813(02)00181-7

20. Peruzzini M, Germani M, Favi C (2012) Shift from PLM to SLM: a method to support business requirements elicitation for service innovation in Product Lifecycle Management. In: Proceedings of IFIP advances in information and communication technology 388 AICT. Springer, New York, pp 111–123, https://doi.org/10.1007/978-3-642-35758-9_10
21. Wiesner S, Peruzzini M, Doumeingts G, Thoben KD (2012) Requirements engineering for servitization in manufacturing service ecosystems (MSEE). In: 4th CIRP IPS2 conference, Japan
22. Peruzzini M, Marilungo E, Germani M (2014a) A QFD-based methodology to support product-service design in manufacturing industry. In: Proceedings of 2014 international conference on engineering, technology and innovation: engineering responsible innovation in products and services ICE 2014, Bergamo, Italy, 23–25 June pp 1–7. https://doi.org/10.1109/ice.2014.6871572
23. Peruzzini M, Marilungo E, Germani M (2015) Structured requirements elicitation for product-service system. Int J Agile Syst Manag 8(3/4):189–218
24. Peruzzini M, Marilungo E (2016) User-centred approach for product- service design using virtual mock-ups. In: Proceedings 14th international design conference, DESIGN 2016, Cavtat, Dubrovnik, Croatia, 16–19 May 2016, in DS 84, pp 1805–1814
25. Mengoni M, Peruzzini M (2016) How to support the design of user-oriented product-related services. Proceedings 18th international conference on human-computer interaction, HCI international 2016, Toronto, Canada, 17–22 July 2016, in distributed, ambient and pervasive interactions, Lecture Notes in Computer Science, vol 9749, pp 103–110. https://doi.org/10.1007/978-3-319-39862-4_10
26. Nilsson P, Fagerström B (2006) Managing stakeholder requirements in a product modelling system. Comput Ind 57(2):167–177. https://doi.org/10.1016/j.compind.2005.06.003
27. Elgh F (2007) Modelling and management of manufacturing requirements in design automation systems. In: Loureiro G, Curran R (eds) Complex Syst Concurr Eng. Springer, London, pp 321–328
28. Rouse WB, Sage AP (2009) Handbook of systems engineering and management, 2nd edn. Wiley Series in Systems Engineering and Management. Wiley, Hoboken, NJ
29. Boehm B, Basili VR (2001) Top 10 list [software development]. Computer 34(1):135–137. https://doi.org/10.1109/2.962984
30. Nuseibeh B, Easterbrook S (2000) Requirements engineering. In: Finkelstein A (ed) ICSE'00 Proceedings of the conference on the future of software engineering, Limerick, Ireland, pp 35–46. https://doi.org/10.1145/336512.336523
31. Baxter D, Gao J, Case K, Harding J, Young B, Cochrane S, Dani S (2008) A framework to integrate design knowledge reuse and requirements management in engineering design. Robot Comput-Integr Manuf 24(4):585–593. https://doi.org/10.1016/j.rcim.2007.07.010
32. Laporti V, Borges MRS, Braganholo VP (2007) A collaborative approach to requirements elicitation. In: 11th international conference on computer supported cooperative work in design, Melbourne, Australia, pp 734–739. https://doi.org/10.1109/cscwd.2007.4281527
33. Azadegan A, Papamichail KN, Sampaio P (2013) Applying collaborative process design to user requirements elicitation: a case study. Comput Ind 64(7):798–812. https://doi.org/10.1016/j.compind.2013.05.001
34. Goedkoop MJ, van Halen CJG, Te Riele HRM, Rommens PJM (1999) Product service systems—ecological and economic basics
35. Baines TS, Lightfoot HW, Evans S, Neely A, Greenough R, Peppard J, Roy R, Shehab E, Braganza A, Tiwari A, Alcock JR, Angus JP, Bastl M, Cousens A, Irving P, Johnson M, Kingston J, Lockett H, Martinez V, Michele P, Tranfield D, Walton IM, Wilson H (2007) State-of-the-art in product-service systems. Proc Inst Mech Eng, Part B: J Eng Manuf 221(10):1543–1552. https://doi.org/10.1243/09544054JEM858
36. Meier H, Roy R, Seliger G (2010) Industrial product-service systems—IPS2 CIRP annals. Manuf Technol 59(2):607–627. https://doi.org/10.1016/j.cirp.2010.05.004
37. Pahl G, Beitz W (2007) Konstruktionslehre: Grundlagen erfolgreicher Produktentwicklung; Methoden und Anwendung, 7th edn. Springer-Lehrbuch

38. van Husen C (2007) Anforderungsanalyse für produktbegleitende Dienstleistungen. IPA-IAO-Forschung und Praxis, Nr. 458. Jost-Jetter, Heimsheim
39. Berkovich M, Leimeister JM, Krcmar H (2011) Requirements engineering für product service systems. Wirtschaftsinf 53(6):357–370. https://doi.org/10.1007/s11576-011-0301-3
40. Ehrlenspiel K (2007) Integrierte Produktentwicklung: Denkabläufe, Methodeneinsatz, Zusammenarbeit, 3rd edn. Hanser, München
41. Ulrich KT, Eppinger SD (2012) Product design and development, 5th edn. McGraw-Hill, New York
42. Ahrens G (2000) Das Erfassen und Handhaben von Produktanforderungen: Methodische Voraussetzungen und Anwendung in der Praxis
43. Lindemann U (2009) Methodische Entwicklung technischer Produkte. Springer, Berlin, Heidelberg
44. Liu X, Raorane S, Leu MC (2007) A web-based intelligent collaborative system for engineering design. In: Li WD, McMahon C, Ong SK, Nee AYC (eds) Collaborative product design and manufacturing methodologies and applications. Springer Series in Advanced Manufacturing. Springer, London, pp. 37–58
45. Murthy DNP, Rausand M, Østerås T (2008) Product reliability: specification and performance. Springer Series in Reliability Engineering. Springer, London
46. Berkovich M, Leimeister JM, Krcmar H (2009) Suitability of product development methods for hybrid products as bundles of classic products, software and service elements. In: ASME 2009 international design engineering technical conferences and computers and information in engineering conference, San Diego, California, USA, August 30–September 2, pp 885–894. https://doi.org/10.1115/detc2009-86939
47. Bullinger HJ, Scheer AW, Schneider K (2006) Service Engineering: Entwicklung und Gestaltung innovativer Dienstleistungen, 2nd edn. Springer, Berlin
48. Bullinger HJ, Schreiner P (2006) Service Engineering: Ein Rahmenkonzept für die systematische Entwicklung von Dienstleistungen. In: Bullinger HJ, Scheer AW (eds) Service engineering. Springer, Berlin/Heidelberg, pp 53–84
49. Spath D, Demuß L (2006) Entwicklung hybrider Produkte - Gestaltung materieller und immaterieller Leistungsbündel. In: Bullinger HJ, Scheer AW (eds) Service engineering. Springer, Berlin/Heidelberg, pp 463–502
50. Ramaswamy R (1996) Design and management of service processes. Engineering process improvement series. Addison-Wesley Pub. Co., Reading, MA
51. Edvardsson B, Olsson J (1996) Key concepts for new service development. Serv Ind J 16(2):140–164. https://doi.org/10.1080/02642069600000019
52. Frietzsche U, Maleri R (2006) Dienstleistungsproduktion. In: Bullinger H-J, Scheer A-W (eds) Service engineering. Springer, Berlin/Heidelberg, pp 195–225
53. Kersten W, Kern EM, Zink T (2006) Collaborative service engineering. In: Bullinger HJ, Scheer AW (eds) Service engineering. Springer, Berlin/Heidelberg, pp 341–357. https://doi.org/10.1007/3-540-29473-2_14
54. Pohl K (2008) Requirements Engineering: Grundlagen, Prinzipien, Techniken, 2nd edn. Dpunkt-Verl, Heidelberg
55. Kotonya G, Sommerville I (1998) Requeriments engineering processes and techniques. Wiley, New York
56. van Lamsweerde A (2009) Requirements engineering: from system goals to UML models to software specifications. Wiley, Chichester, England, Hoboken, NJ
57. Hull E, Jackson K, Dick J (2005) Requirements engineering, 2nd edn. Springer, London
58. Aurum A, Wohlin C (2005) Engineering and managing software requirements. Springer, Berlin, London
59. Lanubile F (2009) Collaboration in distributed software development. In: Hutchison D, Kanade T, Kittler J, Kleinberg JM, Mattern F, Mitchell JC, Naor M, Nierstrasz O, Pandu Rangan C, Steffen B, Sudan M, Terzopoulos D, Tygar D, Vardi MY, Weikum G, de Lucia A, Ferrucci F (eds) Software engineering. Lecture Notes in Computer Science. Springer, Berlin, Heidelberg, pp 174–193

60. Whitehead J (2007) Collaboration in software engineering: a roadmap. In: Future of software engineering, Minneapolis, MN, USA, pp 214–225. https://doi.org/10.1109/fose.2007.4
61. Li Z, Rahman QA, Ferrari R, Madhavji NH (2009) Does requirements clustering lead to modular design? In: Hutchison D, Kanade T, Kittler J, Kleinberg JM, Mattern F, Mitchell JC, Naor M, Nierstrasz O, Pandu Rangan C, Steffen B, Sudan M, Terzopoulos D, Tygar D, Vardi MY, Weikum G, Glinz M, Heymans P (eds) Requirements engineering: foundation for software quality. Lecture Notes in Computer Science. Springer, Berlin, Heidelberg, pp 233–239
62. Lindahl M, Sundin E, Sakao T, Shimomura Y (2007) Integrated product and service engineering versus design for environment—a comparison and evaluation of advantages and disadvantages. In: Takata S, Umeda Y (eds) Advances in life cycle engineering for sustainable manufacturing businesses. Springer, London, pp 137–142
63. Aurich JC, Schweitzer E, Fuchs C (2007) Life cycle management of industrial product-service systems. In: Takata S, Umeda Y (eds) Advances in life cycle engineering for sustainable manufacturing businesses. Springer, London, pp 171–176
64. Elnadi M, Shehab E (2015) Main enablers and factors for successful implementation of lean in product-service systems. Int J Agile Syst Manag 8(3–4):332–354
65. Böhmann T, Langer P, Schermann M (2008) Systematische Überführung von kundenspezifischen IT-Lösungen in integrierte Produkt-Dienstleistungsbausteine mit der SCORE-Methode Wirtschaftsinf 50(3):196–207. https://doi.org/10.1365/s11576-008-0047-8
66. Vasantha GVA, Roy R, Lelah A, Brissaud D (2012) A review of product–service systems design methodologies. J Eng Des 23(9):635–659. https://doi.org/10.1080/09544828.2011.639712
67. Maussang N, Sakao T, Zwolinski P, Brissaud D, (2007) A model for designing product-service systems using functional analysis and agent based model. In: 16th international conference on engineering design (ICED'07), Paris, France
68. Morelli N (2006) Developing new product service systems (PSS): methodologies and operational tools. J Clean Prod 14(17):1495–1501. https://doi.org/10.1016/j.jclepro.2006.01.023
69. Freitag M, Kremer D, Hirsch M, Zelm M (2013) An approach to standardise a service life cycle management. In: Zelm M, van Sinderen M, Ferraira Pires L, Doumeingts G (eds) Enterprise interoperability. Wiley, Chichester, pp 115–126
70. Scheithauer G, Kett H, Kaiser J, Hackner S, Hu H, Wirtz G (2010) Business modeling for service engineering. In: Shin SY, Ossowski S, Schumacher M, Palakal MJ, Hung CC (eds) the 2010 ACM symposium, Sierre, Switzerland, p 118. https://doi.org/10.1145/1774088.1774113
71. Schweitzer E, Fiekers C, Möhrer J (2010) Realisierung investiver Produkt-Service Systeme. In: Aurich JC, Clement MH (eds) Produkt-Service Systeme. Springer, Berlin, Heidelberg, pp 95–116
72. Mont O (2002) Clarifying the concept of product-service system. J Clean Prod 10(3):237–245. https://doi.org/10.1016/S0959-6526(01)00039-7
73. Lindow K, Müller P, Stark R (2011) NEW job roles in global engineering-from education to industrial deployment. In: Proceedings of the 18th international conference on engineering design (ICED 11), impacting society through engineering design, vol 8: design education. Lyngby/Copenhagen, Denmark, pp 205–215
74. Abramovici M, Aidi Y, Jin F, Göbel JC (2012) Lifecycle management von Hybriden Leistungsbündeln. In: Meier H, Uhlmann E (eds) Integrierte Industrielle Sach- und Dienstleistungen. Springer, Berlin, Heidelberg, pp 265–284
75. Müller P, Stark RR, Fraunhofer IP (2014). Integrated engineering of products and services: layer-based development methodology for product-service systems. Fraunhofer Verlag
76. Clarkson J, Eckert C (2005) Design process improvement: a review of current practice. Springer, London
77. Larsson A, Ericson Å, Larsson T, Isaksson O, Bertoni M (2010) Engineering 2.0: exploring lightweight technologies for the virtual enterprise. In: Randall D, Salembier P (eds) From CSCW to web 2.0: European developments in collaborative design. Computer supported cooperative work. Springer, London, pp 173–191

78. Chirumalla K (2013) Managing knowledge for product-service system innovation: the role of web 2.0 technologies. Res Technol Manag 56(2):45–53. https://doi.org/10.5437/08956308x5602045
79. Nemoto Y, Akasaka F, Shimomura Y (2015) A framework for managing and utilizing product–service system design knowledge. Prod Plan Control 26(14–15):1278–1289. https://doi.org/10.1080/09537287.2015.1033493
80. Cedergren SI, Elfving SW, Eriksson J, Parida V (2012) Analysis of the industrial product-service systems (IPS2) literature: a systematic review. In: 2012 IEEE 6th international conference on management of innovation & technology (ICMIT 2012), Bali, Indonesia, pp 733–740. https://doi.org/10.1109/icmit.2012.6225897
81. Zhang D, Hu D, Xu Y, Zhang H (2012) A framework for design knowledge management and reuse for product-service systems in construction machinery industry. Comput Ind 63(4):328–337. https://doi.org/10.1016/j.compind.2012.02.008
82. Zhu H, Gao J, Cai Q (2015) A product-service system using requirement analysis and knowledge management technologies. Kybernetes 44(5):823–842. https://doi.org/10.1108/K-11-2014-0244
83. Wewior J (2015) Role-play based assessment of IPS2-specific intellectual capital. Proc CIRP 30:415–420. https://doi.org/10.1016/j.procir.2015.02.127
84. Bertoni M (2010) Bottom-up knowledge sharing in PSS design. A classification framework. In: Marjanović D (eds) Design 2010: DS 60: proceedings of the 11th international design conference, Dubrovnik, Croatia, Zagreb, pp 1461–1470
85. Romero D, Rabelo RJ, Molina A (2012) On the management of virtual enterprise's inheritance between virtual manufacturing & service enterprises: supporting "dynamic" product-service business ecosystems. In: 2012 18th international ICE conference on engineering, technology and innovation (ICE), Munich, Germany, pp 1–11. https://doi.org/10.1109/ice.2012.6297695
86. Stjepandić J, Verhagen WJC, Liese H, Bermell-Garcia P (2015) Knowledge-based engineering, In: Stjepandic J et al. (eds) Concurrent engineering in the 21st century: foundations, developments and challenges. Springer International, Switzerland, pp 255–286
87. Baxter D, Roy R, Doultsinou A, Gao J, Kalta M (2009) A knowledge management framework to support product-service systems design. Int J Comput Integr Manuf 22(12):1073–1088. https://doi.org/10.1080/09511920903207464
88. Kremer D, Leyh J (2010) Mit Technologietreiber-Rollen neue Technologien schneller in Produkte umsetzen B. Dworschak, & A. Karapidis, Professional Traing Facts, pp 49–74
89. Nergard H, Ericson A (2012) Changes in present product design—opportunities for industrial oriented research. In: 2012 IEEE 3rd international conference on cognitive infocommunications (CogInfoCom), Kosice, Slovakia, pp 499–503. https://doi.org/10.1109/coginfocom.2012.6422032
90. Mouritsen J, Larsen HT (2005) The 2nd wave of knowledge management: the management control of knowledge resources through intellectual capital information. Manag Account Res 16(3):371–394. https://doi.org/10.1016/j.mar.2005.06.006
91. Bell S (2006) Lean enterprise systems: using IT for continuous improvement. Wiley Series in Systems Engineering and Management. Wiley-Interscience, Hoboken, NJ
92. van den Ende J, Frederiksen L, Prencipe A (2015) The front end of innovation: organizing search for ideas. J Prod Innov Manag 32(4):482–487. https://doi.org/10.1111/jpim.12213
93. Budak C, Agrawal D, El Abbadi A (2011) Structural trend analysis for online social networks. Proc VLDB Endow 4(10):646–656. https://doi.org/10.14778/2021017.2021022
94. McGuinness DL, van Harmelen F (2004) OWL web ontology language overview. W3C Recommendation 10(10)
95. Nawroth C, Schmedding M, Brocks H, Kaufmann M, Fuchs M, Hemmje M (2015) Towards cloud-based knowledge capturing based on natural language processing. Procedia Comput Sci 68(2015):206–216
96. Bertoni M, Larsson AC (2010) Coping with the knowledge sharing barriers in product service systems design. In: Horváth I, Mandorli F, Rusak Z (eds) Tools and methods of competitive engineering: proceedings of the eighth international symposium on tools and methods of

competitive engineering—TMCE 2010, 12–16 April 2008, Ancona, Italy. Delft University of Technology, Delft, Netherlands, pp 903–914
97. Nonaka I, Toyama R, Konno N (2000) SECI, Ba and leadership: a unified model of dynamic knowledge creation. Long Range Plan 33(1):5–34. https://doi.org/10.1016/S0024-6301(99)00115-6
98. Gries B, Restrepo J (2011) KPI measurement in engineering design: a case study. In: Proceeding of international conference on engineering design, ICED11, vol 1, pp 531–537
99. Deming WE (1992) Quality, productivity and competitive position, MitCenter for Advanced Engineering Study, place
100. Adams WM (2006) The future of sustainability: re-thinking environment and development in the twenty-first century. In: Technical report, IUCN renowned thinkers meeting
101. Dombrowski U, Schmidtchen K, Ebentreich D (2013) Ballanced key performance indicators in product development. Int J Mater, Mech Manuf 1(1):27–31
102. Abramovici M, Jin F, Dang HB (2013) An indicator framework for monitoring IPS2 in the use phase. In: Product-service integration for sustainable solutions, pp 311–322. https://doi.org/10.1007/978-3-642-30820-8_27
103. Mourtzis D, Doukas M, Fotia S (2015) Performance indicators for the evaluation of product-service systems design: a review. In: Proceeding of IFIP WG 5.7 international conference, Tokyo, Japan, vol 460. https://doi.org/10.1007/978-3-319-22759-7_68
104. Kerzner HR (2013) Project management metrics, KPIs, and dashboards: a guide to measuring and monitoring project performance. Proj Manag J 43(2):102. https://doi.org/10.1002/pmj.21263
105. McAloone TC, Mougaard K, Restrepo J, Knudsen S (2010) Eco-innovation in the value chain. In: Proceedings of international design conference, Dubrovinik, Croatia
106. Aurich JC, Mannweiler E, Schweitzer E (2010) How to design and offer services successfully. Proc CIRP J Manuf Sci Technol 2(3):136–143. https://doi.org/10.1016/j.cirpj.2010.03.002
107. Jeswiet J (2009) A definition for life cycle engineering. In: Proceedings of 36th international seminar on manufacturing systems, Saarbrucken, Germany
108. ISO 14040:2006 (2006) Environmental management—life cycle assessment—principles and framework
109. Woodward DG (1997) Life cycle costing—theory, information acquisition and application. J Proj Manag 15(6):335–344
110. Weidema B (2006) The integration of economic and social aspects in life cycle impact assessment. Int J Life Cycle Assess 11(1):89–96. https://doi.org/10.1065/lca2006.04.016
111. Peruzzini M, Marilungo E, Germani M (2013b) Product-service sustainability assessment in virtual manufacturing enterprises. IFIP International Federation for Information Processing AICT 408, Proceedings 14th IFIP WG 5.5 working conference on virtual enterprises, PRO-VE 2013, Dresden (Germany), 30 September–2 October 2013, Camarinha-Matos LM, Scherer RJ (eds), pp 13–21. ISSN 1868-4238, ISBN 978-3-642-40542-6. https://doi.org/10.1007/978-3-642-40543-3
112. Kwak M, Kim H (2013) Economic and environmental impacts of product service lifetime: a life-cycle perspective. In: Proceedings of 5th CIRP international conference on industrial product-service systems, Bochum, Germany. https://doi.org/10.1007/978-3-642-30820-8_16
113. Wognum N, Bil C, Elgh F, Peruzzini M, Stjepandić J, Verhagen WJC (2019) Transdisciplinary systems engineering: implications, challenges and research agenda. Int J Agile Syst Manag 12(1):58–89

Chapter 11
Systems Engineering for Machining

John P. T. Mo and Songlin Ding

Abstract Machining is the traditional product shaping process by removing materials from a block of original materials. Practically, the machining process itself has not changed much in the last couple of centuries but the accessories around the process have improved significantly, like data logging features in modern computer numerically controlled machines. The machining process is a system, the components of which should be considered as independent units which work harmonously with other systems in the enterprise. In this chapter a systems approach is adopted to examine methods and techniques that can improve five key performance indicators of the machining system, i.e. sustainability, accuracy, efficiency, precision and reliability. In particular, High Speed Machining, tool breakage prevention, thin wall deflection, tool geometry and chatter monitoring are studied in relation to the five performance indicators, respectively. Application of these techniques has produced good machining outcomes showing strategic development direction leading to better performance of the machining system.

Keywords Machining · High speed machining · Titanium alloys · Polycrystalline diamond tools · Finite element analysis · Online chatter monitoring

11.1 Introduction

Machining is the traditional manufacturing process that makes products from a block of material by removing some parts of it. This manufacturing process has been around for centuries. A lot of research and development in machining processes have been reported since the turn of last century. Practically, the machining process itself has not changed much but the accessories around the process have improved significantly. For example, modern computer numerically controlled (CNC) machines have many data logging features that help the operator to diagnose machining problems. The future trend of machining engineering is going to move towards more difficult tasks, in particular, machining of exotic materials which are regarded as difficult to work

J. P. T. Mo (✉) · S. Ding
Manufacturing Engineering, RMIT University, Melbourne, Australia
e-mail: john.mo@rmit.edu.au

J. Stjepandić et al. (eds.), *Systems Engineering in Research and Industrial Practice*,
https://doi.org/10.1007/978-3-030-33312-6_11

with. Due to customer preferences, manufacturers have to find ways to complete these manufacturing tasks.

The machining process is a system. Therefore, components of a machining system should not be considered separately, thereby ignoring the system's properties. Currently, the component approach is a common manufacturing practice in today's environments, where different manufacturers produce the various components of the machining system, but no one seems to be responsible for system coherency and thus for its sustainability. Low efficiency, subpar quality, and tool failure are direct consequences of such an approach [1].

In this chapter a systems approach is adopted to examine methods and techniques that can improve five key performance indicators of the machining system, i.e. sustainability, accuracy, efficiency, precision and reliability. In particular, High Speed Machining, tool breakage prevention, thin wall deflection, tool geometry and chatter monitoring are studied in relation to the five performance indicators, respectively. Application of these techniques has produced good machining outcomes showing strategic development direction leading to better performance of the machining system.

The outline of the chapter is as follows. Section 11.2 defines the scope of machining system and the method of measuring its performance. Section 11.3 examines different dimensions of machining process performance. Section 11.4 elaborates strategies that can be used in an important machining process: high speed machining (HSM). Section 11.5 explores the methods to prevent a critical problem in machining: tool breakage. Section 11.6 analyses the issues in machining along thin walls of the workpiece where deflection of the material is excessive, causing uneven thickness when the cutting forces are removed. Having considered the mechanics around the machining process, Sect. 11.7 examines the geometrical shape of cutting tools. In Sect. 11.8, the phenomenon of chatter during machining is explained. The section outlines a method to monitor this detrimental phenomenon so as to prevent it before its starts so that the machining process can continue without disruption. Section 11.9 gives concluding remarks.

11.2 The Machining System

The machining system consists of the components schematically shown in Fig. 11.1: machine, holder, tool, fixture, part, control, metalworking fluid and regime. Component and their interdependences have impact on the performance of the machining system which must be optimized as a whole.

An example is the High Performance Virtual Machining System which consists of a machining process simulation kernel, machine tool controller kernel and graphical system to display the results and interact with the user. The scientific details of the simulation algorithms for each module have been published in the literature. The digital modeling of such an integrated manufacturing system allows simulation and optimization of machining operations in a virtual environment ahead of costly

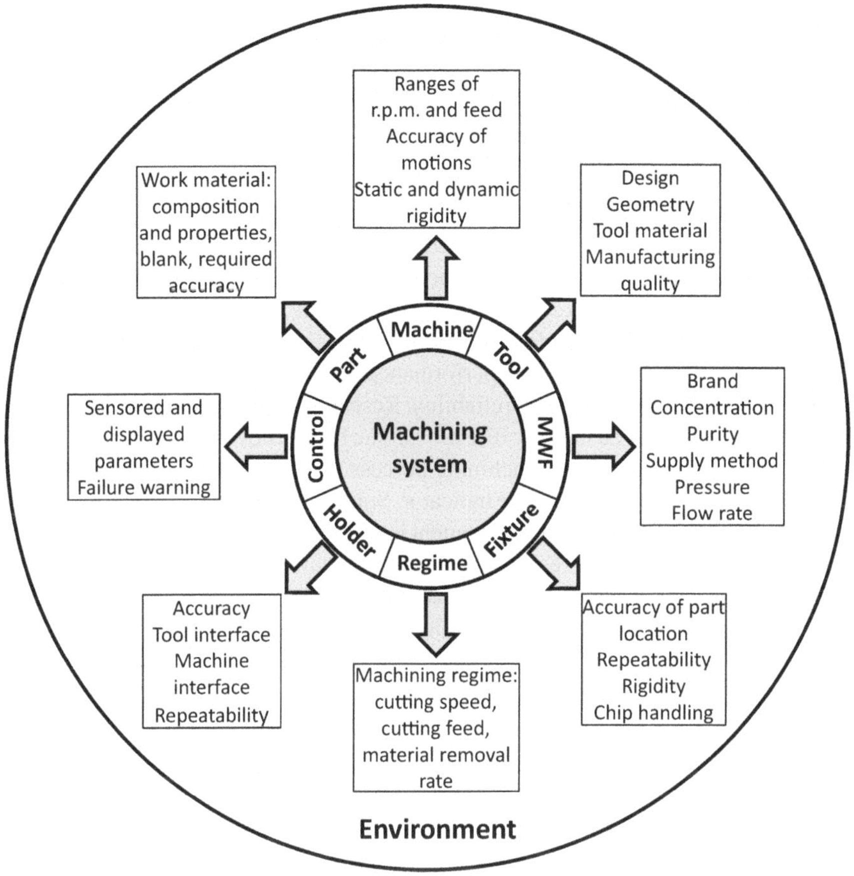

Fig. 11.1 Schematic representation of the machining system [1]

physical trials. The digital models are connected to real time machine tool monitoring and process control systems to further improve the productivity and accuracy of developed models [2].

The systems consideration is supported by the implementation of AR in a CNC machining environment to achieve an in situ CNC machining simulation system, namely, the ARCNC system. A virtual workpiece is rendered onto the worktable of a real CNC machine, and a virtual cutter is registered with the real cutter moving according to given NC codes [3]. During experiments, the ARCNC system renders a virtual workpiece being machined according to the movement of the real cutter, as well as the machining conditions estimated by the physical simulation methods proposed. The user is allowed to alter the parameters in the given NC codes during the experiments and to observe the new machining process using the updated codes. Simple collision detection has also been integrated in the system so that a warning message is rendered to the user when a collision is detected between the cutter and a virtual fixture.

11.2.1 Key Performance Indicators

Most past researches on machining focused on the techniques of the process. To make a good product, the "system" in which the process occurs should be managed well. This chapter examines five key performance indicators of the machining "system" that are essential to be managed so that desirable outcomes from the machining process can be achieved. These five key performance indicators (KPIs) are sustainability, accuracy, efficiency, precision and reliability. Research on each of these performance indicators can be explored in many directions. The researches outlined in this chapter focus on certain aspects of the machining process that can lead to improvements of the performance as measured by the indicator. Some good machining outcomes have been achieved but the extent of improvement is still insufficient to cover the complete system of machining. More researches along these investigation paths are required to identify processes that can push the quality of the system to better performance in the five key performance. In the next section some challenges of KPIs are indicated.

11.3 The Challenges of Key Performances

Key performances are measured from the outcomes of the machining system. To reach the target set by the manufacturer, the machining system needs to be planned, monitored and controlled in certain ways. More specifically, engineering analysis aiming at improving any of the key performance indicators should be designed to influence the machining system towards meeting the target. The KPIs mentioned in Sect. 11.2.1 are discussed below.

11.3.1 Efficiency

In any manufacturing environment, efficiency drives the cost of production and hence the profitability of the manufacturing organization. Efficiency in machining operation can be improved by changing system parameters. One of the common methods is to increase the speed of machining.

The concept of High speed machining (HSM) was first suggested by Salomon in 1924 [4, 5]. Owing to the high feed rate and high rotating speed, the cutting principle of HSM is different from that of conventional machining. For example, usually as the cutting speed increases, the amount of cutting heat increases quickly. However, when the cutting speed exceeds a certain threshold value, less cutting heat would be generated. This new discovery contradicts traditional knowledge in metal cutting and lays the foundation for HSM.

With the advance of material, control and manufacturing technologies over the past century, the environment for the application of HSM has changed significant compared to that of ninety years ago. The spindle speed could reach 80,000 rpm and the cutting speed was up to 8000 m/min [6]. The technology is not only used to achieve high material removal rate in roughing, but is also utilized to obtain high surface quality in finishing so as to reduce the following up bench work or Electrical Discharge Machining (EDM).

Owing to the ultra-high feed rate and spindle speed, the tool path strategies applied in conventional CNC machining cannot be effectively used in HSM. Different from conventional CNC machining in which tool path is not a critical factor, tool path patterns play a key role in the successful application of a HSM system. Tool paths, which are the trajectories the cutter, have to be smooth and the cutting load should be constant. Without an appropriate tool path strategy HSM cannot be realized even though the CNC machine tool employed has advanced HSM potentials. In practice, industry pays more attention to the hardware, like advanced servo systems and balanced tool holders, instead.

Research to improve the efficiency of a machining system is aimed at developing a tool path generation method or a CAM system to serve HSM more effectively. It is one of the important factors in realizing the successful application of an advanced HSM system and will be explored in the HSM section of this chapter.

11.3.2 Sustainability

A sustainable machining system should complete cutting tasks with specified quality consistently. One of the major problems in machining is tool breakage during machining. When the tool breaks, it will ruin the workpiece (causing bad marks) and delays the operation (change of tool as well disrupting the cutting program). Higher costs and lower quality are consequences. This problem is more serious when machining titanium parts from blocks of titanium material. Titanium and its alloys are non-ferrous metals with excellent corrosion resistance, fatigue properties, and high strength-to-weight ratios, as well as good ductility and have good ability in harsh environments [7]. The specific weight of titanium is approximately two thirds of steel and higher than aluminum. This property reduces the weight of products without loss of strength and achieves lower energy consumption [8].

However, Titanium has strong chemical reactivity with almost all tool materials available at the cutting temperature (>500 °C) [9]. High affinity of titanium alloys contributes partially to the hardening of titanium and its alloys in addition to the strain hardening [10]. Titanium tends to ignite during machining and sparks have been observed during cutting in some experiments [11].

The metallurgical characteristics of titanium such as low thermal conductivity, low young modulus, as well as high chemical dependency to cutting tool material

Fig. 11.2 Low quality surface of titanium part after occurrence of chatter

make it more difficult to cut than other metals due to interaction between the cutting tool and the titanium materials [12]. As a result, the high cutting force between work piece and cutting tool causes the cutting tool to wear out quickly. More serious problems such as low quality surface such as Fig. 11.2 and tool failure can in during machining resulting expensive rejects.

Research in tool life is aimed at finding methods of controlling the system so that breakage becomes predictable and can be prevented. By predicting the life of cutting tool, the worn tool can be replaced before the next machining cycle (in which tool breakage is expected) starts. Sustainability of the machining system can then be maintained and productivity can be optimised.

11.3.3 Accuracy

Advancement of CNC machining technology is enabling more complex shaped components to be made in one piece to replace inefficient assembly of parts into structures [13]. These components have the characteristics of a thin-wall monolithic part.

Thin-wall machining of monolithic parts allows for higher quality and precise parts in less time, thus impacting business including inventory and Just-In-Time (JIT) manufacturing.

Due to poor stiffness of thin-wall parts, deformation is more likely to occur during machining, which results in dimensional errors. In current industry practice, the resulting errors are usually compensated through one or more of the following techniques: (i) using a repetitive feeding and final 'float' cut to bring the machined surface within tolerance; (ii) manual calibration to determine 'tolerable' machining conditions; and (iii) a lengthy and expensive trial and error numerical control validation process [14].

Accuracy of machined components is one of the most critical considerations for many manufacturers, especially in the aerospace industry where most of the parts use a thin-walled structure [15]. Research of machining accuracy is aimed not only at setting on-machine system parameters, but also at predicting deformation due to forces generated in the machining system.

11.3.4 Precision

The ability to cut exact dimensions requires more than just compensating forces in the machining system. It also depends on the ability of the cutting tool to cut at precisely the location that is planned by the tool path. However, precision of cutting is determined not only by the hardness of cutting tool. The shape of the tool at the point of cutting dictates how the material breaks away from the metal block. To a large extent, the material deformation and cracking process in machining depends on the "sharpness" of the tool [16]. Research has shown that some regions of the cutter suffered from higher than detected temperature due to material diffusion and chemical reaction on the cutting point of the tool. Results from chip morphology also illustrated that serration frequency changed on each single chip. The shape of cutting tool plays a critical role in determining the precision of the machining system.

In order to improve precision of machining, some companies use polycrystalline diamond (PCD) tools to machine titanium blocks. The main advantage is that PCD tools are much harder than normal tungsten carbide tools and hence theoretically it will last a lot longer without deterioration of sharpness.

Polycrystalline diamond (PCD) is a composite of diamond particles sintered together with a metallic binder. Since its development by General Electric (GE) in the 1950s, PCD has found widespread application in numerous machining and wear part applications owing to its outstanding wear resistance and unique ability to machine difficult-to-cut materials.

Research to improve precision performance in the machining system is aimed at investigating the effect of cutting tool shaping features at the cutting point on material. This investigation involves analysis of the tool geometry and its effect on the cutting process.

11.3.5 Reliability

Machining is the outcome of dynamic interactions among all components of the machining system, including the machine structure, the spindle, cutting tool, workpiece, clamping system and the system's foundation. One of the detrimental interactions that often ends with bad product quality is vibrations in the machining system. In practice, as long as the effect of vibration on the machining outcome is within a tolerable limit (as indicated by the quality of the machined surface), the machining system is regarded as reliable. Therefore, to counteract the effect of vibration, machine tools are designed with high rigidity, e.g., heavy base, strong axis structure, tight pre-tension of mechanisms. However, a phenomenon known as chatter exists irrespective of how heavy, strong, or tight the system has been designed [17].

Chatter is a complex phenomenon characterized by unstable, chaotic motions of the tool and by strong anomalous fluctuations of cutting forces which cause uncontrollable vibration in the machining system. Due to its nature, no machine tool design can eliminate the possibility of its occurrence. The situation becomes more serious in the milling of titanium alloys because of their low Young modulus and extended elastic behaviour.

Research on machining system reliability focuses on various chatter prevention systems and presents an online chatter detection system based on the analysis of cutting forces, which is one of the integrated modules of a multi-sensor chatter detection system consisting acoustic and acceleration sensors. The signals are transformed to another mathematical domain. By computing a key parameter as an indicator of tendency to chatter, the online detection system will be able to predict the onset of chatter and informs the control system to adjust machining parameters accordingly to prevent chattering.

11.4 High Speed Machining for Machining Efficiency

High Speed Machining (HSM) puts stringent demand on the machining system because it requires high moving speeds of all movable parts in a machine tool. It embraces high feed rate, high spindle speed as well as high acceleration and deceleration rates which are achieved through the application of advanced control and servo systems [18, 19].

In high speed finish machining which is used to achieve high quality surfaces, the strategy is normally to remove materials with very shallow cuts and small side step distances between adjacent tool paths so to reduce cutting force and avoid excessive deflection of the cutting tool. The cutting depths usually do not exceed 0.2/0.2 mm (ae/ap). Since the theoretical roughness of the machined surface is determined by the step-over distance in combination with tool radius, the surface finish can be improved significantly by taking smaller step-overs. This, in turn, will reduce the necessity of conducting subsequent polishing, grinding, bench work, and EDM machining, which are labor intensive and time consuming. Below, some challenges of HSM are addressed.

11.4.1 Challenges of HSM

High speed in terms of high feed rate and spindle speed is not the only requirement of HSM. An important factor that may affect the machining efficiency and the final quality of workpieces in NC programming is the pattern of the tool path in addition to the selection of a suitable tool and the setting up of proper cutting parameters.

To cater for different shapes of workpieces, the cutting tool may be expected to move in any direction. If there are sharp changes in the moving direction, in order to maintain machining accuracy, the machine has to decelerate before the change and accelerate again after it passes it. However, frequent deceleration and acceleration in NC machining reduces the average machining speed, and the forces required to quickly decelerate or accelerate moving components may increase tool position errors, increase wear of moving parts of machine tools and decrease motion repeatability. Depending on the type of controllers and machine tools employed, errors of different natures may emerge.

To control the feed rate more accurately and reduce possible machining errors, modern CNC controllers are normally equipped with a "look ahead" function, with which the controller is able to pre-calculate the overall feeding speed hundreds of program blocks before the actual cutting position is reached so that the acceleration and deceleration could be arranged ahead of time to reduce the acceleration force and the possible overshot and undercut. The average feed rate is thus always lower than the programmed one and the machine can never achieve the programmed feed rate as the adaptive control system can decrease feed rate automatically before the tool reaches the cutting position and resume the previous one after the moving direction has been changed over. For systems without "look ahead" function, the force caused by the sharp change in moving directions may cause errors in the machining process.

To increase machining accuracy and reduce cutting force, the tool used in HSM is usually small. The large force caused by the abrupt change in material removal rate or sudden local overload may break or shorten the life of the small tool. One example is when cutting walls with solid carbide tools where the cutting edge is the side rather than the tip of the tools. These tools run at high speed and work best when taking deep cuts with a small stepover. This means that the tool only has a small angle of engagement. The risk of tool damage increases as soon as the angle of engagement increases. To minimize tool damage the angle of engagement should be minimized and the cutting load should be maintained as constant as possible.

The sharp change in moving directions is one of the main causes of acceleration and uneven cutting force. To reduce errors and minimize feed rate losses, conventional tool paths or paths that have sharp changes of feeding directions can no longer be used for HSM. Smoother tool path patterns with new requirements are required. Cutting strategies must be changed for the successful application of HSM.

11.4.2 Smooth Engagement and Retraction Tool Path Strategy

Feeding into material at full depth with fast feed rate incurs large cutting forces to the cutting tool. This force may mar the surface and shorten tool life. In HSM it is important to have control over the way the tool enters and leaves the workpiece to minimize the cutting force. At the starting stage of cutting an effective way of minimizing cutting force is to make the tool cuts into the workpiece smoothly. Conventional methods of plunging straight down into a part could cause the tool to snap. Proper engagement and retraction (lead in and lead out or entry/exit) is in the tangent direction of the profile or the surface to be machined.

The tangent engagement and retraction may occur on the horizontal plane, the vertical plane as well as the surface normal plane. For HSM of closed regions the tool has to cut into the workpiece from the top of the part. To achieve smooth tool engagement and ease the load up, the tool may contact the part in single ramping, zigzag ramping or a helical motion, as illustrated in Fig. 11.3a–c. These continuous motions reduce the possible scratches on steep surfaces of the closed regions and avoid sudden load being applied on the tool. They are especially suitable for high speed machining of cavity, pocket or slot when entry must be accomplished from the top of the closed area.

The parameters used to define a helical tool path include the radius of the helix, lead in angle and height of the helix. As for the ramp, the defining parameter is the tilt angle which determines the cutting direction (Fig. 11.3d).

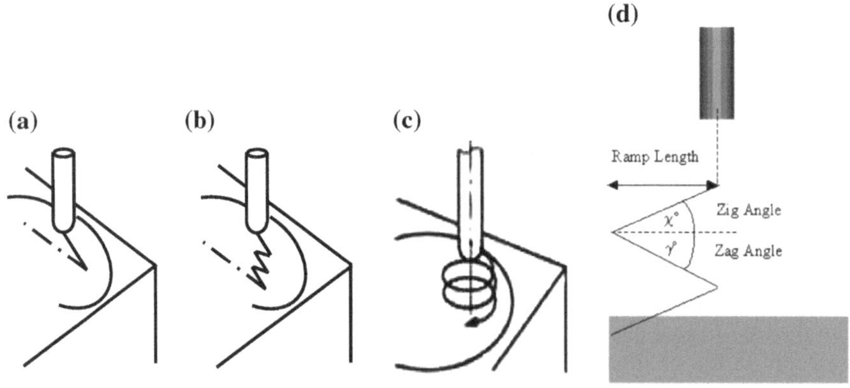

Fig. 11.3 **a** Single ramp entry, **b** zigzag ramp entry, **c** helical entry, **d** parameters to define a ramp entry

11.4.3 Smooth Connections Between Adjacent Tool Paths Strategy

Iso-planar machining is predominantly used in conventional metal cutting [20]. It provides the ability of continuously machining multi-surface (compound or composite) models. However, as the connection between consecutive passes is usually short, it tends to produce sharp step-over movement, which makes it not suitable for HSM. To take advantage of this machining strategy, improvements can be made by connecting two adjacent tool paths with an arc or a segment of curve that is tangent to the two consecutive passes so that smooth transition could be achieved. As shown in Fig. 11.4a, the radius of the transitional arc is determined by the distance between two neighboring passes. The transition of cutting direction is tangent to both passes, so smooth motion is ensured.

This strategy is proved effective when the step-over is large enough and the cutting is conducted at a moderate feed rate. However, when the side step is small or the feeding speed is particularly high, which is quite common in high speed finish machining, transiting arcs with larger radii can be used to generate smooth transitions. The simple rounded move is too sharp and the original fluid transition becomes non-smooth. To deal with this type of cases, more advanced connection strategies of using 3D arcs other than 2D curves are developed recently. As shown in Fig. 11.4b, the transit bridge is a 3D arc. In this way, the radius of the transitional arc is no longer restricted by the stopovers, and a reasonable value can be assigned to it. For example, when the step-over is 0.5 mm, the radius of the 2D arc could be only 0.25, but that of a 3D arc could be of any value, for example 3 mm. With this large radius, smooth transition motion is no longer a problem.

"Golf club" is a special 3D connection method. As shown in Fig. 11.4c, the "golf club" is a simple 3D loop that has the geometric shape of a golf club. It can be seen clearly from the figure that the momentum of the high speed tool can be maintained with this shape.

In addition to the smooth connection between adjacent paths in the formation of zigzag tool paths, smooth transition between layers or loops of tool paths is necessary as well to ensure smooth tool motions in "layer machining" or "contour

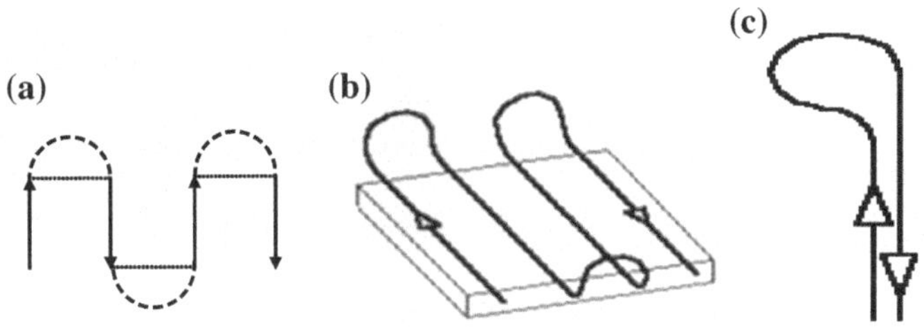

Fig. 11.4 **a** 2D Arc connections in zigzag tool paths, **b** transition with 3D arcs, **c** golf-club connections [2]

offset machining". The smooth transition could be a single arc, spiral curve, "S" shape, helical and looped curves.

11.4.4 Rounded Corners in Tool Paths Strategy

Abrupt change in moving directions may result in the controller to reduce feed rate in order to avoid possible overshot and undercut, or may cause large acceleration force at the changing points which is detrimental to the cutter and machine tool if the controller has not the "look ahead" function. To maintain consistent feed rate and constant cutting load, sharp corners in tool path should be smoothed. When a tool cuts along a straight line with a constant radial depth of cut, the load on the tool is constant. However, when the cutting tool enters a concave corner or curve, it engages more material and the load on the tool may be increased. Therefore, from the perspective of constant cutting load, sharp corners in tool paths should be eliminated as well [21].

Figures 11.5a shows the corners after being rounded, not limited to the corners exist in the tool path for material removing. To obtain smooth tool motions in the overall machining process, sharp corners in rapid feeding (or air cutting) should be rounded as well. In addition to these directly smoothing methods, there are some other types of advanced rounding algorithms in practice. In these algorithms, rather than adding the user defined rounding angles at sharp corners, overshot loops are utilized to absorb the momentums of the fast moving tool. Figure 11.5b illustrates this type of rounding method applied in CAM systems like Cimatron and Mastercam.

Spiral tool path, race line tool path, trochoidal tool path, curvilinear tool path, constant load tool path and NURBS (Non-Uniform Rational B-Splines) tool path, etc. are examples of these new developments.

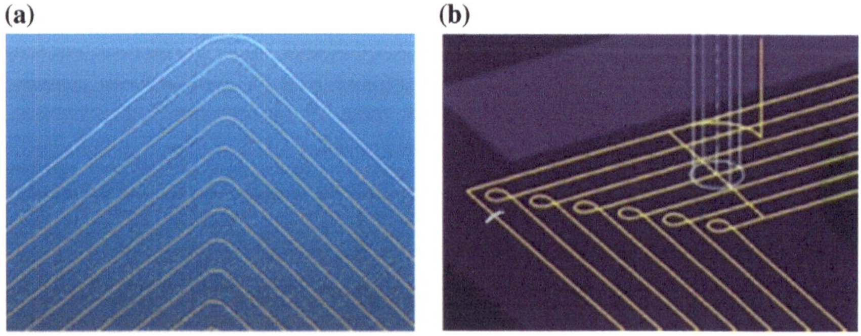

Fig. 11.5 **a** Rounded tool paths, **b** looped tool path at corners [22]

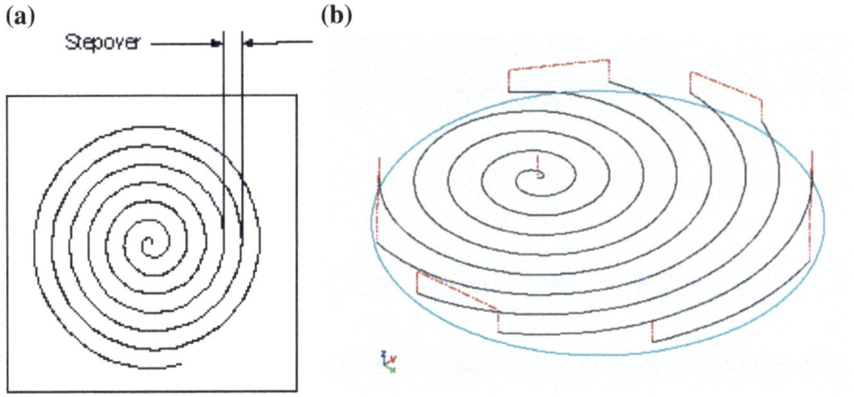

Fig. 11.6 **a** Step over in spiral tool path, **b** mixed mode [23]

11.4.5 Spiral Tool Paths

Applying arcs and loops to corners is only an improvement on conventional tool path patterns so to make them applicable in HSM. It works very well when the tool load variation is less severe. However, HSM is much more than just taking out sharp corners and rounding them off. When parts become more complicated and machining requirements become more aggressive, this approach may not be effective and more advanced tool path patterns catering to HSM are needed. This stimulates the development of completely new tool paths from first principles.

The characteristic of spiral tool path uses a spiral curve as tool path to ensure smooth tool motions. With this type of tool path, there is no lifting or downing motion on the milling path that may affect the surface quality. It is especially applicable for cutting workpiece with features of circles or round shapes.

Spiral tool path can be generated by projecting an initial spiral curve defined on a plane (Fig. 11.6a) to the workpiece surface. Geometrically the initial spiral is defined by the center point, maximum radius and stopovers on the plane. Figure 11.6b illustrates the tool paths with user-defined off-centered entry points. Considering the spiral direction, the cutting may be performed clockwise or counter-clockwise. Likewise, considering the cutting direction, operations may start on the inside boundary and cut outward, or start on the outside boundary and cut inward. The difference in machining directions results different tool paths.

11.4.6 Curvilinear Tool Paths Strategy

Curvilinear tool path follows a gradually evolving spiral curve. As shown in Fig. 11.7, this type of tool path shares some similarities with above spiral tool path. By using

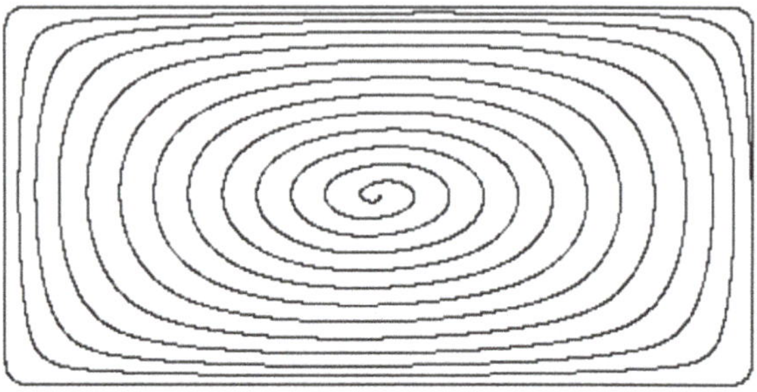

Fig. 11.7 Curvilinear tool paths [24]

a user-specified maximum width of cut, the curve morphs from having low, nearly-constant curvature at the pocket center to one that is geometrically correct at the pocket boundary. However, unlike spiral tool path, the generation of such a curve was achieved mathematically by solving a scalar elliptic partial differential equation (PDE) boundary value problem on an entire pocket.

11.4.7 Trochoidal Tool Paths Strategy

Trochoid is the trajectory of a point at a distance l from the center of a circle of radius r rolling on a fixed line. Geometrically, it is the locus of a point fixed to a curve which rolls on another curve without slipping. As shown in Fig. 11.8a, the curve becomes smoother with the increase of l/r.

Trochoidal cutting is a milling method that uses trochoidal curves as tool paths. There are two obvious advantages with this strategy:

1. The cutting tool moves in a cycled circular looping pattern while the center of the circle moves along a path. This produces smooth tool motion even at corners and allows a consistent feed rate to be maintained throughout the machining process.
2. Since the machining is carried out in a cycled motion, the tool cut into workpiece when it moves forward in each cycle and cool down when it moves backward,

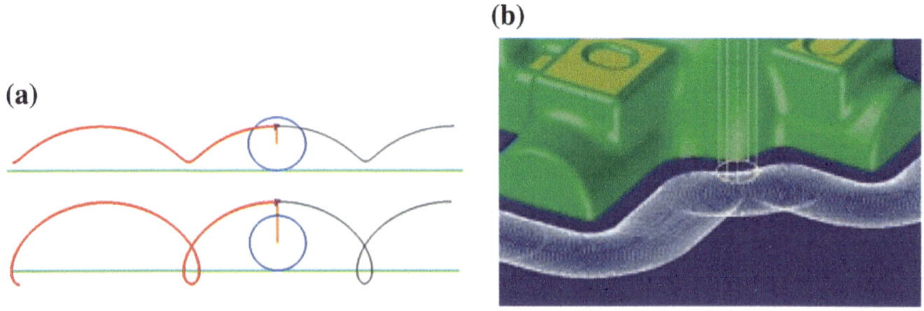

Fig. 11.8 **a** Trochoidal curve, **b** trochoidal milling in Cimatron [22]

this enables chips to be expelled in an easy manner. The cutting edge could be in contact with the material through only about 5% of the cutter's revolution. Therefore, the tool is never buried in materials, full-width cut is avoided and longer tool life can be achieved.

Trochoidal machining is well suited for high speed machining of open pockets, slots and grooves. It is very effective in controlling excessive stepover to prevent the tool from breaking when it is fully embedded in the cut [25]. Trochoidal cut pattern may have a larger width and a smaller path width (Fig. 11.8b).

Trochoidal strategy minimizes sharp changes in tool moving directions, and the small-radius, overlapping circles relieve the tool from burying in materials. However, there exist some other disadvantages:

1. The extra circular motions offset some benefits of fast feed by extending the total length of tool path.
2. Because there are no cutting actions when it moves backward in the cycle, based on a stricter criterion, Trochoidal is not a method of constant cutting load.

11.4.8 Constant Cutting Load Tool Paths Strategy

In the machining of free-form surfaces geometries of workpieces may vary widely. If feed rate and parameters such as cutting depth, stepover and spindle speed are all constant, varying chip loads may be formed due to the unevenly distributed stock materials. This may result in large variations of cutting forces that are detrimental to the cutting process. As shown in Fig. 11.9a, when the tool approaches corners with a constant speed, the chip thickness as well as the tool engagement angle (TEA), which is the central angle describes the portion of the tool buried in materials, increases dramatically. From Fig. 11.9b, it can be seen that half-width cut becomes full-width

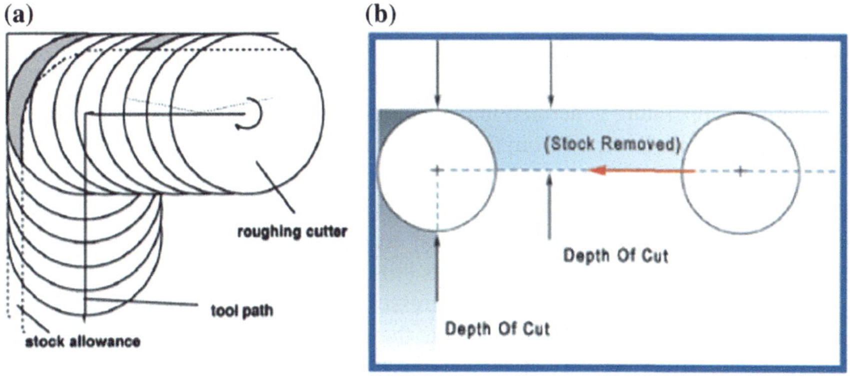

Fig. 11.9 **a** Change in chip thickness when cutting corners, **b** full-width cut in the corner [26]

as the tool enters the corner. These excess materials left from previous machining process cause large cutting force which increases the risk of the breakage of a fast moving cutting tool.

In order to obtain constant cutting force, the material removal rate should be maintained constant as much as possible. “Constant cutting load” tool path generation is such a technology that optimizes the feed rate and spindle speed based on volume of materials removed so to keep the chip load and cutting force constant. The CAD/CAM system analyzes and detects machining conditions automatically. When more materials are removed and the chip load comes greater, the feed rate is adjusted slower; when less material removed, it will return to higher feed rate. To accomplish this task, it is important for the CAM system to understand what changes in surface geometry have taken place in current and previous machining processes. This is often referred as the “knowledge of machining process”, “knowledge of stock remaining (KSR)” or “In Process Workpiece (IPW). With this knowledge, after analyzing the in-process state of the workpiece, the system can determine where the machine should make its next cut and what the appropriate feed rate is to keep the cutting load constant.

11.5 Tool Life Prediction for Machining Sustainability

Although there have been great advances in the development of cutting tool materials which have significantly improved the machinability of a large number of metallic materials, including cast irons, steels and some high temperature alloys such as nickel-based alloys, no equivalent development has been made for cutting titanium alloys due primarily to their peculiar characteristics. Ezugwu and Wang [27] found that the straight tungsten carbide (WC/Co) cutting tools continue to maintain their superiority in almost all machining processes of titanium alloys, whilst chemical vapour deposition coated carbides and ceramics have not replaced cemented carbides due to their reactivity with titanium and their relatively low fracture toughness as well as the poor thermal conductivity of most ceramics. There are researches in special machining methods, such as rotary cutting and the use of ledge tools, which showed some success in the machining of titanium alloys [28].

The high temperature generated in machining titanium is the principal reason for the rapid wear of tools. Apart from the poor thermal conductivity, the contact length between the chip and the tool is less than one-third the contact length of steel with the same feed rate and depth of cut [29]. This implies that high cutting temperature and high stress are simultaneously concentrated near the cutting edge (within 0.5 mm). The temperature zone of 700 °C comes as close as 0.1 mm from the cutting edge. Both the high temperature and the high stresses developed at the cutting edge may cause plastic deformation and/or accelerate the wear of the tools.

11.5.1 Finding Parameters for Machining Titanium Alloys

To find the optimum setting for Titanium machining, a series of experiments are designed to provide the data for analysis. These experiments are designed to support practical industrial applications by using (1) commercial cutting tools purchased from the market; (2) normal cooling media—coolant; (3) existing machining equipment. The aim is to maximise utilization of the cutting tool in its tool life. Four cutting parameters can be controlled without increasing investment of the system:

1. Feed rate F (mm/min)
2. Spindle speed S (RPM)
3. Cutting depth in axis direction Ap (mm)
4. Cutting depth in radial direction Ae (mm)

Using Taguchi approach, 27 runs of experiments are set to determine the best combination of machining parameters as shown in Table 11.1. According to the Taguchi-based experiment design, a L9 (34) orthogonal arrays were determined.

To create a statistical significance without the need for substantial number of experiments, the runs are randomised as shown in Table 11.2.

The cutting tests were done on a mid-size 4-axis CNC machining centre. The material was a Titanium block with dimension 150 × 100 × 20 mm. A total of 16 work pieces were used in the tests. Both sides (top and bottom) of the workpiece were used to save material giving 32 cutting experiments. In order to standardize the

Table 11.1 Factors investigated in titanium machining and their levels

Factors	Level 1	Level 2	Level 3
V_c—speed (m/min)	40	42.5	45
F_z—chip load/tooth (mm/tooth)	0.03	0.05	0.07
A_p—depth of cut (mm)	2	3	4
A_e—side cut (mm)	0.25 D	0.5 D	0.75 D

Table 11.2 Factors investigated in titanium machining and their levels

Run	V_c	F_z	A_p	A_e
1	1	1	1	1
2	1	2	2	2
3	1	3	3	3
4	2	1	2	3
5	2	2	3	1
6	2	3	1	2
7	3	1	3	2
8	3	2	1	3
9	3	3	2	1

outcome, a flat end mill of diameter 6 mm was used. The normal coolant for steel machining was used due to its availability. All cuttings were cut on climb milling and linear motions.

11.5.2 Tool Life Predication

Having determined the machining parameters for high speed Titanium machining, a tool failure model for the predication of tool life can be developed. We assume:

1. Tool wear and damage is directly caused by cutting forces occurring at interactions between tool and workpiece, and tool and chips;
2. Force variation in air cutting is not considered for the calculation of Root Mean Square (RMS) in milling process;
3. The machining environment is constant.

Although the theories of tool failure are very complicated and many factors contribute to the tool break, the root cause leads to tool break is the cutting force added on the cutter. Therefore, it is the most straightforward and effective method to analysis the cutting force on the cutter and its evolution trend so make judgement when the tool will break.

By monitoring the variation of cutting forces (RMS values) constantly in nine runs of the designed experiments, a set of reference values were obtained. The first reference value is Mean Change Force (MCF). MCF is defined as the mean value of the force variation in the tool lifespan. The second reference value is mean gradient of cutting forces (RMS values) over time. The nine groups of cutting tests total 27.8 h of machining were conducted to analyse the cutting force for the generation of a predication model. The general descriptive statistical characteristics of *X* forces and gradients are listed in Table 11.3.

The first row (Max) in Table 11.3 indicates the maximum forces measured before each tool broke. The mean variation is measured as Root Mean Square (RMS) of the cutting forces. The RMS for all tools is 277.76 N. This provides an indicator of the magnitude of force that the tool will be subjected to over its life. The other important factor resulting in tool breakage is the rate of change of force. This can be measured by average force gradient in row 5 of Table 11.3. Since the actual force gradient changes during the cutting process, detection of breakage is detected from setting a warning level at the upper quartile of the force gradient. This tool life prediction model is illustrated using a test on tool no. 6 (Fig. 11.10).

It can be seen in Fig. 11.10 that there are no extreme high forces in the lifespan of tool 6. When the force change reaches 260 N, which is bigger than 75% of the mean change force (i.e. 277 N) and at the same time, the gradient reaches about 6 N per minute, which is considered as more than third quartile of the total sampling gradients, and much higher than average gradient (0.7 N/min), then, it can be predicted that the tool is about to break and tool replacement becomes necessary.

Table 11.3 Descriptive characteristics of X forces and gradient

Tool number	1	2	3	4	5	6	7	8	9
Max X force (N)	471.75	402.88	433.73	482.21	701.48	588.06	548.13	468.79	569.12
Min X force (N)	138.86	33.11	177.71	101.05	552.52	251.13	277.71	291.89	342.36
Range of X force (N)	332.89	369.78	256.02	381.17	148.95	336.93	270.42	176.91	226.76
Av. X force (N)	241.39	250.19	229.28	191.79	625.77	373.25	343.75	349.27	429.40
Av. Grad. (N/min)	1.10	1.43	10.97	1.69	2.91	0.73	1.81	6.50	0.32
Lower quartile	−3.99	−12.10	−37.81	−3.18	−105.41	−3.44	−5.40	−18.59	−4.79
Upper quartile	5.40	19.25	52.24	4.71	115.98	4.73	9.35	23.23	5.68

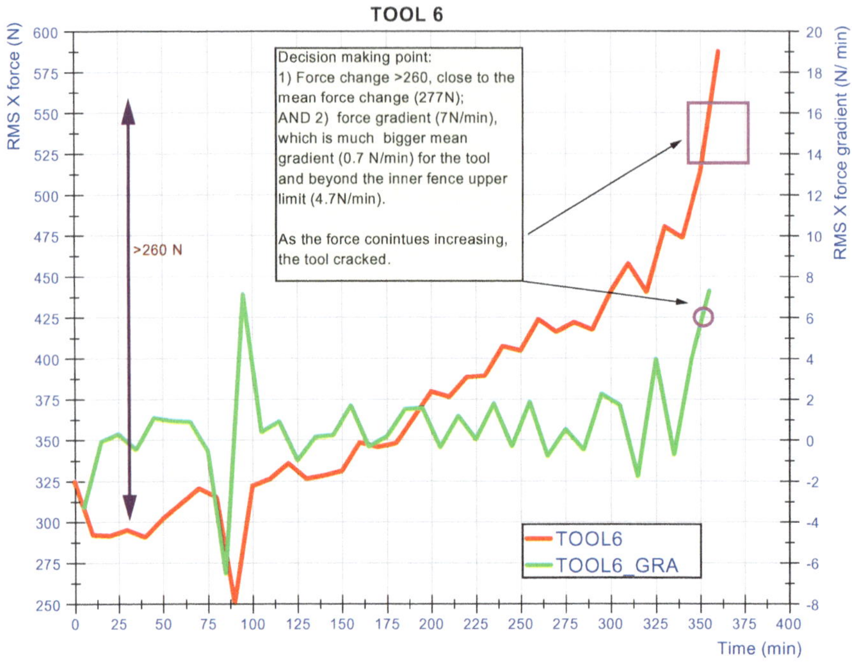

Fig. 11.10 *X* force, gradient and decision making point (tool 6)

Based on these experiments, the extreme cutting situation is identified by determining the extreme cutting force value (Fex). If the cutting force reaches Fex, the interruption control process will be activated. In this process, the time limit (Tex) is set to check whether the cutting force remains equal to or above Fex during the time of Tex. If it is so, it indicates that a tool replacement is necessary.

11.6 Deflection Compensation for Machining Accuracy

Different methods of estimating deflection of thin wall during machining have been reported [30–35]. These methods have specific application constraints which are difficult to adapt to other uses. An integrated methodology was developed by Izamshah et al. [36] consists of a machining load computational model from the machining parameters, a feature based geometry model, a deflection analysis model and an NC machining verification model. The advantages of the method are: fast design-analysis loop and the flexibility to create complex finite element models while maintaining association with the master design, thereby avoiding the time-consuming and error-prone transfer of geometry.

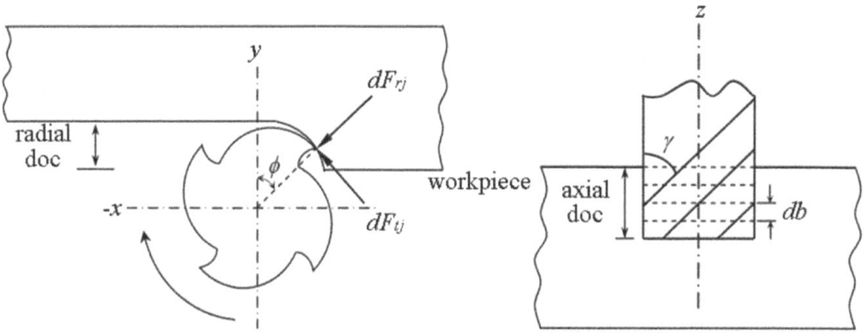

Fig. 11.11 Cutting force model for helical endmill

11.6.1 The Finite Element Machining Model

This model has incorporated the assumption that cutting is done with a helical endmill tool. As shown in Fig. 11.11, the machining loads acting on a helical flute endmill are equally discretized into a finite number of elements along the tool axis [37].

The tangential, radial and axial forces in x, y, z cartesian directions during cutting can be derived from Fig. 11.11 by considering interaction of the tool with the material. The instantaneous cutting forces acting on the whole endmill can then be obtained, which are used as the input for FEA to compute the deflection of the workpiece.

The structure of the thin-wall workpiece is modelled with the three-dimensional twenty-node parabolic hexahedron solid element. The parabolic hexahedron solid element is preferred since the wall is very thin and the change in structural properties of the wall due to material removed is very important for accurate prediction of the wall deflections [38]. Figure 11.12 shows the thin-wall component model for deflection calculations. The initial wall thickness t_{i} is reduced to t_{c} at the transient zone where the cutter flutes enter and exit the material in the milling process. The displacements of the whole structural component are obtained by normal three dimensional nodal displacement equations applied to each finite element.

11.6.2 Statistical Analysis

However, the time required to run the FEA is extremely long. Practically, the system needs to create a new finite element mesh in every simulation step. Experience shows that it takes a couple of weeks for a simulation run of a cut along a length of 200 mm to complete. To develop the deflection model, multiple regression technique is used to perform the statistical analysis to determine the correlation between a criterion variable, part deflection and a combination of predictor variables, namely speed, feed rate, radial depth of cut, wall thickness, wall height and wall length [39]. It can be used to analyse data from any of the major quantitative research designs such as

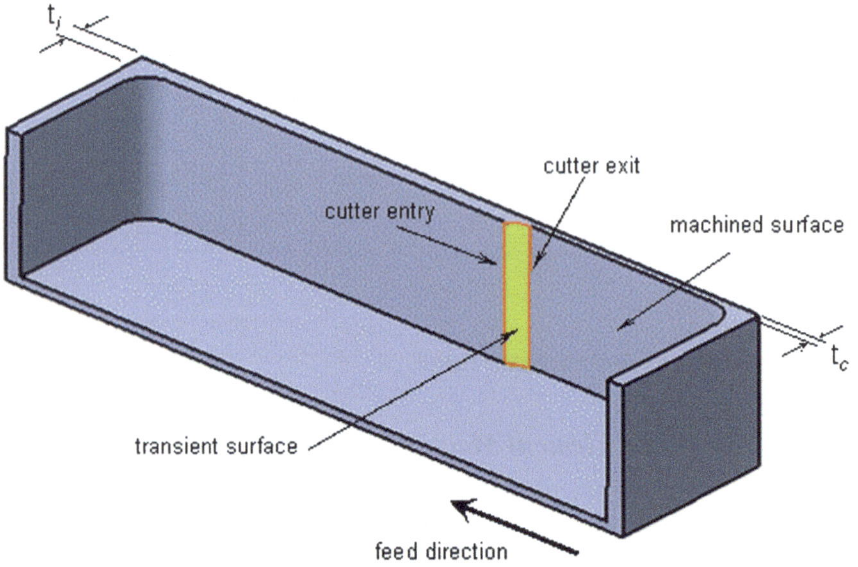

Fig. 11.12 Modelling the thin-wall component

causal-comparative, correctional and experimental. This method is also able to handle interval, ordinal, or categorical data and provide estimates both of the magnitude and statistical significance of the relationship between variables. The multiple regression model can be expressed as:

$$y = \beta_0 + \beta_S S + \beta_F F + \beta_C C + \beta_T T + \beta_H H + \beta_L L$$

where:

- y = displacement (μm)
- S = Speed (rpm)
- F = Feed rate (mmpt)
- C = Radial depth of cut (mm)
- T = Workpiece thickness (mm)
- H = Workpiece height (mm)
- L = Workpiece length (mm)

To validate the hypotheses, experiments were performed with a combination of six variables as shown in Table 11.4.

Figure 11.13 shows the displacement values for three sensors, comparing simulation and experiment. The cutter feed step is set at 30 equally spaced locations at one side of the wall along the feed direction. Other machining parameters are wall height 17 mm, axial depth of cut 15 mm, radial depth of cut 0.3 mm and the workpiece material is titanium alloy.

Figure 11.14 shows the instantaneous predicted and measured force F_y for one

Table 11.4 Design of experiment for the multiple regression analysis

	Level 1	Level 2	Level 3
Speed (rpm)	4244	4509	4774
Feed rate (mmpt)	0.02	0.05	0.08
Radial depth of cut (mm)	0.1	0.2	0.3
WP thickness (mm)	1.5	2	2.5
WP height (mm)	5	10	15
WP length (mm)	60	90	120
Upper quartile	5.40	19.25	52.24

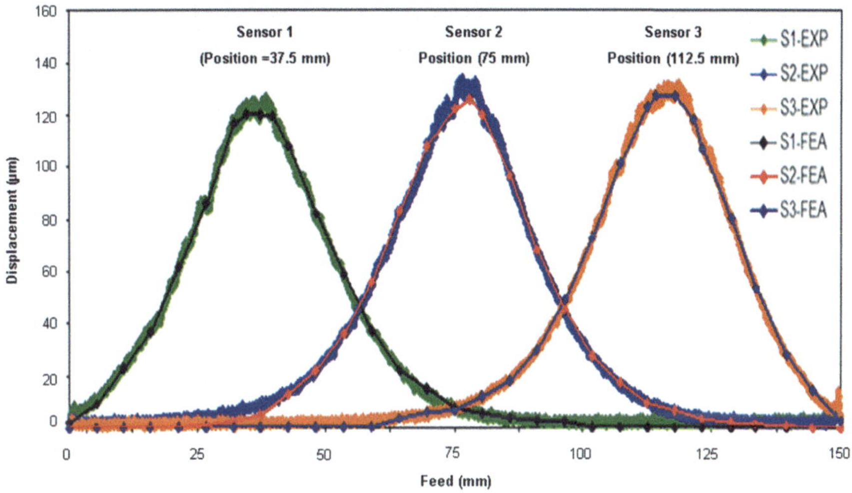

Fig. 11.13 Forces detected

cutter revolution. As it can be clearly observed, both the values of predicted and measured force are in a good agreement. The calculated machining loads are used as an input for the FEA to calculate the deflection of the workpiece during machining.

The experiments showed that the displacement obtained by simulation closely matches the displacement measured in the experiment. The agreement value between predicted and measured results is between 80.3 and 99.9%. The criterion variable is calculated at five different locations along the workpiece feed direction (i.e. D1 = 0, D2 = L/4, D3 = L/2, D4 = 3L/4 and D5 = L). Based on the validity of the assumptions, ANOVA was used for the regression analysis. From the ANOVA analysis, the R square values obtained for displacement at D1, D2, D3, D4 and D5 were 92.3%, 86.2%, 87.6%, 85.9% and 90.7% respectively, which indicated a high correlation coefficient between the dependent variable and the predicted value. All these evidences showed a strong linear relationship between the predictor variables (S, F, C, T, H and L) and the predicted variables (Table 11.5).

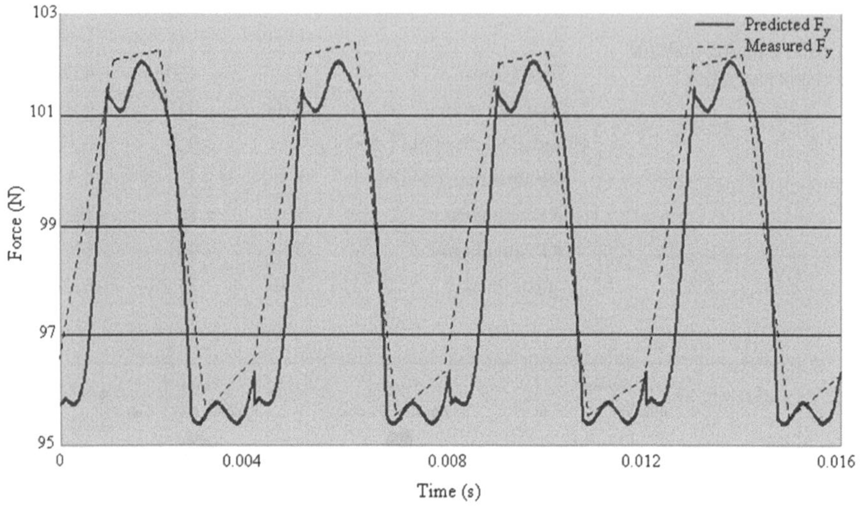

Fig. 11.14 Calculated cutting force for one cutter revolution

Table 11.5 The model coefficients

	D1	D2	D3	D4	D5
Constant	0.0005	2.052	−0.633	1.552	0.03621
S	0.00003505	−0.000102	0.000657	0.000049	0.00002688
F	1.7070	26.624	31.672	28.746	1.5672
C	0.39878	5.952	5.503	5.618	0.29444
T	−0.10834	−3.3362	−3.5850	−3.4204	−0.09039
H	0.004670	0.47683	0.54291	0.48117	0.002717
L	0.0000339	0.010356	0.001730	0.009648	−0.0000469

11.7 Tool Geometry for Machining Precision

The study of tool geometry involves understanding the mechanism of different features on the cutting tool. Flank wear reflects the friction between the workpiece surface and the flank face of the cutting tool. Figure 11.15 shows the worn areas on the flank face of a PCD tool after machining for 10 min. The images (50×, Alicona EdgeMaster) show that flank wear on the tool surface was caused by the abrasion between PCD flank face and the surface of the workpiece. It can be seen in Fig. 11.15a–d that tool wear developed from a narrow area around the cutting edge into a large triangular area.

SEM images of regions near the tool nose show more detailed information about the flank wear (Fig. 11.16). It can be seen that material adhered to the worn flank face, and some of them was removed by abrasion and attrition, fresh PCD surface was

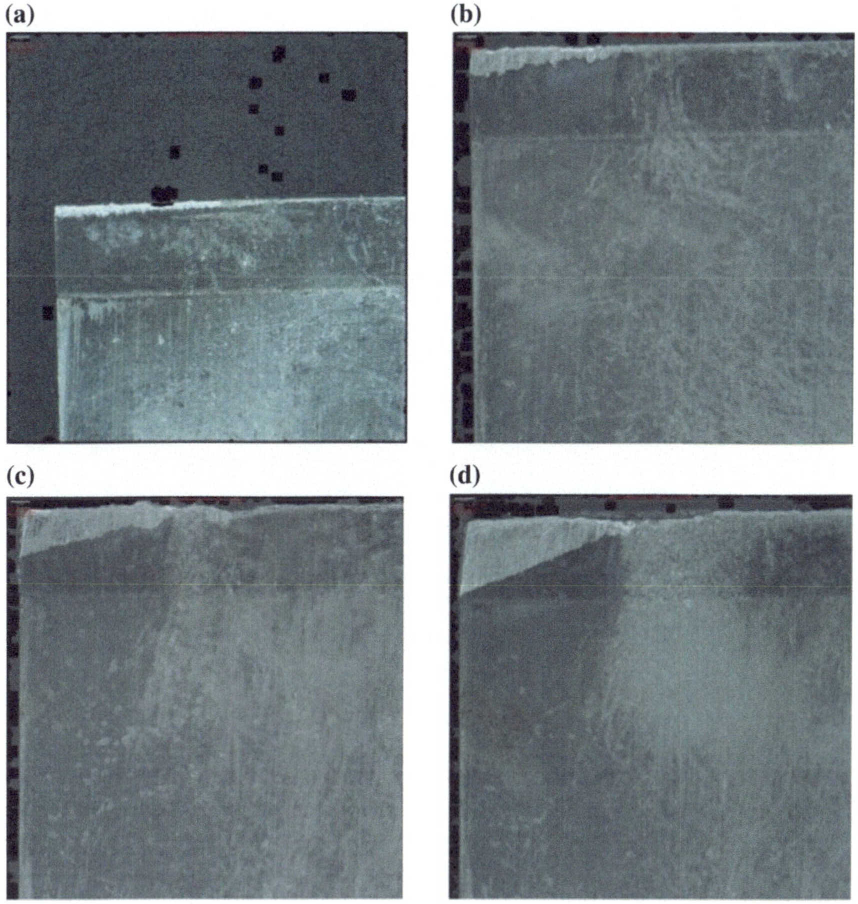

Fig. 11.15 Intermittent images of flank wear on a PCD tool after machining of **a** 1 min, **b** 3 min, **c** 6 min, **d** 10 min

exposed and led to further wear process. Results of energy dispersive X-ray spectrum (EDS) analysis show that the accumulated material is Titanium alloy. This indicates that serious diffusive-abrasive wear happened in this area. The adhesion of Titanium was the main factor that contributed to flank wear. According to the theory developed by Bhaumik et al. [40], adhesion happened frequently when there was chemical affinity between workpiece and tool surface. This caused further material diffusion between the work piece and cutting tools. High temperature and high compressive stress were generated between the surface of workpiece and flank surface, which accelerated the diffusion and attrition process. Although the temperature on flank face was lower than the temperature on clearance face, abrasion on flank face was constant and lasted longer, the constant and longtime contact provided sufficient time for diffusive-abrasive reaction. Furthermore, PCD is sensitive to high temperature

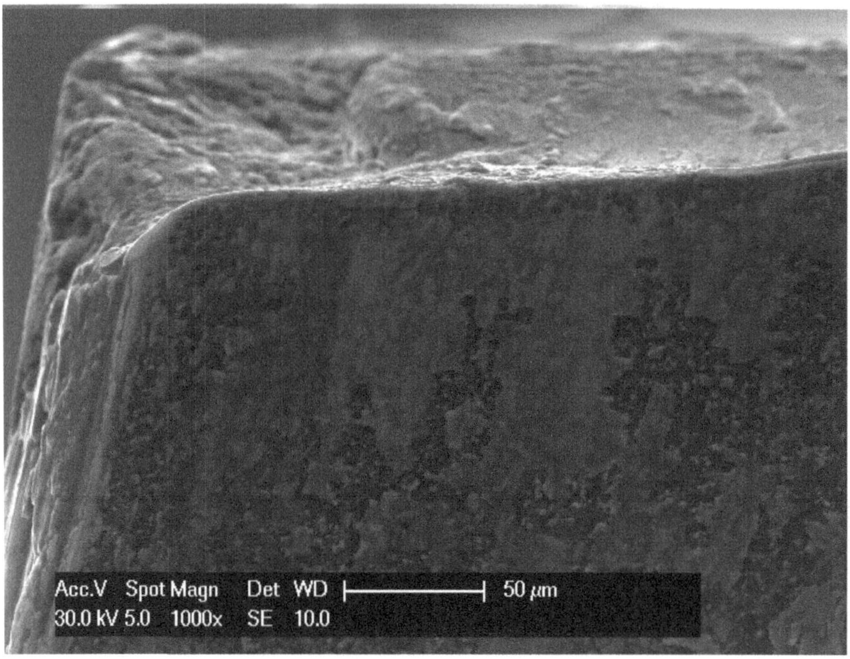

Fig. 11.16 SEM images (1000×) of rake flank face after 10-min turning

and around 36% of its hardness can be reduced when the temperature rises from 300 to 500 K. Even though coolant was applied in the machining, the instantaneous temperature near the cutting edge and tool nose was high enough to weaken the diamond structure on the flank face. As a result, flank wear was formed and developed by the continuous cycle of adhesion, diffusion and the removal of this layer.

11.7.1 Tool Nose Wear Analysis

As the landing area of flank wear, the rate of nose wear is generally higher than flank wear. To be specific, the worn area extended nearly to the boundary between PCD layer and the WC substrate (Fig. 11.17).

In Fig. 11.17, notching near tool nose is found along the secondary cutting edge. On a cutting tool, tool nose suffers the maximum cutting temperature which might cause thermal softening of material if it is high enough. Sreejith et al. [41] proved that notching near tool nose was produced by the oxidation wear combined with high temperature. According to the result of EDS analysis at the notching area (Fig. 11.18), titanium and oxygen present at worn area. This indicates that chemical reaction between titanium alloy, carbon and oxygen contributed to notching around the tool

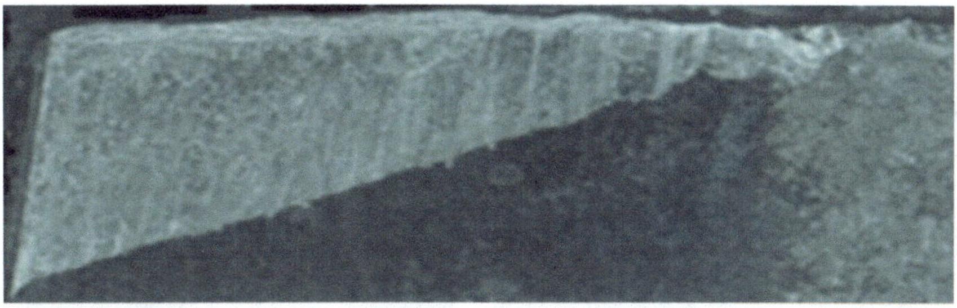

Fig. 11.17 Enlarged images of flank face after 10-min turning (50×, "Alicona", EdgeMaster)

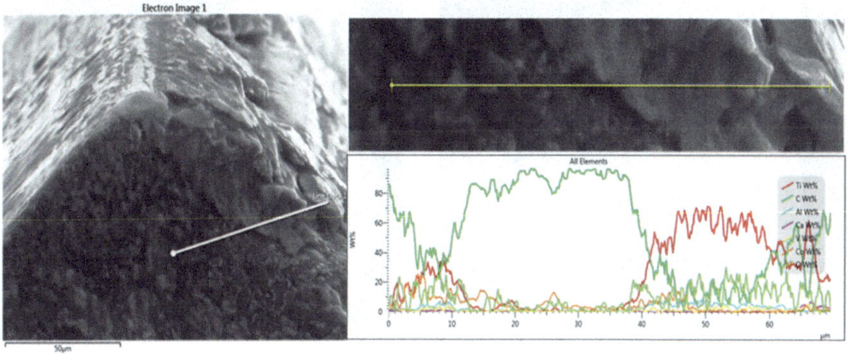

Fig. 11.18 EDS analysis of elements composited on worn clearance surface of conventionally ground insert after 10-min turning

nose. Also, this phenomenon proved that the temperature at the tool nose was very high although coolant was applied in the turning process.

11.7.2 Crater Wear Analysis

It is known that PCD has high resistance to crater wear under normal cooling conditions. However, tool wear was found after 10-min turning on the rake face of the PCD tool. Figure 11.19 illustrates the tool wear on the rake surface and cutting edge of the PCD insert. The cutting edge becomes blunt owing to the loss of tool material on the rake face near the cutting edge (Fig. 11.19d). It can be seen in the SEM images that severe crater wear occurred in the 10-min turning process: there is titanium adhesion near cutting edge and a "hollow area" near the worn cutting edge. It has been proven that this type of wear is generated by chemical diffusion at tool/chip interface [42]. It is thermally activated and developed by removing the adhesive material by plucking action. Because the rate of chemical diffusion and adhesion depends on the temperature at tool/chip interface, it is reasonable to assume that the temperature at

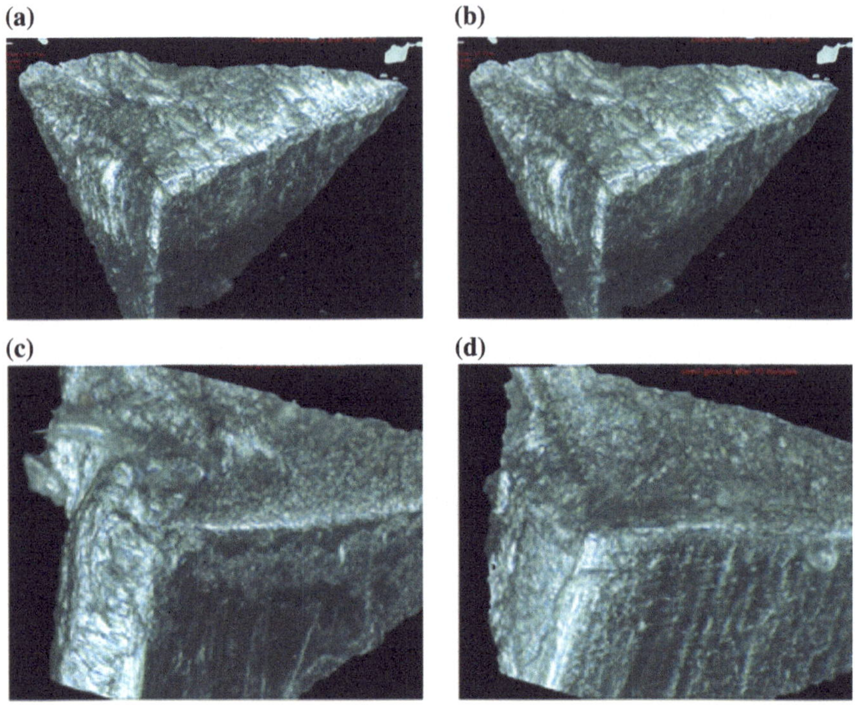

Fig. 11.19 Wear on tool nose and cutting edge after each step of turning experiment after **a** 1 min, **b** 3 min, **c** 6 min, **d** 10 min

the worn area was higher and the chemical reaction was severe in turning Titanium with PCD tools.

11.8 Chatter Prediction for Machining Reliability

Researches to determine reliability of the cutting process have been reported [43, 44]. Some methods need expensive equipment or highly qualified professional employees making it not feasible to use in industry [45]. Many researches have been done to develop methods detecting chatter off-line or in lab conditions to provide stability lobes diagram [46], or methods on how to prevent the process from chatter, but they are slow and time consuming [47].

There are three basic types of vibration in milling process: free vibration, forced vibration and self-excited vibration (chatter). In particular, chatter is much more destructive than other two types of vibration. It occurs as a result of an interaction between the dynamics of the machine tool and the workpiece. The onset of chatter may cause abnormal tool wear or tool breakage, damage of both tooling structure and spindle bearings, poor surface roughness and poor dimensional accuracy of the workpiece.

Chatter is characterized by unstable relative motions between cutting tool and the workpiece. Stephenson and Agapiou [48] explained that variations in the cutting forces, dry friction, built up edge, metallurgical vibrations in workpiece material, as well as regenerative effects are factors that produce chatter during machining of titanium. When it occurs, it induces strong cutting forces and vibration [49].

Koohestani et al. [50] applied phase-space reconstruction method to analyse and predict chatter. Reconstruction of phase space is a method to identify vibration and instability from time series [51]. It does not have the issue of other methods as described above. It does not require clearing of noise from the input signals. It can adapt to individual cutting parameters and dynamics of machining process. The technique is applicable for both deterministic and chaotic systems. Hence, it has great potential of using as an on-line chatter prediction method.

The phase space reconstruction method consists of re-writing the equations as a system of differential equations that are first-order in time and by introducing new variables at embedded time. The original variables at time (t) and new variables at time ($t + \tau$) form a vector in the phase space. The solution becomes a curve in the phase space which is parameterized by time. The differential equation is reformulated as a geometrical description of the curve, as a differential equation in terms of the phase space variables only, without the original time parameterization [52].

Delay-time method is used for the time series in order to reconstruct the phase space of the dynamical system [53]. The transition of single variable $X(t)$ is related to other variables which corporate with the original variable and information are contained in the history of the time series and its delays that are coordinated for the new vector, where τ is referred to the time delay and m is embedding dimension.

$$X_t = \left\{x_t, x_{t+\tau}, x_{t+2\tau}, \ldots, x_{t+(m-1)\tau}\right\}$$

The new vector time series is a multiple of the sampling interval used. The embedding dimension (m) is considered as the sufficient dimension for recovering the object without distorting any of its topological properties, thus it may be different from the true dimension of the space where this object lies [54]. Based on Takens theorem if m is large enough, the attractors that are reconstructed by delay-time embedding have similar mathematical properties as the original system, if provides an image similar to the original system [55].

It is required to record vibration between workpiece and cutting tool as is the main reason of chatter. This vibrations are occurred in three axes and can be recorded during the process time and provide suitable time series along X, Y, and Z axis in order to use in reconstruction of phase space of cutting process. In order to ensure the quality of data, the experiment has been setup to record vibration from the spindle, where the signal capture is closest to the source of the vibration, the tool. Figure 11.20 shows signals from a typical experiment.

The delay parameter can be any delay time but not all values produce a useful outcome for analysis. The optimum time delay value of τ can be found by plotting

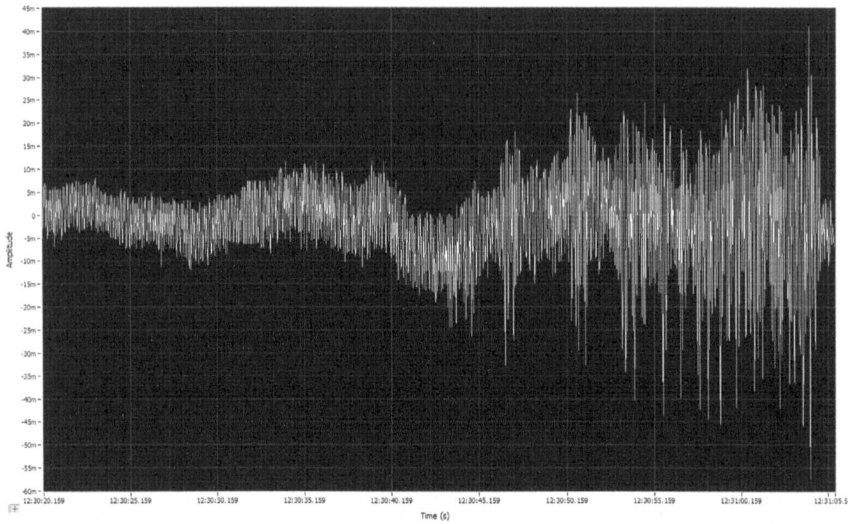

Fig. 11.20 Time series recorded by cutting parameters No. 1

the autocorrelation factor for different *k*.

$$C_{(k)} = \frac{\sum_{i=0}^{n-1}\left(X_i - \overline{X}\right)\left(X_{i+k} - \overline{X}\right)}{\sum_{i=0}^{n-1}\left(X_i - \overline{X}\right)^2}$$

where $-1 < C(k) \leq 1$ and is the average of series, k is the delay in number of samples in the series $X_n = \{x_0, x_1, x_2, \ldots, x_{n-1}\}$. Based on the behavior of the correlation coefficient, time delay can be considered. The first zero can be an optimum point to consider as time delay τ.

In order to apply phase space reconstruction method, the optimum time lag is required to be found from the autocorrelation between $Y(t)$ and $Y(t+\tau)$. Figure 11.21

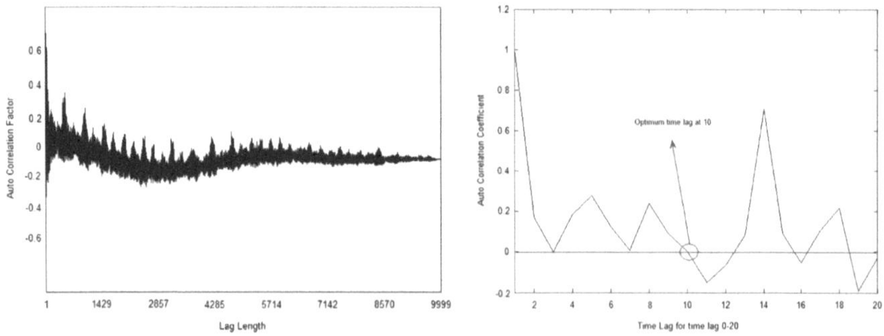

Fig. 11.21 Determination of optimum time delay by autocorrelation method

shows the plot of the autocorrelation vs. different values of k. The value of k, when the plot crosses the x-axis, is approximately 10 (Fig. 11.20b).

Once the optimum time delay is determined, different sections of the phase space reconstructed map are plotted. The method is to divide the time signals in equal periods of time each 10 s and with overlap of 2 s. the Poincarè maps for each period was plotted in Cartesian coordinate.

Figure 11.22a shows the Poincarè map at the period of 0–10 s, when the milling process is in stable conditions. This map can be used as reference for other step.

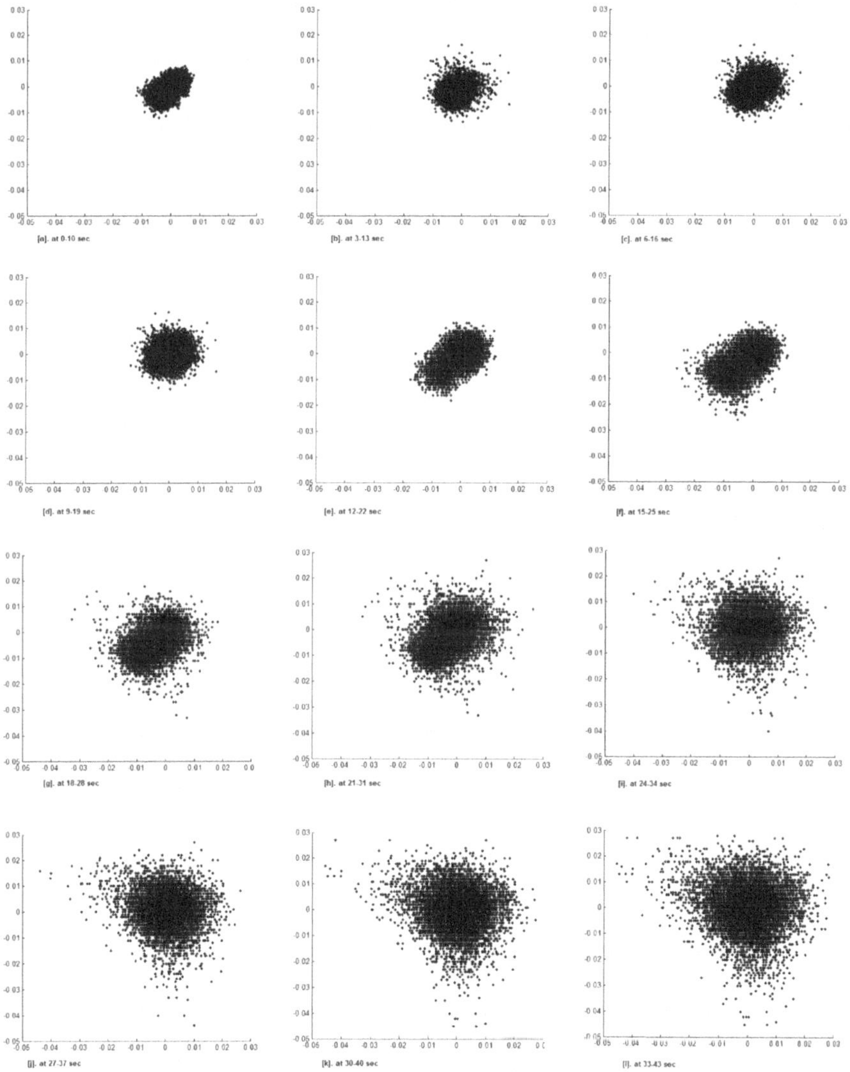

Fig. 11.22 The progress of instability during the cutting process, using Poincarè maps

Data points are so dense in the map and the shape is similar to a circle. The area of data point has in minimum size, as well as standard deviation (SD). The standard deviation of the section is 0.004 and the area is around 3.14.

Figure 11.22b is the next Poincarè map in the next period of 3–13 s. The system is in stable condition and the area of data point is extended 26%. The SD coefficient of this section is 0.004. However, data points are not as dense as the reference image and it shows that is going to be expanded.

The start of instability is time period 15–25 s (Fig. 11.22f). The expansion of maps has continued smoothly in average of 20% in each step. Data point area was extended almost 120% at the onset of instability. By onset of chatter, 5 s time is to enter completely into the instable phase or chatter.

The cutting process was instable at time period 24–34 s (Fig. 11.22i) completely. The data points are extended five times comparing with the reference or stable map which it was seven times bigger at the end of process. SD factor at this period of time has reached to 0.009.

It is clear from the above that the SD factor is a measure of the onset of chatter. Detection of chatter is visualized by the Poincare maps which are progressively developed from time domain signals.

11.9 Conclusion

This chapter has reviewed the 5 key performances in a machining system. These 5 key performances are: efficiency, sustainability, accuracy, precision and reliability. For each of these performance requirements, the machining system should be enhanced with appropriate technologies and findings.

High speed machining is the primary method to improve efficiency and increase productivity in the machining system. In the application of HSM, one of the important components which are often neglected is the tool path strategy. The successful application of HSM can only be realized in conjunction with proper CAM functions. Without a suitable tool path strategy, HSM technology includes new spindles and tool clippers, new control theories and advanced servo systems cannot be used to their full potentials. As a result, an enormous productivity advantage may be missed.

Tool life prediction contributes to maintaining sustainability of the machining system, in particular, Titanium alloys are increasingly used in applications requiring high strength-weight ratio. However, Titanium alloys are extremely difficult to machine owing to the low thermal conductivity and high chemical reactivity with many cutting tool materials. To improve cutting sustainability, strategies to determine when the tool will break is essential so as to optimize cutting costs.

Associated with new CNC machining equipment is the increasing trend of unitised monolithic machining resulting many thin-wall components. Due to machining forces, many thin walls are distorted if not carefully machined. This affects the accuracy of the machining system. A finite element analysis (FEA) machining model has been developed to predict the distortion or deflection of the part during end milling process and the result can be used to compensate the machining path to maintain a straight thin wall.

Machining with accuracy may not immediately produce workpiece with exact dimensions. Tool wear in the machining can affect cutting point precision and subsequently precision of the machining process. Flank wear of PCD tools was caused by chemical diffusion and abrasion between tool surface and workpiece in turning Titanium alloy which was activated by the high temperature at tool/workpiece interface. To improve cutting precision, the appropriate tool geometry for specific cutting requirements should be applied.

Reliability of machining depends on the reliability of all components. The phenomenon of chatter causes unstable, chaotic motions of the tool and bad surface finish. An online chatter detection system based on state-space analysis of cutting forces is able to predict the onset of chatter and informs the control system to adjust machining parameters accordingly to prevent chattering.

References

1. Astakhov VP (2017) Improving sustainability of machining operation as a system endeavor. In: Davim JP (ed) Sustainable machining. Springer International Publishing Switzerland, pp 1–30
2. Altintas Y (2016) Virtual high performance machining. Procedia CIRP 46:372–378
3. Zhang J, Ong SK, Nee AYC (2012) Design and development of an in situ machining simulation system using augmented reality technology. Procedia CIRP 3:185–190
4. King RI (1985) Handbook of high-speed machining technology. Chapman and Hall, New York
5. Ding SL, Mo J, Yang D (2010) HSM strategies of CAD/CAM systems—part I tool path generation. Key Eng Mater 426–427:520–524
6. Byrne G, Dornfeld D, Denkena B (2003) Advancing cutting technology. Ann CIRP 52(2):483–507
7. Dandekar CR, Shin YC, Barnes J (2010) Machinability improvement of titanium alloy (Ti–6Al–4V) via LAM and hybrid machining. Int J Mach Tools Manuf 50(2):174–182
8. Boyer R (2010) Attributes, characteristics, and applications of titanium and its alloys. JOM 62(5):21–24
9. Brinksmeier E, Lucca DA, Walter A (2004) Chemical aspects of machining processes. CIRP Ann 53(2):685–699
10. Ezugwu EO, da Silva RB, Bonney J, Machado AR (2005) The effect of argon-enriched environment in high-speed machining of titanium alloy. Tribol Trans 48(1):18–23
11. Tonshoff HK, Winkler J, Gey C (1999) Machining of light metals. Materialwissenschaft und Werkstofftechnik 30(7):401–417
12. Ítalo Sette Antonialli A, Eduardo DA, Pederiva R (2010) Vibration analysis of cutting force in titanium alloy milling. Int J Mach Tools Manuf 50(1):65–74
13. Marinac D (2000) Tool path strategies for high speed machining. Mod Mach Shop 72(9):104–110
14. Haron CHC, Ginting A, Arshad H (2007) Performance of alloyed uncoated and CVD-coated carbide tools in dry milling of titanium alloy Ti-6242S. J Mater Process Technol 185:77–82
15. Izamshah RAR, Mo JPT, Ding S (2011) Finite element analysis of machining thin-wall parts. J Key Eng Mater 458:283–288
16. Wan M, Zhang WH, Qin GH, Wang ZP (2008) Strategies for error prediction and error control in peripheral milling of thin-walled workpiece. Int J Mach Tools Manuf 48:1366–1374

17. Quintana G, Ciurana J (2011) Chatter in machining processes: a review. Int J Mach Tools Manuf 51(5):363–376
18. Urbanski JP, Koshy P, Dewes RC, Aspinwall DK (2000) High speed machining of moulds and dies for net shape manufacture. Mater Des 21:395–402
19. Field R, Beard T (1996) High speed machining of dies and molds. Modern Mach Shop 69(6):76–83
20. Ding S, Mannan MA, Poo AN, Yang DCH, Han Z (2003) Adaptive iso-planar tool path generation for machining of free-form surfaces. Comput-Aided Des 35(2):141–153
21. Ding SL, Mo JPT, Yang D (2010) HSM strategies of CAD/CAM systems—part II industry applications. Key Eng Mater 426–427:559–563
22. Marinac D (2017) Tool path strategies for high speed machining. Available from http://www.mmsonline.com/articles/tool-path-strategies-for-high-speed-machining, 9 Sept 2017
23. Esprit CAM software online help, Available from http://www.espritcam.com/support/overview, 9 Sept 2017
24. Bieterman M (2001) Curvilinear tool paths for pocket machining. Seminar on Industrial Problems, Institute for Mathematics and its Applications (IMA), University of Minnesota, 16 Mar 2001
25. Ding S, Yang D, Han Z (2005) Boundary-conformed machining of turbine blades. Proc Inst Mech Eng Part B: J Eng Manuf 219(3):255–263
26. Fallböhmer P, Rodríguez CA, Özel T, Altan T (2000) High-speed machining of cast iron and alloy steels for die and mold manufacturing. J Mater Process Technol 98(1):104–115
27. Ezugwu EO, Wang ZM (1997) Titanium alloys and their machinability a review. J Mater Process Technol 68:262–274
28. Nabhani F (2001) Machining of aerospace titanium alloys. Robot Comput Integr Manuf 17:99–106
29. Shivpuri R, Hua J, Mittall P, Srivastava AK (2002) Microstructure-mechanics interactions in modeling chip segmentation during titanium machining. CIRP Ann 51(1):71–74
30. Budak E, Altintas Y (1994) Peripheral milling conditions for improved dimensional accuracy. Int J Mach Tools Manuf 34:907–918
31. Kline WA, DeVor RE, Shareef IA (1982) The prediction of surface accuracy in end milling, ASME. J Eng Ind 104:272–278
32. Elbestawi MA, Sagherian R (1991) Dynamics modelling for the prediction of surface errors in the milling of thin-walled sections. J Mater Process Technol 25:215–228
33. Sutherland JW, DeVor RE (1986) An improved method for cutting force and surface error prediction in flexile end milling system. ASME J Eng Ind 108:269–279
34. Tsai JS, Liao CL (1999) Finite element modelling of static surface errors in the peripheral milling of thin-walled workpiece. J Mater Process Technol 94:235–246
35. Ratchev S, Huang W, Liu S, Becker AA (2004) Milling error prediction and compensation in machining of low-rigidity parts. Int J Mach Tools Manuf 44:1629–1641
36. Izamshah RRA, Mo JPT, Ding S (2012) Hybrid deflection prediction on machining thin-wall monolithic aerospace components. J Eng Manuf 226(4):592–605
37. Gradisek J, Kalveram M, Weinert K (2004) Mechanistic identification of specific force coefficients for a general end mill. Int J Mach Tools Manuf 44:401–414
38. Chen W, Xue J, Tang D, Chen H, Qu S (2009) Deformation prediction and error compensation in multilayer milling process for thin-walled parts. Int J Mach Tools Manuf 49:859–864
39. Wan M, Zhang WH, Tan G, Qin GH (2007) New cutting force modelling approach for flat end mill. Chin J Aeronaut 20:282–288
40. Bhaumik SK, Divakar C, Singh AK (1995) Machining Ti6Al4V alloy with a wBN-cBN composite tool. Mater Des 16(4):221–226
41. Sreejith PS, Krishnamurthy R, Malhotra SK (2000) Evaluation of PCD tool performance during machining of carbon/phenolic ablative composites. J Mater Process Technol 104(1):53–58
42. Liang L, Liu X, Li X, Li YY (2015) Wear mechanisms of WC–10Ni3Al carbide tool in dry turning of Ti6Al4V. Int J Refract Metal Hard Mater 48:272–285

43. Gradišek J, Kalveram M, Insperger T et al (2005) On stability prediction for milling. Int J Mach Tools Manuf 45(7–8):769–781
44. Ding Y, Zhu L, Zhang X et al (2010) A full-discretization method for prediction of milling stability. Int J Mach Tools Manuf 50(5):502–509
45. Totis G (2009) RCPM-A new method for robust chatter prediction in milling. Int J Mach Tools Manuf 49(3–4):273–284
46. Quintana G, Ciurana J, Ferrer I et al (2009) Sound mapping for identification of stability lobe diagrams in milling processes. Int J Mach Tools Manuf 49(3–4):203–211
47. Grabec I, Gradišek J, Govekar E (1999) A new method for chatter detection in turning. CIRP Ann Manuf Technol 48(1):29–32
48. Stephenson DA, Agapiou JS (2006) Metal cutting theory and practice. CRC Taylor & Francis
49. Wan M, Wang Y-T, Zhang W-H, Yang Y, Dang J-W (2011) Prediction of chatter stability for multiple-delay milling system under different cutting force models. Int J Mach Tools Manuf 51(4):281–295
50. Koohestani A, Mo JPT, Yang S (2014) Stability prediction of titanium milling with data driven reconstruction of phase-space. Mach Sci Technol Int J 18(1):78–98. https://doi.org/10.1080/10910344.2014.863638
51. Rusinek R, Warminski J (2009) Attractor reconstruction of self-excited mechanical systems. Chaos Solitons Fractals 40(1):172–182
52. Kautz R (2011) Chaos: the science of predictable random motion. Oxford University Press, New York. ISBN 978-0-19-959457-3
53. Wu CL, Chau KW (2010) Data-driven models for monthly streamflow time series prediction. Eng Appl Artif Intell 23(8):1350–1367
54. Shang P, Na X, Kamae S (2009) Chaotic analysis of time series in the sediment transport phenomenon. Chaos Solitons Fractals 41(1):368–379
55. Kodba S, Perc M, Marhl M (2005) Detecting chaos from a time series. Eur J Phys 26(1):205

Chapter 12
Technology Nationalization in the Space Sector: The Brazilian Perspective

Timo Wekerle, Luís Gonzaga Trabasso and Luís E. V. Loures da Costa

Abstract Brazil as an emerging country needs to catch up with technology to extend its position on the international market, especially in the space sector. The Technology Nationalization Framework (TNF) is a strategy for nationalization and industrialization of high technology products. The TNF is meant to assure that strategic technologies, that are currently lacking, will be designed, produced, and operated in Brazil as long as needed, without the risk of export bans or unavailability of components. The framework is based on reengineering with subsequent transfer to the national industry. The strategy starts with the identification of strategic technologies in relation to technologies already present in Brazil. For the nationalization process of these technologies a decision-making process is needed taking into account available resources and competencies. In this chapter the TNF will be introduced and explained, while also a pilot project is described in which the TNF strategy is applied.

Keywords Technology nationalization framework · Design for autonomy · Design for X · Space technology

12.1 Introduction

Brazil as an emerging country needs a technological catch-up process to assure and extend its position on international markets. Astronautics is a base for technological, economic and military potential of a nation. Particularly in the space sector Brazil needs to catch-up with technologies.

This Chapter introduces a strategy for nationalization and industrialization of high technology products called Technology Nationalization Framework (TNF). The

T. Wekerle (✉) · L. G. Trabasso · L. E. V. L. da Costa
Aeronautics Institute of Technology (ITA), Praça Marechal Eduardo Gomes, 50, CEP 12, 228-900
São José dos Campos, SP, Brazil
e-mail: t.wekerle@gmx.de

Present Address:
T. Wekerle
Airbus Operations GmbH, Kreetslag 10, Hamburg 21129, Germany

J. Stjepandić et al. (eds.), *Systems Engineering in Research and Industrial Practice*,
https://doi.org/10.1007/978-3-030-33312-6_12

objective of the TNF is to assure that certain strategic technologies, which are currently lacking and are only available abroad, are converted into national products that can be designed, produced and operated in Brazil for a defined period of time at a minimum risk of being dependent on export bans or unavailability of components.

The Technology Nationalization Framework is a strategy for nationalization of foreign technology based on reengineering with subsequent transfer into the national industry. It is a comprehensive systematic strategy that begins with the identification of strategic technologies in Brazil and continues with support for the decision-making process for nationalization of these technologies. An evaluation of feasibility of development of national domain and subsequently its coordination and cooperation helps to stimulate the best use of resources and competencies available in Brazil.

Design for Autonomy, an integrated product development tool within the TNF strategy, is a decision and design supporting tool that copes with high complexity and generates alternative views for a robust national design. The application of the TNF fosters innovation and competitiveness in Brazil and ensures non-dependence of strategic technologies. This can be achieved by a balance between completely domestic/national development with intrinsically high cost, lead time and risk, and blind implementation of technology transfer with risk of failure due to never-ending-projects, incomplete transfer, more expensive solutions and/or higher vulnerability to embargoes. In a pilot project the TNF is successfully being applied in the Brazilian space sector.

The outline of the chapter is as follows. In Sect. 12.1, integrated product development and systems engineering are briefly discussed as well as their mutual relationships. In Sect. 12.2, the Technology Nationalization Framework is introduced. In Sect. 12.3, the TNF model is presented with its eight steps. In Sect. 12.4, the processes of each of the eight steps are presented, followed by thoughts about follow-up in Sect. 12.5. Implementation and integration issues are briefly described in Sect. 12.6. In Sect. 12.7, application of the processes to a pilot project is presented. Section 12.8 contains a discussion on implications for Aerospace Engineering and other sectors, including abroad, for each of the eight steps of the model. Limitations of the approach are also discussed. The chapter ends with Sect. 12.9 with concluding remarks and an outlook for further work.

12.2 Integrated Product Development and Systems Engineering

You might have heard the classical sentence once: "Oh, that's the name then. I've been doing that for ages… I only did not know that was the name". We believe that this sentence would apply to a dialogue between a mechanical/industrial engineer and an electric/electronic engineer. The latter tells the former all about Systems Engineering. The former listen mindfully and then s/he replies…

In this introduction we give an overview of the terms Integrated Product Development (IPD) and Systems Engineering (SE), showing common ground and differences.

12.2.1 Integrated Product Development

According to Pessôa and Trabasso [1], Integrated Product Development is a product design and development approach where the requirements of all technical areas related to the product lifecycle are considered, weighed, discussed, and balanced at the conceptual phase of the product development process. As a result, the outcome from integrated product development is a product, which is designed not only to function, but also to be easily and cheaply manufactured, assembled, tested, maintained, and recycled. IPD, therefore, expands the horizon of the product evaluation by taking into account all the technical areas and phases the product goes through during its lifecycle. Product requirements are the engineering expressions to make the product lifecycle areas to be active protagonists at the conceptual design phase of the product development as illustrated in Fig. 12.1.

It is usual to have conflicts among engineering requirements: a product configuration that is easy to assemble might be difficult to disassemble. Consequently, it is expected to have a number of design trade-offs to be solved within IPD. The final product configuration yielded by an IPD program is a balanced solution that accommodates—the best possible way—all the requirements posed by the technical areas of the product lifecycle.

There are, essentially, two main resources required to implement the IPD approach, namely, a multifunctional design team and IPD tools.

IPD design team: The mission of the IPD team is to assure that the requirements of all product development phases are evenly represented in the IPD's conceptual

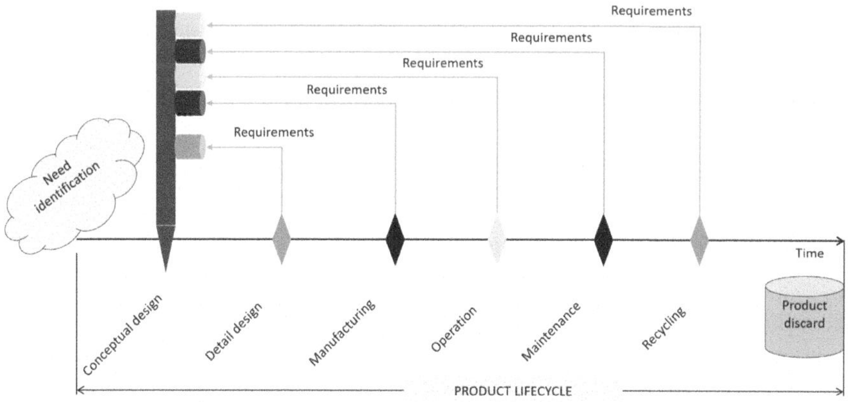

Fig. 12.1 IPD—the engineering requirements role

design phase. All people from the IPD design team should be committed to obtaining the best possible balanced results for the product, even if that means giving away some of his/her technical area expectations. In a typical PD team meeting, nobody leaves either "100% happy" or "100% unhappy." It is the role of the project leader to ensure the team's focus on the mission and achieve a balanced result.

IPD design tools: A great number of design tools are available to promote the integration of the technical areas of the product development, such as Design for Manufacturing (DFM), Design for Assembly (DFA), Design for Recycling (DFR), Design for Service (DFS), Design for Packing (DFP), Design for E-Business (DFEB), Design for Automation (DFAut), and so forth. All of them are related to some phase of the product life cycle and have one characteristic in common: an attempt to integrate the requirements of their product life cycle phase into the conceptual design phase of the product development process. It's worth stressing the words "attempt to integrate" because all the representatives of product life cycle phases will try to do the same—to advocate their cause. It is quite possible that the DFM product option conflicts with that of DFS, thus raising an engineering tradeoff whose solution might partially fulfill both areas.

DFX versus DTX: the literature also presents DTX design tools associated to IPD such as DTC—Design to Cost, DTW—Design to Weight, DTCG—Design to Center of Gravity and so forth. Pessôa and Trabasso [1] pose that the DTX design tools are related to a specific type of design variables named Integrative Design Variables (IDV). According to the authors, costs, weight, center of gravity, and net electric power are examples of integrative design variables. The characteristics of these variables are the following:

There is a target value associated with them within a specific product development. Examples: the cost of an aircraft cannot be greater than $14.5 M; the maximum weight of a robot end effector is 80 kg; the net power of a satellite is 2300 W.

These variables are affected by almost all design decisions. Examples: the choice of a single component impacts cost, weight, center of gravity, and perhaps net electric power if the component requires it for operation.

It is easy to grasp the concept around integrative design variables. Design people do not need to be lectured about them as their understanding is quite straightforward. Examples of design variables that do not meet this characteristic of IDV are: aerodynamic drag, wear, and stiffness.

12.2.2 Systems Engineering

According to the System Engineering Body of Knowledge [2], there are three types of systems engineering, namely, (1) Product Systems Engineering—PSE, (2) Enterprise Systems Engineering—ESE and Service Systems Engineering—SSE. As a matter of comparison with IPD, only the first type is addressed here. PSE is the traditional systems engineering focused on the design of physical systems consisting of hardware and software. ISO/IEC/IEEE 15288 [3] define systems engineering as the interdisciplinary approach governing the total technical and managerial effort required to transform a set of stakeholder needs, expectations, and constraints into a solution and to support that solution throughout its life. INCOSE [4] adds that systems engineering focuses on defining needs and required functionality early in the development cycle, documenting requirements, and performing the design synthesis and system validation while considering the complete problem (operations, cost and schedule, performance, training and support, manufacturing, and disposal). The United States Department of Defense (DOD) [5] complements that systems engineering transforms needs and requirements into a product as well as generates information for decision makers and provides input for next levels of development (adding value and more detail with each level). DOD [5] describes other important characteristics of systems engineering: it is a top-down, comprehensive, iterative and recursive effort. Mar [6] states that most systems engineering efforts are based on the hierarchical decomposition of the system into its parts. These characteristics help to describe the chronology of the systems engineering effort as beginning at the highest hierarchical level and going down to lower levels by adding value and more detail with the execution of the processes and the passage of iterations and recursions. The Vee model depicted in Fig. 12.2 exhibits that a systems engineering effort starts from the highest level (i.e. the system being the object of the systems engineering effort) and flows down to lower levels that appear by decomposition; then the lower levels are integrated and verified to realize the higher levels (Chap. 2). in contrast, Transdisciplinary engineering (TE) is focused on solving ill-defined and society-relevant problems, like sustainability and environmental problems [7, 8].

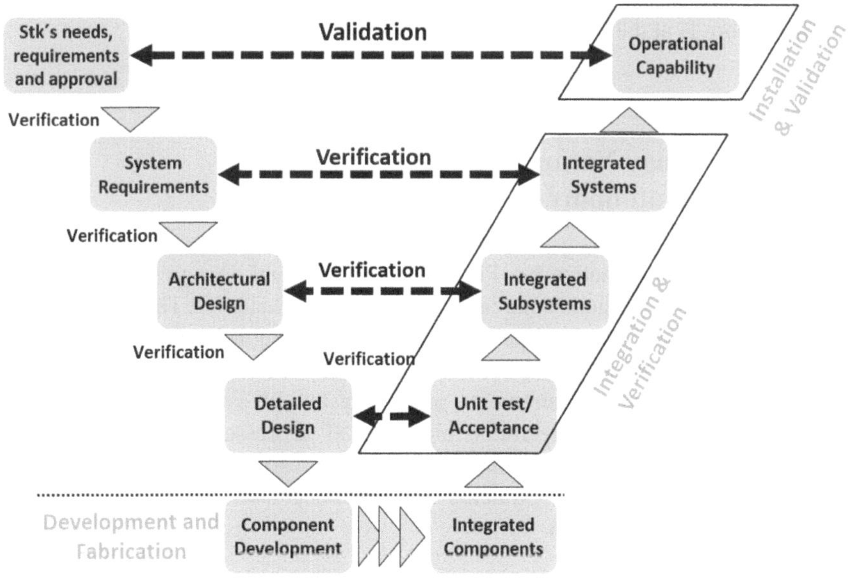

Fig. 12.2 The V-model of systems engineering, after Stevens et al. [9]

12.2.3 Common Ground and Differences Between IPD and SE

By analyzing the IPD and SE concepts and definitions given above, one is forced to reckon that there are a number of similarities between them, namely [10]:

- **Inputs and Outputs**: both take as inputs the requirements of the stakeholders. SE consolidated the word "stakeholder" while the original equivalent word for IPD was "customer". Example: the QFD—Quality Function Deployment—IPD method has one field named the Voice of Customer. The deliverable (overall output) of IPD is a product or service, that could be extended to systems, while the SE counterpart is a system, that could be shrinked to a product.
- **Focus on product/system lifecycle**: both IPD and SE focus on the overall lifecycle of products or systems.
- **Focus on early stages of the design process**: both IPD and SE reckon that is easier and cheaper to include product/system lifecycle requirements at the early stages of the design process, specifically, the conceptual design phase.
- **Deployment/decomposition**: both IPD and SE prescribe the top-down approach to design: the overall problem is broke-down into sub problems, sub solutions are found to them, then the overall solution is presented by integrating the sub solutions.
- **Teams and design tools**: IPD explicitly requires multifunctional teams to carry out the product development process. It also prescribes design tools to assist the design team to accomplish its task. SE implicitly signalizes the need of a such a team.
- **Integration**: is at the heart of both IPD and SE.

12.3 Technology Nationalization Framework and DfAutonomy

The Technology Nationalization Framework (TNF) and Design for Autonomy (DFAutonomy) are very good application examples of SE and IDP. The reader has the opportunity to follow general aspects of SE and IPD being instantiated by practical and down-to-earth problems and solutions. The baseline for TNF and DFAutonomy are customer/stakeholder requirements, product/system lifecycle, early design stage actions and most of all, the integration mindset. TNF addresses broader aspects than DFAutonomy: this can also be found in the relative position between SE and IPD: the organizational aspects are explicitly taken into account in SE and implicitly in IPD (design team structuring).

12.3.1 Reengineering: Reverse and Forward Engineering

Forward engineering is the traditional process of moving from high-level abstractions and logical, implementation-independent designs to the physical implementation of a system [9]. The term "forward" is necessary to implement in order to distinguish this process from reverse engineering. Reverse engineering can be seen as a process of analyzing an existing system to identify its components and their interrelationships and to investigate how it works to redesign or produce a copy without access to the design from which it was originally produced. Reengineering, also known as renovation or reclamation, is the examination and alteration of a subject system to reconstitute it in a new form and the subsequent implementation of the new form [10]. Reengineering is a process of modifying the internal mechanisms of a system without changing the functionality. In Fig. 12.3 the relationship between forward-, reverse- and reengineering is depicted. The process of reengineering initiates with reverse engineering, an analysis of the original product that includes design recovery originating from the implementation phase and the design phase, restructuring the requirements of the system (data-to-data) and the design (graphical and functional). Once the new and modified requirements are defined, the second part of the reengineering process, the forward engineering continues the process with a development in order to achieve a new system. In the last step of reengineering, restructuring and redocumentation is applied where redocumentation is the creation or revision of existing documentation, based on the original system.

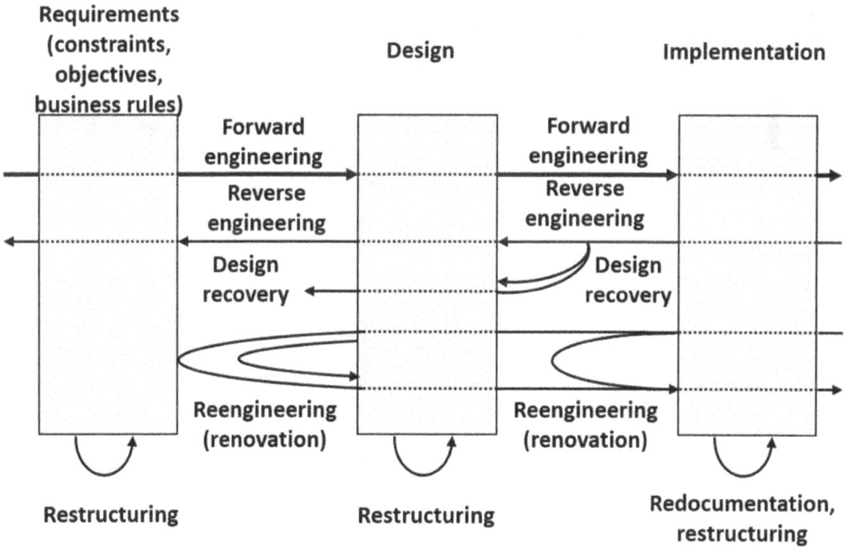

Fig. 12.3 Relationship between the terms of forward-, reverse- and reengineering represented by life-cycle phases [11]

12.3.2 Technology Nationalization Framework

The Technology Nationalization Framework is a strategy for nationalization of foreign technologies realized by reengineering with subsequent transfer into the national industry and industrialization/commercialization. The term nationalization can be rendered by the meaning of making something distinctively national or giving a national character to something.

The TNF is a comprehensive systematic strategy for integrated product development that starts with the identification of strategic technologies for the Brazilian space program and continues with support for the decision-making process for nationalization of these technologies by an evaluation of feasibility of development of a national domain and subsequently its coordination and cooperation, incentivizing the best use of resources and competencies available in Brazil.

The Design for Autonomy tool within the TNF strategy allows managing high complex products by reengineering strategic technologies applied, embedded or implemented within a product. It includes product analysis of critical elements in order to avoid entering in a fatal spiral of total nationalization (100% of product produced in Brazil) and represents a balanced way of reverse engineering that provides observations beyond those perceived by the original designer, creating an innovative scenario for straightforward engineering to prevent errors, save time and money and add value to the new national product. Design for Autonomy is a decision and design-supporting tool that copes with high complexity and generates alternative views for a robust national design.

The political and economic environment of the Brazilian space sector is considered throughout the TNF strategy, leading to a maturation of nationalized product within the executive organizations prior to the transfer to the national industry for successive commercialization.

The core objective of the Technology Nationalization Framework is to assure that certain lacking strategic technologies only available abroad are converted into a national product that can be designed, produced and operated in Brazil for a defined period of time at a minimum risk of being dependent on export bans or unavailability of components.

An estimated comparison of qualitative knowledge acquisition using TNF/Design for Autonomy with classical learning and acquisition curves is plotted in Fig. 12.4. (1) Starting a technology development from scratch is represented by the S-shaped learning curve, a sigmoid function. The initial phase with slow increase of knowledge is attributed to the lack of know-how at the beginning of a project. With a certain amount of experience and expertise the knowledge gain is approaching an exponential growth rate, but then declines in a negative acceleration phase through the time-spending qualification of the product/process until it reaches the state of technological domain. (2) The knowledge gain for acquisition is represented by a flatter S-shaped curve. Eventually, the technology will stay a black box and technological domain will not be achieved. The accomplishment of operability/use of the acquired technology is represented by a cross at the end of the acquisition curve. (3) The advance of

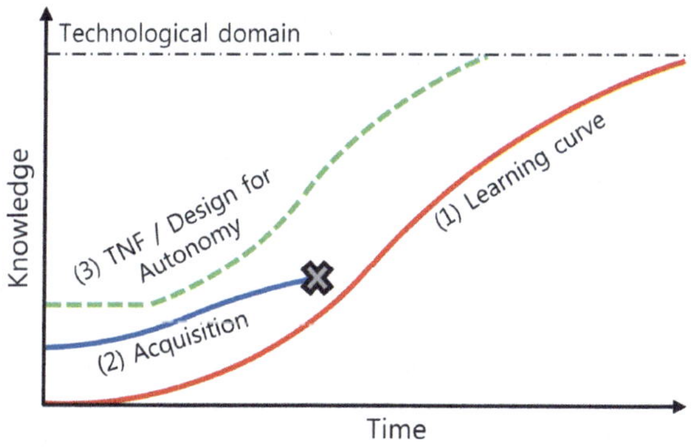

Fig. 12.4 Qualitative knowledge acquisition over time due to (1) proper learning; (2) acquisition of technology; and (3) TNF/design for autonomy

knowledge for TNF/Design for Autonomy right at the beginning can be explained through reverse engineering with the use of original needs/mission/concept from the technology transfer or that can be utilized as baseline configuration. By the time, the given knowledge/data, available and accessible from the very beginning, is being evaluated and comprehended, which is represented by the horizontal section in green. Thereafter, careful adaptations, adjustments and improvements are required, the process of forward engineering, which leads to the characteristic shape of the learning curve, making the initial phase with slow increase of knowledge obsolete.

12.3.3 TNF Guidelines

For the development of new DFX tools, the DFX shell [12–14] suggests to define so-called DFX Guidelines. For developing the Technology Nationalization Framework, the TNF Guidelines are derived in the following. These principles or imperatives can be organized into three categories of key characteristics, namely focus and flexibility, functionality, and operability and are detailed in Table 12.1.

Focus and flexibility imperatives shall be included in order to achieve the right balance between functionality and operability; Functional imperatives shall be included in order to define the purpose and utility of the Technology Nationalization Framework and the Design for Autonomy tool; and Operability imperatives shall be included in order to ease the use the TNF/DfAutonomy and to fulfill its functions effectively.

12.4 TNF Model

The Technology Nationalization Framework model is composed of eight steps in which each step is comprised by a function modeling methodology based on IDEF0. The IDEF0 functional modeling is a member of the IDEF modeling language for

Table 12.1 Imperatives of TNF guidelines

Focus and flexibility imperatives
The target product sector for the TNF shall be the Brazilian space sector
The business process of the DFX tool within the TNF shall be the variable *A* for Autonomy, resulting in Design for Autonomy
The Technology Nationalization Framework shall cover the whole product development life cycle up to verification and validation of the product. The stage of product development process of the Design for Autonomy application shall be the conceptual phase up to the detailed design phase
The Technology Nationalization Framework shall be used as a tool for reengineering, comprising reverse and forward engineering. The Design for Autonomy tool shall focus on the decision and design-making process for the design of the product
Functional imperatives
The Technology Nationalization Framework shall permit comparison to the original and alternative technologies and tracing of changed requirements and architectural/system design
The Technology Nationalization Framework shall predict what-if effects in case of unavailability of components or possible export controls and shall provide design alternatives
The Design for Autonomy tool shall include means for comparing the design alternatives and may include backup alternatives
The Technology Nationalization Framework shall incentive design improvements, innovation and added value for the national product
The Technology Nationalization Framework shall be implemented in an iterative way and shall be applied in multiple cycles
Operability imperatives
The application of the Technology Nationalization Framework shall be pragmatic and easy to apply
The Technology Nationalization Framework shall have a systematic procedure to follow
The collection and presentation of data shall be simple and for further processing traceable
The Technology Nationalization Framework shall create visible and measurable benefits. The different steps of the TNF shall include performance measurements in order to evaluate the pretended outcome of the respective step (where applicable; measurable quantity or quality)
The Technology Nationalization Framework shall encourage innovation and creativity, rather than impose restrictions

software engineering, which has been designed to graphically model and represent decisions, actions, and activities of an organization or system [15]. The syntax used is illustrated in Fig. 12.5. Each function or activity is placed in a box, identified with a number at the bottom right. Inputs are represented by arrows entering the left side, outputs by arrows exiting the right side. Control/management is represented by arrows entering the top of the box and mechanisms/processes by arrows entering from the bottom of the box. In contrast to the standard IDEF0 model, the executors of the function/activity are represented by arrows entering from the top, together with the control of the function/activity. From the bottom, the mechanisms/processes enter, not the performer of these mechanisms.

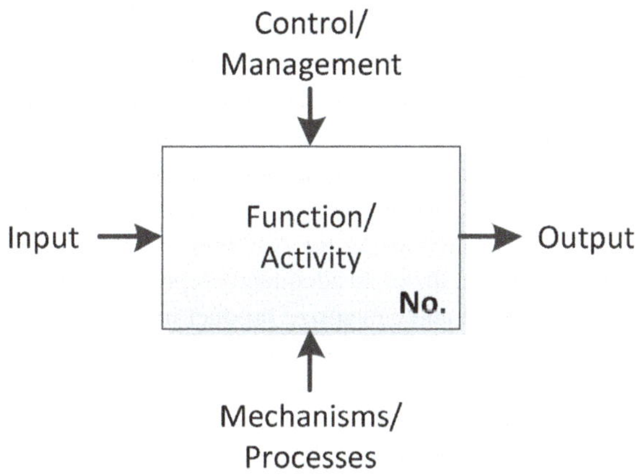

Fig. 12.5 Syntax of modified IDEF0 model for TNF

The Technology Nationalization Framework, illustrated in Fig. 12.6, requires a minimum of eight steps in order to be functional and obtain operability for the nationalization and subsequent industrialization of technology for the Brazilian space sector.

The first step is required to define the technologies necessary to obtain and to assure non-dependency and feasibility of a Brazilian space program within the different space sectors of launch vehicles, space segment, applications, and ground segment.

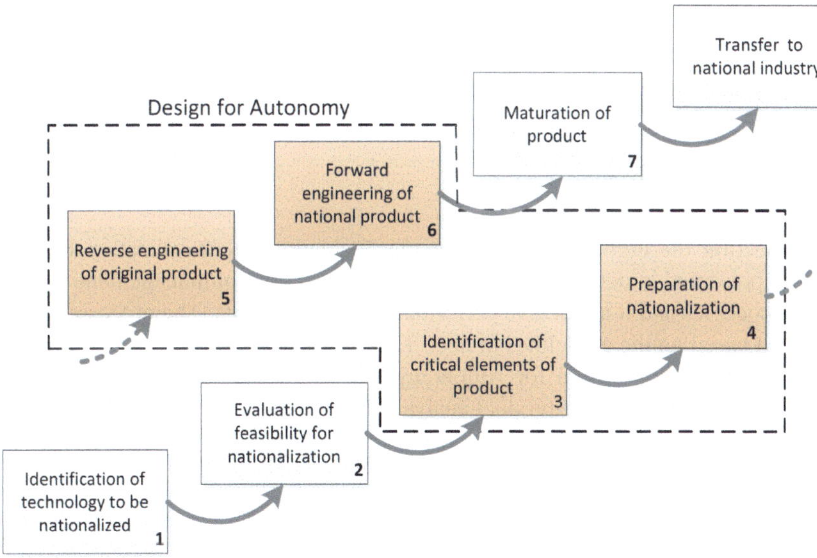

Fig. 12.6 Model of technology nationalization framework consisting of eight steps

The second step is necessary to evaluate the feasibility to nationalize a certain technology, defined in the first step. This step is fundamental for the decision if the task of nationalization should continue or not. With this step, unnecessary waste of resources can be prevented in an early phase of nationalization. Furthermore, the transition from a desired technology to a specified product to be nationalized is achieved.

The third step, the start of the Design for Autonomy tool, is required in order to analyze the product to be nationalized, to adequately represent the product for further design decisions and to collect and categorize product information.

The acquired data from the third step is necessary to prepare the nationalization, the fourth step, where the national industry and research sector are analyzed and the environment for nationalization is prepared. This step is indispensable for successful reengineering, since an inappropriate environment may cause unnecessary delays and/or cost overruns.

Having a prepared technology and a prepared environment, reverse engineering of the original product is accomplished in the fifth step and subsequently forward engineering in the sixth step in order to obtain a national product.

Due to the Brazilian political and industrial environment with the Brazilian Institute of Aeronautics and Space (DCTA/IAE) and the National Institute for Space Research (INPE) as executive organizations, and the Brazilian industry as their supplier, the seventh step of product maturation within the executive organizations is implemented. This step shall ease the challenges and prevent setbacks of the national industry with the development of immature technologies and shall encourage the interest of the industry to obtain a mature technology in order to transform it into a commercial product.

The eighth step is the transfer of the product to the national industry in order to release resources of the executive organizations.

12.5 TNF Processes

In this Section the eight activities from the Technology Nationalization Framework are presented in detail, including inputs, outputs, control/management, mechanisms/processes and if applicable performance measurements. The processes were developed based on the guidelines in Sect. 12.3.

The 3rd–6th step, Design for Autonomy, begins with a subsection where the product is being modeled for product analysis that is used for process description.

12.5.1 Identification of Strategic Technologies to Be Nationalized

The first step, the identification of strategic technologies that need to be nationalized is a high level activity for decision-makers in business and politics. The process is illustrated in Fig. 12.7. Coordination of and cooperation on technologies have to be concluded by all stakeholders, namely executive organizations, space industry and space agency in order to ensure the best use of resources and competencies available in Brazil. The planning has to contain a strategy with short, medium and long term vision. Besides the current demands, needs for future developments have to be anticipated and a strategy implemented to ensure that the right technology is at the right maturity at the right time.

The inputs for the first step are the product breakdown structures of the principal products of the Brazilian space program, organized within the different segments of launch vehicles, space segment (satellites and re-entry systems), applications (operation of payloads and data processing), and ground segment (control and monitoring). The subdivision into systems and further classification by level of integration into equipment, building blocks and Electrical, Electronic and Electro-mechanical (EEE) components, mechanical parts and materials is adopted for the product breakdown structures from the ESA Generic Product Tree [16].

The output are certain technologies identified within the product breakdown structures in the level of equipment, building blocks or EEE components, mechanical parts and materials, that need to be nationalized in order to achieve the goals, prevent failures and accomplish the missions of the Brazilian space program.

A triad of the executive organizations (DCTA/IAE and INPE), the space industry, (Aerospace Industries Association of Brazil (AIAB)), and the national space agency (AEB) is envisioned by the authors to define the necessary technologies for the Brazilian space program and to manage the identification of technologies to be

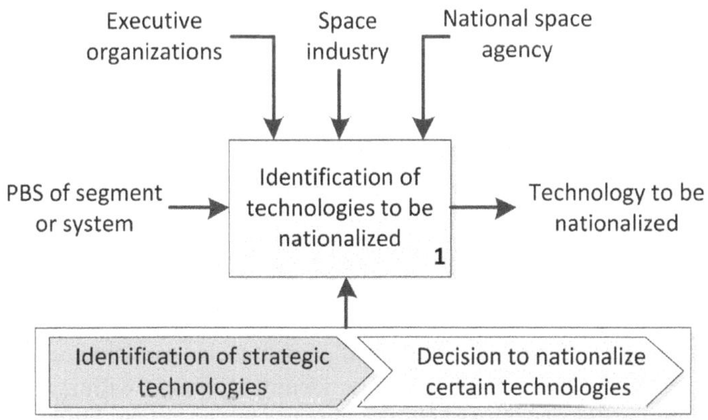

Fig. 12.7 TNF process of 1st step: identification of technologies to be nationalized

nationalized in order to avoid deficiencies or duplications of necessary technologies. Certainly, in the Brazilian case, this triad shall be led by AEB. It could be implemented by a commission for identification and management of technologies, comprising all three stakeholders. This commission should define a strategy with short, medium-and long-term vision in order to attend the current demands and to anticipate the needs for technologies for future developments.

The identification of strategic technologies is an iterative task that shall be repeated in defined time frames and shall include the following metrics.

A technology will be defined as *strategic* if the products of the PBS on the level of equipment, building blocks and EEE components, mechanical parts and materials

1. are not available from any Brazilian source and the unrestricted availability from non-Brazilian suppliers cannot be assured, and
2. have a clearly identified function and performance target.

At each iteration of identification of strategic technologies, the elements of the PBS have to be re-evaluated and advances or setbacks in non-dependency, thus the free, unrestricted access to required space technologies can be measured, qualitatively and/or quantitatively. This measurement can be seen as a benchmark for the state-of-the-art of the Brazilian space sector. This performance indicator is more objective and conclusive than a simple declaration that certain systems (launch vehicles, satellites etc.) have been completed/are operational and vague statements for forecasting the finalization of future systems.

Furthermore, this step can contribute to consistent future investments and medium and longterm priority choices. Petroni et al. [17] examine the basic strategic orientations of some of the world's main space agencies (Brazilian, French, European, Japanese, Indian and Russian agencies), and conclude that these agencies tend to express an intention to invest in many different areas and fields in a way that is often ambiguous and rather inconsistent, without revealing the real priorities determined by their stakeholders.

12.5.2 Evaluation of Feasibility for Nationalization

The second step, the evaluation of feasibility for nationalization of a strategic technology, is illustrated in Fig. 12.8. Once a strategic technology is identified and decided to be nationalized, the technology recipient, the transferee, is evaluated within three categories: Technological capabilities; Infrastructure; and Technology Readiness Level. Deficits in one or more of these areas may suggest investments in the respective areas or may lead to the conclusion that at this specific moment it is not feasible to nationalize that technology. If there are sufficient competencies available, a filter will be applied: A technology transferor has to be identified who is willing to hand over the technology to the transferee and bi-national contracts signed that include this technology transfer. Several methods and mechanisms for technology transfer may be applied.

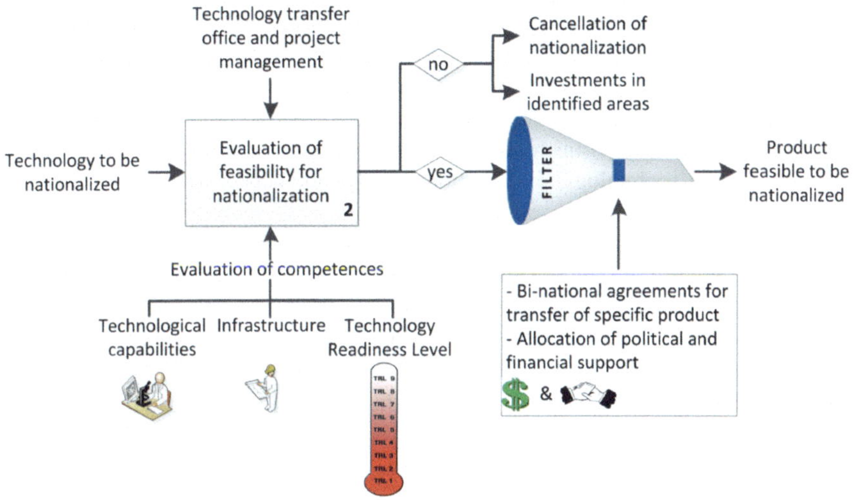

Fig. 12.8 TNF process of 2nd step: evaluation of feasibility for nationalization

Common examples are (1) Partnerships and strategic alliances between research institutes and industry or universities; (2) Intergovernmental cooperation; (3) Direct foreign investment by purchasing technology via joint ventures with international companies; (4) Licensing of systems and subsystems; and (5) Acquisition of turnkey projects. Compliance with the regulations of intellectual property rights according to the World Intellectual Property Organization (WIPO) standards, recommendations and guidelines [18] is a key prerequisite.

A positive outcome of this step will assure the feasibility of nationalization of a certain technology applied, embedded or implemented into a product. The input for this step is the defined strategic technology to be nationalized and the output is a specified product with the defined technology applied, embedded or implemented.

The project management/product owner of the system (launch vehicle, satellite etc.) together with a so-called technology transfer office (TTO) may execute the evaluation of feasibility for nationalization of a certain technology. The term TTO is intentionally introduced to emphasize the new challenges to be addressed. A potential candidate for the tasks of a TTO in Brazil would be the Technological Innovation Centers (NITs). The interplay between the technical side where a technology is transformed into a product to be acquired from a different country and the negotiations with business and legal questions of the contractual partners has to be conducted by the TTO at the best possible conditions for the transferee.

The evaluation of technological capabilities can be supported by a tool for mapping of human resources in the Brazilian research sector, developed by the Space Technology Observatory (OTE) from the Brazilian Center for Strategic Studies and Management (CGEE) [19]. This tool searches and classifies human resources according to their scientific and technical production based on information from the CV-Lattes system from the Brazilian National Research Council (CNPq). It identifies professionals and their networks of cooperation and expertise through co-authorships and semantic analysis of the CVs of the researchers. This tool may help to give an overview of the state-of-the-art in the Brazilian research sector in a specific topic and to identify available human resources in Brazil for possible cooperation.

There are currently no tools for identification of available infrastructure of Brazilian Scientific and Technological Institutions (ICTs), and hence, specific inquiries are necessary.

The technology readiness level of the technology to be nationalized has to be evaluated both within Brazil and outside, in order to identify the current status of the technology in Brazil compared to the readiness of the technology worldwide. This gives an important additional parameter for the decision-making process.

If the evaluation results reveal insufficient competencies that make a nationalization substantially difficult or impossible, the possibilities of cancellation of the nationalization attempt or recommendations for investments in order to advance in the respective competence area are introduced.

If the TTO and the project management consider a certain technology feasible to be nationalized, a filter is applied. A transferor for the product/technology has to be identified and bi-national contracts for the transfer of technology assigned. The legal questions have to be included in the contracts and financial and political support for a national development assured. Only if these conditions are met, the product/technology will be feasible to be nationalized.

Annotation: The 2nd step is case specific and cannot guarantee completeness nor to be exhaustive, since every problem has to be approached separately. Various methods for bi-national negotiations and agreements may be pursued and the evaluation of competences and control and management is only a representative example. The TTO is a suggestion from the authors to reinforce technology transfer and to reduce the burden of the project manager.

12.5.3 Design for Autonomy—3rd to 6th Step of TNF

The Design for Autonomy tool is an integrated product development instrument and a new member of the DFX family, developed on the basis of the DFX shell [12]. It is introduced in Wekerle et al. [20] and detailed in an extended paper by Wekerle et al. [21]. The term *autonomy* can be rendered as self-rule or self-determination and is used in this context as freedom from external control or influence.

The first activity of the product development process, the 3rd step, contains modeling of the product for further analysis including the identification of critical elements of the product. The nationalization of the product is being prepared in the 4th step. Having a prepared product/technology and a prepared environment for nationalization, reengineering is carried out in order to obtain a national product (5th and 6th step).

The process of reengineering is illustrated in Fig. 12.9, initiating with reverse engineering, an analysis of the original product that includes design recovery originating

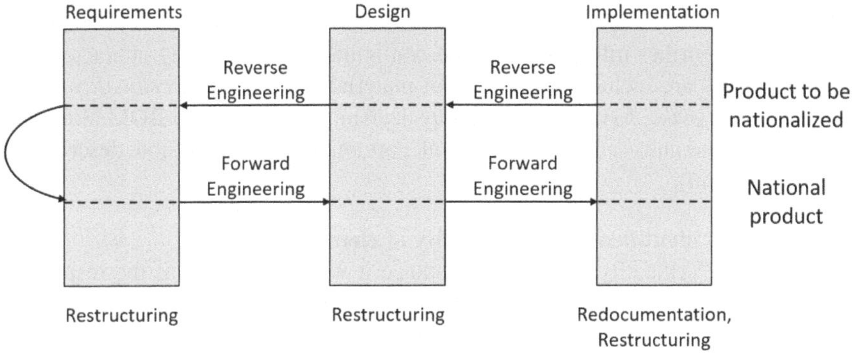

Fig. 12.9 Design for Autonomy tool for reengineering including the interrelations between the terms of forward-, reverse- and reengineering represented by life-cycle phases, after Chikofsky and Cross [11]

from the implementation phase and the design phase, restructuring the requirements of the system (data-to-data) and the design (graphical and functional). Once the new and modified requirements have been defined, the second part of the reengineering process, the forward engineering continues the process with own development in order to achieve a national product. The term "forward" is necessary to implement to distinguish this process from reverse engineering.

12.5.3.1 Identification of Critical Elements

The activity in the 3rd step is determined to identify critical elements of the product and to create action lists in case of unavailability of components. Not identified critical elements may hinder or impede the product development or may result in delays or excessive cost. This activity is the initial step of the Design for Autonomy tool, breaking down a complex product into manageable elements that are being identified.

Modeling for product analysis

In order to model the product, Huang [12] suggested to determine three general categories of product information, namely, composition, configuration, and characteristics.

The Product Breakdown Structure (PBS), a technical tree, which is a structured representation of all various elements of a system [22], represents the composition of what the product consists of. According to the American National Standards Institute (ANSI)/Electronic Industries Association (EIA) [23], the PBS is a hierarchical structure of the complete set of physical systems and subsystems including operational system, training system, development support, production support, etc. which identifies the configuration items. The PBS hierarchically sorts the physical components of the respective product top-down into manageable elements.

The configuration is also partly included in the PBS, defining the relation between the elements. Further information of the configuration and the key characteristics of the elements are included in the bill of materials (BOM). The *criticality* of the elements and a *make, buy or make and buy* decision is added to the BOM alongside with the standard entries (hierarchical level, part number (PN), revision, description, quantity and unit).

Mechanism of identification of criticality of elements

To identify the criticality, in the first instance it will be assessed if the respective element has a potential to be critical for the development of a national product. For this purpose, the questionnaire in Table 12.2 was developed, based on the TRA Deskbook from DOD [24]. This questionnaire defines (A) if the element is relevant for the product, and (B) if this element requires development of technology. A *potentially critical element* is identified only if the respective element obtains at least one 'yes' in both categories (A) and (B).

In the second instance, the *potentially critical elements* are used to determine their criticality. Therefore, the flow chart in Fig. 12.10 was developed which is based on the InsightTec tool from CGEE. The input requires the following information of an element: manufacturer, manufacturing country, majority shareholder of manufacturing company, and export restrictions from

1. respective national institutions, e.g. EAR, ITAR and OFAC for US goods or Federal Office for Economic Affairs and Export Control (BAFA) for German goods, and
2. multilateral export control regimes, e.g. MTCR or the European Council Regulation No. 428/2009 [26].

Four different categories of criticality can be obtained from the evaluation:

Table 12.2 Identification of potentially critical elements, after United States [24] and CGEE [25]

Questionnaire for identification of potentially critical elements		Yes	No
Analysis of relevance	(A.1) The system to be developed depends on this critical element in order to achieve the operational requirements?		
	(A.2) The present limitations of understanding of this element may cause risk in the project time schedule or may introduce risk due to excessive cost?		
	(A.3) Constrains/restrictions for acquisition are existent or foreseeable?		
Development of technology	(B.1) This element is innovative in Brazil?		
	(B.2) This element is the finding of a modified technology?		
	(B.3) This element is used in a different environment than it was originally projected for?		

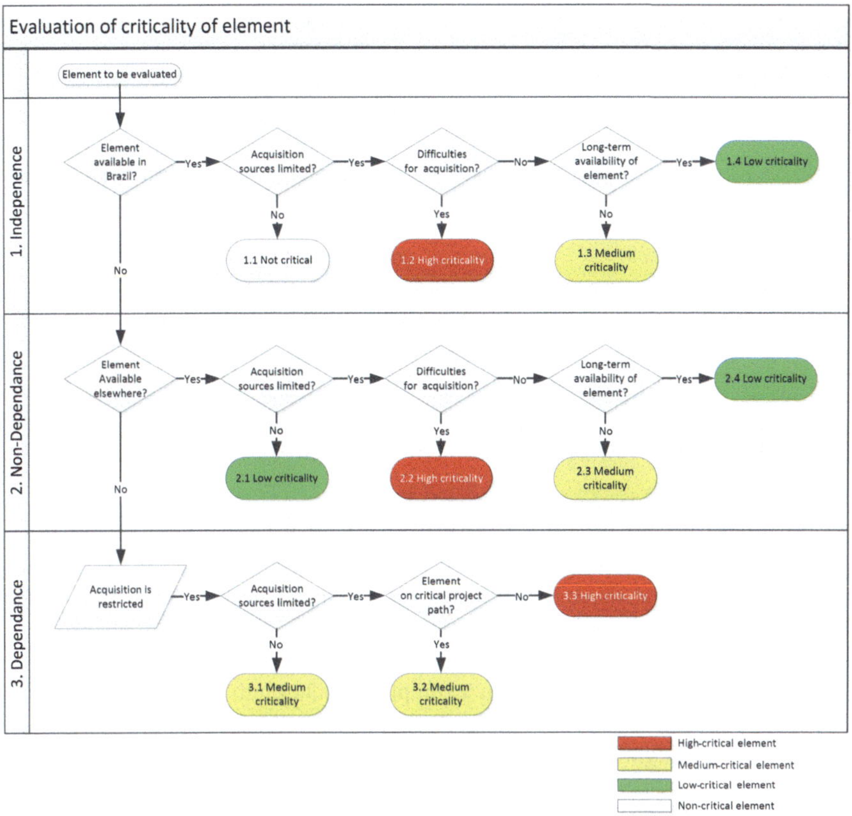

Fig. 12.10 Determination of criticality of elements, after European Space Agency [16] and CGEE [25]

- Non-critical elements: No short or long term restrictions for acquisition or production of element in Brazil; sufficient alternatives available.
- Low-critical elements: Longterm availability for a specific element with limited acquisition resources in Brazil or unlimited acquisition resources out of Brazil assured.
- Medium-critical elements: No long term availability assured and uncertainty of future acquisitions or unlimited acquisition resources for a restricted element in foreign country. Furthermore, an element on critical project path may be classified as medium-critical element.
- High-critical elements: Restricted access or difficulties in acquisition and availability for identified element.

The availability of elements is characterized by three different stages, namely:

1. Independence—The required technology is/was developed and the element is produced in Brazil,

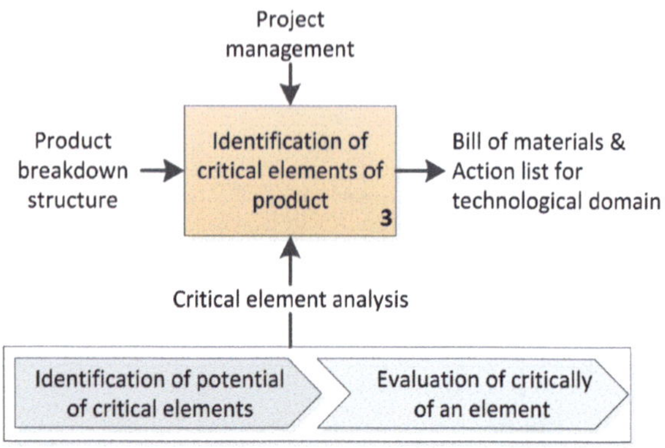

Fig. 12.11 TNF process of 3rd step: identification of critical elements of product

2. Non-dependence—Brazil has free, unrestricted access to the element and its technology, and
3. Dependence—Brazil has restricted access for acquisition of the element.

Annotation: The definitions used herein have been adapted from EC-ESA-EDA workshops on critical space technologies for European strategic non-dependence [27].

The activity of identification of critical elements is depicted in Fig. 12.11.

The input requires breaking down a complex product into manageable elements through a PBS. The output is a completely filed out BOM based on this PBS, which determines the configuration and the key characteristics of the elements. The respective criticality of elements leads to the following means for further handling:

- High-critical elements require the generation of an action list. This action list is case specific and aims at developing technological domain, utilizing alternatives or circumventing the use of that element. This may be accomplished by a morphological box for example. The action list shall be traceable and include a time envelope for troubleshooting.
- Medium-critical elements require careful observation of availability and if possible potential alternatives.
- Low-critical elements may be kept in mind for possible changes of criticality and revision of analysis of criticality.
- Non-critical elements can be disregarded for further analysis.

The execution of this activity, the control and management may be realized by the project coordinator or project manager in charge/designated for project coordination. The processes are intentionally kept simple and pragmatic to assure low time effort. Due to the comprehensive, reproducible processes of this step, a superior, supervisor, funding agency etc. may inspect and audit the decisions.

12.5.3.2 Preparation of Nationalization

The fourth step represented in Fig. 12.12, the preparation of nationalization, involves specific preparations for the product itself and preparations of the setting in order to generate an adequate and prepared environment for the nationalization. The input requires general information about the product/technology itself and the BOM obtained for the 3rd step. The output is a prepared environment with an adequate infrastructure and the creation of organizational structures including a team buildup with mission, commitment, skills and time envelope.

The product specific preparation provides insight into the national and international industry for the product and its elements. Furthermore, a research of possible patents avoids the violation of international laws and the national and international research review gives an overview of the state-of-the-art.

The preparation of the product setting includes the creation of an organizational structure for the Product Development Process (PDP) and the modification and adaptation of the necessary infrastructure.

The preparation of nationalization is an ongoing process during the whole product development. Efforts especially for the modification and adaption of the infrastructure should be proceed until the end of the product development in order to ensure the *status quo* of a prepared environment.

For the present framework, CGEE provided access and support of their InsightData and InsightNet tools which assisted in the preparation for technology development in Brazil. InsightNet is capable to identify available human resources in Brazil working on a specific technology. This tool can be used for team buildup, to foster scientific or technological collaborations, and to assess scientific and technological capabilities in Brazil. The InsightData tool gathers and provides information

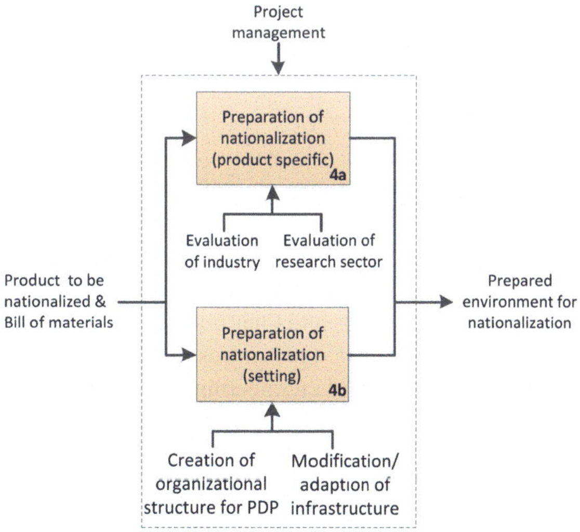

Fig. 12.12 TNF process of 4th step: preparation of nationalization

that allows the assessment of the technological state-of-the-art. It collects, processes and analyzes data from about 4.000 sources, including scientific and technological journals, patent databases, news magazines sites.

(4a) Product specific preparation for nationalization
The product specific preparation gives insights into the national and international industry and research sector of the product and its technologies. Furthermore, interest and capability of the national industry for future industrialization and commercialization of the product or parts of it is evaluated.

(4b) Preparation of the setting for the nationalization/product development
Development projects require a separate organizational structure from the basic organization since the basic organizational pattern may not fit for the required tasks to be done for a successful product development. The organizational structure and team buildup for the product development shall be carried out as a project. A project, according to Andreasen and Hein [28], has the following characteristic properties:

- It runs for a defined period of time.
- It has a defined work force, the project group.
- It is executed within well-defined resource limits.
- It spans an entire organization; team members must work in a different manner than that defined by the basic organization.
- It is interdisciplinary.
- It is characterized by development, and innovative products are expected to be created.
- Since it is normally important for the basic organization, thus, it is controlled and dependent on top management.

The existing infrastructure and facilities of the respective executive organization have to be modified and/or adapted for the product development in order to be prepared for the reverse engineering and subsequently forward engineering of the product.

12.5.3.3 Reverse Engineering of Original Product

Reverse engineering of the original product, the first activity of reengineering is presented in Fig. 12.13. The goal is to analyze and examine the product in order to identify its components and their interrelationships and to obtain the know-how and know-why. This knowledge is necessary in order to continue reengineering with the second activity, forward engineering to redesign a national product in the 6th step.

The outputs of the previous steps provide a technology ready to be nationalized and a prepared environment, which is required as the input for the 5th step. The output of this step is learned knowledge (know-how and know-why) from the technology. This activity is executed by the development team and managed by the project coordinator/manager.

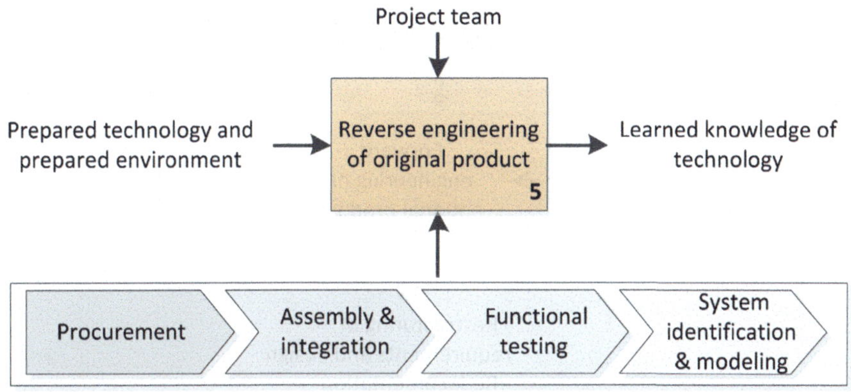

Fig. 12.13 TNF process of 5th step: reverse engineering of original technology

The activities of reverse engineering are case specific, depending on the product to be re-engineered. The following items represent a first idea.

1. Procurement—The original product or at least a large part of it has to be acquired and imported from the transferor according to the bi-national contracts.
2. Assembly and integration—The original product has to be assembled and integrated in the national laboratories with the adequate equipments and infrastructure.
3. Functional and performance testing—Functional testing where appropriate with technical staff and/or support from transferor has to be accomplished in order to train the team and assure save handling and use and to understand form and functionality of components and their interrelationships.
4. System identification and modeling.

Note: The reverse engineering has to be documented adequately to facilitate the forward engineering of a national product. Besides the mentioned examples of methods and processes for technology transfer, training from the transferor could be included into the reverse engineering.

12.5.3.4 Forward Engineering of National Product

The second activity of reengineering, the forward engineering to achieve a national product is depicted in Fig. 12.14. This step leads to the development of a product with national domain that has a minimum risk of being dependent on export bans or unavailability of components. Furthermore, the knowledge of the original technology and the identification of its strengths and weaknesses gives the opportunity for product enhancements, leading to innovation and added value.

The input for the forward engineering of a national product is the documentation obtained from the transferor together with the generated documentation of the

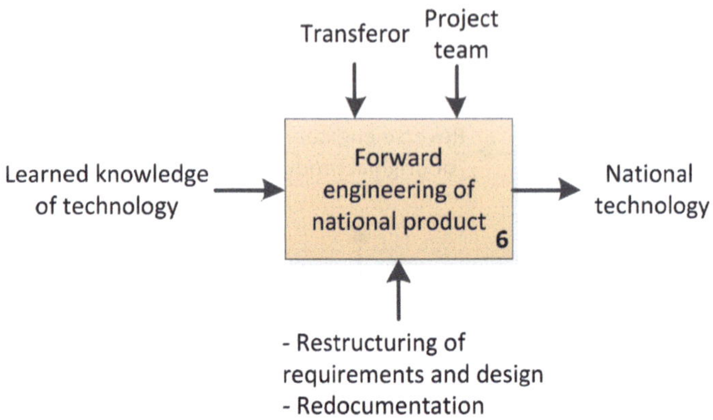

Fig. 12.14 TNF process of 6th step: adaptation for new environment

reverse engineering. The output of the forward engineering is national technology to be transformed into a product. The forward engineering is carried out by the project team, and if possible, supported by the transferor. Depending on the established relationship with the transferor, iterative feedback processes can be established, creating a win-win situation. Control, in terms of intellectual property protection, depending on the arrangements, may be executed by the transferor [29].

The reverse engineering from the 5th step brought knowledge of the design and requirements of the original technology. For the development of a national product, the original requirements get restructured, thus changed, modified or maintained and a new specification is obtained. With this new specification the designs get reviewed and, if necessary, changed or modified for the national product. Redocumentation assures a consistent technical documentation. Therefore, the available documentation of the transferor is reviewed, adapted, changed or modified and converted to the national layout, norms and standards.

The performance of DfAutonomy can be measured during the forward engineering by performance evaluation, distinguished by classification of qualitative and quantitative criteria. At a certain level of product maturity (e.g. preliminary design review), this analysis can be conducted. The baseline configuration of the original product is being compared with the forward engineered national product. A non-exhaustive list of criteria is given as starting point:

Quantitative performance criteria

1. Number of elements which require end-user certificate or military elements.
2. Number of elements that could suffer from import restrictions.
3. Make, buy or make and buy decisions of elements.
4. Ratio of national and imported elements.
5. Lead time reduction due to acquisition of national elements.
6. Cost reduction due to acquisition of national elements (avoiding import taxes and lower labor cost).
7. Test results.

Qualitative performance criteria

1. National product corresponds adequately to adapted requirements.
2. Ability of reconfiguration in case of unavailability of elements.
3. Modularity.

12.5.4 Maturation of Product

The seventh step, the maturation of the national product, is depicted in Fig. 12.15. It is introduced in order to monitor the technology progress throughout the product development. The input for this step is the level of maturity after the forward engineering of the national product (outcome of 6th step). The maturity of a product/technology may be evaluated by ISO 16290:2013(en), Space systems—Definition of the Technology Readiness Levels and their criteria of assessment [30]. Evaluation of the TRL is used to quantify the technology maturity status of an element that is intended to be used in a mission. The assessment ranges from TRL 1, the report and observation of basic principles, up to TRL 9, the status of "flight proven" elements through successful mission operations. The outcome of this step is a product/technology with a higher TRL resulting in an increased interest and feasibility of industrialization and commercialization of the product by the national industry.

The maturation of the product is introduced due to the special political and industrial environment of the Brazilian space sector with its executive organizations DCTA/IAE and INPE and the Brazilian industry as their supplier.

The mechanisms/processes for TRL raising are case specific. In general, the maturation of the product may be accomplished at the executive organization, however, time, cost and risk of maturation can be lowered by the introduction of the so-called triple helix concept [31], a triadic relationship between executive organizations, university and industry, depicted in Fig. 12.16. It is based on the hybridization of all

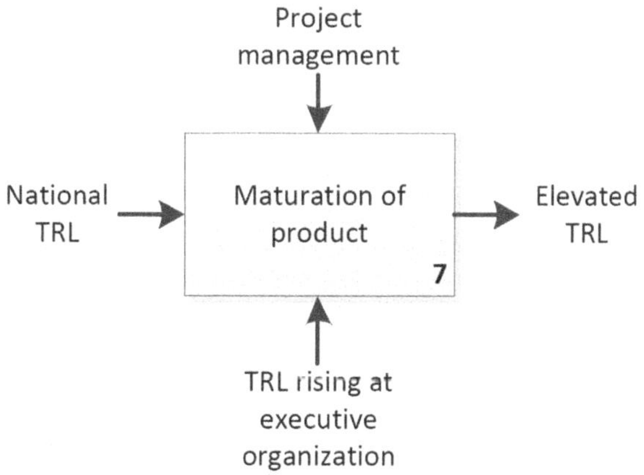

Fig. 12.15 TNF process of 7th step: maturation of product

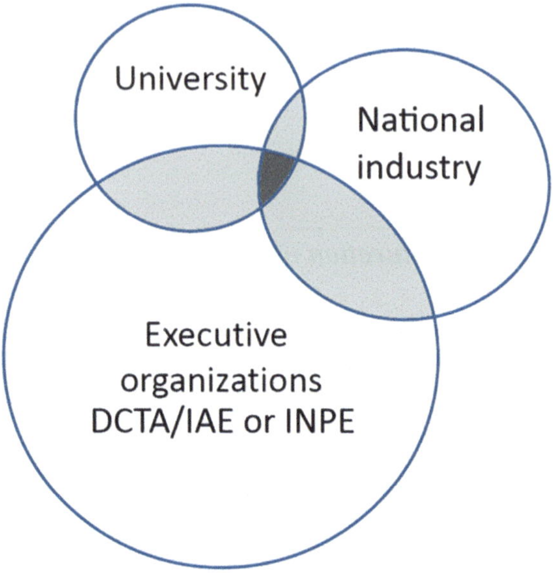

Fig. 12.16 Triple helix concept for maturation of product

institutional spheres and aims at achieving a higher potential for innovation and economic development in a knowledge society. The DCTA campus with university (ITA), the executive organizations (DCTA/IAE and close-by INPE) and the private sector concentrated in the region of São José dos Campos, including the technology parks seem predestined for such an exploration of complex innovation dynamics. The development may use the advantageous combination of infrastructure, expertize and know-how at the executive organizations, basic and applied sciences of universities and the speed and flexibility of the industry. The control and management of the work packages of a defined work breakdown structure for each institutional sphere may be accomplished by the project management, allocated within the executive organization. The legal issues of the intersections of institutions depicted in Fig. 12.16 are not part of the current work and may be considered for future work.

12.5.5 Transfer to National Industry

Once the technology is mature enough for production and commercialization, the technology should be transferred into the industry, the 8th step. This step is essential for

1. to release capacities used for production within the executive organizations and enable the focus on development,
2. to strengthen the national industry by fostering innovation and spin-offs, and
3. to transform and diversify the value chain of space technology in Brazil.

The technological innovation centers (DCTA/NIT and INPE/NIT) of the executive organizations are responsible for generation of the institutional politics for innovation and have, amongst other, the attribute to promote and support the institutional capacity for technological innovation, intellectual property and transfer of technology [32].

12.6 Follow-Up of TNF: Securing of Technological Domain

A follow-up of the Technology Nationalization Framework is required in order create a sustainable space program by maintaining technological domain on strategic technologies already nationalized or existent in Brazil. The follow-up can be based on the product breakdown structures of the different segments of the Brazilian space program (see 1st step) and may be characterized by

1. an iterative identification of strategic technological domain in Brazil,
2. the support of strategic companies by maintaining technical staff trained and avoid bankruptcy or selling abroad of these companies. In case of no or too low demand contracts are necessary for advancements/miniaturization/enhancement/optimization of technology, and
3. legal provisions to assure continuous production of strategic technologies of the Brazilian space sector within the country, with an analog objective as Brazilian law No. 12.598 for the military sector [33].

Brazilian law No. 12.598 defines Strategic Products of Defense (PED) and requires as strategic interest for national defense, amongst others, for national industry

- to maintain the headquarter, administration and industrial establishments within the country,
- to assure that the foreign owners, shareholders etc. may not exceed 2/3 of the votes in a general assembly, and
- to assure continuous productivity within the country.

An example of not sustainable development and maintenance of technology is the production of Hydroxyl-terminated polybutadiene (HTPB), which is used as binder for rocket propellant and was developed in the 1970s by a cooperation of DCTA and Petrobras in Brazil. Petrobras initiated the industrial production in 1982 within the Petroflex unit. With its privatization in 1994 the HTPB plant was acquired by Braskem. In 2008, the production was discontinued from the new owner, the German company Lanxess [34]. In 2014, Avibras Indústria Aeroespacial S.A. received INOVA AERODEFESA subventions of more than 6 million Real for the development of a new HTPB production in Brazil [35]. This example shows the necessity of a sustainable strategy with short, medium and long-term vision for the follow-up of strategic technologies [36].

12.7 Implementation and Integration Issues

The involved countries/institutions/companies have to have bi-lateral cooperation that permit the transfer of material and non-material, components, knowledge, skills, organization, and values from the transferor to the transferee (2nd step of TNF). Since the product is in development or already exists in the country of origin, the phases of conception, mission analysis and feasibility can be reduced in terms of time and cost and the development risk can be decreased. However, several barriers might hinder the implementation of the Technology Nationalization Framework and the integration into a product development process.

Barriers for TNF based on literature and application of TNF on Pilot Project:

- Restrictions and trade-related barriers on transferable technology may impede technology transferor to sign bi-national contracts (Filter of 2nd step).
- Communication and coordination issues between executive organizations and industry: Frequent design changes; Limited visibility; Lack of information and input; and Little knowledge and access for industry to technologies at executive organizations.
  → Strategic partnerships and collaborations required
- Technology maturation issue—Importance of level of technology when industry is brought in:
  - Too early: "TRL valley of death" for product in industry, not enough technological capabilities or insufficient infrastructure for development. This may result in not fully understood tasks being accomplished incompletely, inaccurately or faulty, reducing the added value.
  - Too late: Benefits from industry like flexibility and speed for resolving problems cannot be not utilized adequately. Late industrial involvement may not adequately increase and foster the formation and training of the industry, creating a gap between research and production in the product development life cycle. Potential for innovation and design for manufacturing and assembly are limited for late transfer [37].

  Both scenarios, too early and too late involvement of industry in the development, may result in program stretch-out or abandonment, cost-overruns or not achieved program goals. Careful evaluation of each case is required. In Fig. 12.17 the risk

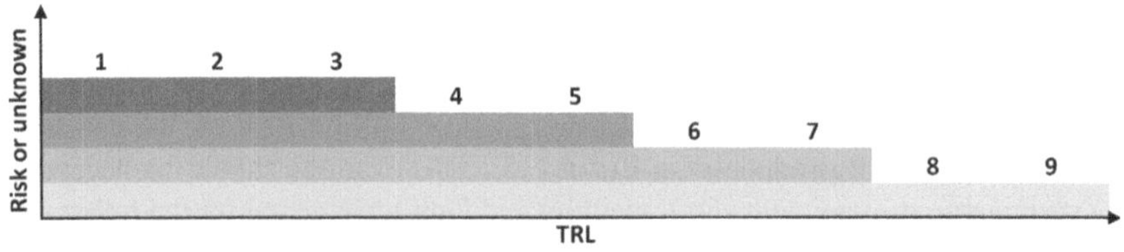

Fig. 12.17 Risk and unknown versus TRL in the transfer of technology to national industry

and unknown outcome of a transfer from the executive organizations to the industry is qualitatively depicted.

- Intellectual property: Property-related negotiations and uncertainties may be one key barrier for technology transfer [32].
- Funding issues: Lack and instability of funding may hinder continuous, efficient product development.
- Qualified human resources issue: Retirements and relocation (especially for military personal) result in loss of expertise in the executive organizations [38]. The lack of projects may lead to loss of technology fundamentals amongst new engineers.
- Cultural Barriers: Research institutes have culture where the focus is on invention and novelty. There are few incentives to engage in technology transfer. Lack of interest in technology transfer on the part of research institute and their technical staff might exist. Motivations such as self realization, completing a task and benefiting research, described by Kremic [39], might not be sufficient. Individual motives must be determined case by case.
- Limited willing partners: No country wants to be philanthropic with space technology because of the investments involved, national security and fear of encouraging competition to its domestic space industry [40].
- Incentives: No guarantee of sales of product combined with a financial and development risk of high technology innovative products in the space sector lead to missing inducement and consequently involvement of industry.
- Acquisition of the product from the company by the executive organizations: Difficulties for purchase of specific product from specific company that cooperated in development.
  → Application of Brazilian Law No. 13.243 [41], instead of utilization of Brazilian Law No. 8.666 [42].
- Poor leadership, retrogressive policies and lack of political support: A solid political and policy backing is required for a successful space sector [40].
- Uncontrollable barriers: Natural or human-made disasters; Stock market crashes; New disruptive technology insertion; National or global market shifts due to trade agreements or disputes between nations; and Cancellation of programs or projects [43].

According to Bach et al. [44], three main characteristics impact the process of technology transfer:

1. The nature of the technologies with their degree of maturity, their degree of diversity and the extent to which they are generic or specific,
2. The nature of the R&D network of participants with the degree of mutual trust and the existence of absorptive capabilities [45], and
3. The nature of the organizational structure of the participants.

12.8 Application of TNF on Pilot Project

Brazilian space activities started in the 1960s with a sounding rocket program, and in 1993 the first Brazilian satellite was inserted into low Earth orbit with a foreign launch vehicle. Unfortunately, attempts to gain autonomous access to space still failed. Key technologies for launch vehicles include the complex guidance and control systems which are most important for proliferation prevention [46]. The challenge of the pilot project for the TNF is to develop the technology for a national TVC system. The technology originates from the German Aerospace Center (DLR) and is transferred via the Brazilian Institute of Aeronautics and Space (DCTA/IAE) into the Brazilian industry which is based on the ongoing German-Brazilian cooperation in the context of aeronautics and space that was initiated in 1971 [47].

The different ways of knowledge acquisition for obtaining technology are modeled in Sect. 12.2, Fig. 12.4. As an example for the knowledge acquisition, the case of TVC actuators for sounding rockets and launch vehicles at DCTA/IAE is illustrated in Fig. 12.18. It shows the three characteristic curves, where the blue curve represents the acquisition of technology, the red curve represents the learning curve of development ab initio and the green curve represents the knowledge acquisition via the Technology Nationalization Framework.

The knowledge of the TVC actuator technology can be composed by three stages:

1. Operations—The knowledge of the technology is sufficient for operating the actuators,
2. Design, manufacturing, identification and testing—The knowledge and capability to design, manufacture and to accomplish system identification and performance testing of own actuators, and
3. Qualification and flight model—The knowledge and ability, to qualify and to produce flight models of actuators.

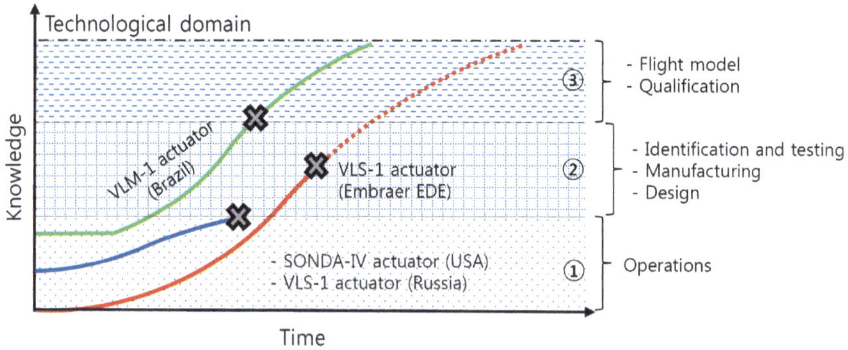

Fig. 12.18 Knowledge acquisition over time of TVC actuators at DCTA/IAE due to acquisition of technology (blue curve), proper learning (red curve) and technology nationalization framework (green curve); the black crosses represent the status; three different stages of knowledge are introduced until reaching technological domain

Fig. 12.19 Left: original actuator from technology transferor; middle: reverse engineering of original technology; right: forward engineered actuator EM

Figure 12.9 presented the approach of reengineering of the DfAutonomy tool, which is composed of Reverse- and Forward Engineering. In Fig. 12.19, the original actuator EM from the technology transferor is depicted on the left side, in the middle reverse engineered CAD model of original technology is plotted and the picture on the right side shows the national forward engineered actuator EM on a testbench. A qualification model is currently under development.

12.9 Discussion

According to Wouters and Hansen [48], the term dependence is a multifaceted phenomenon with possible meanings of

1. determined or conditioned by another,
2. relying on another for support, and
3. subject to another's jurisdiction.

In order to escape and/or lessen technological dependence, the Brazilian experience in the space sector of the TNF may be extrapolated and adapted to other sectors and countries.

12.9.1 Aerospace Sector

Space faring nations like United States, Russia, Japan or China have developed a high level of priority for full independence of technologies. Each country made its own policies. The United States for example have strict procurement rules for government satellites and launch vehicles for preventing dependence on non-US suppliers. Japan only undertakes a space mission if it can be assured that it is able to launch the spacecraft. Japanese governmental satellites have never been launched by foreign launch service providers. The complex programmatic coordination and harmonization of the various technology programs and roadmaps have first been implemented in the European Space Technology Master Plan (ESTMP) [16], which comprises an international organization with 22 Member States.

Autonomic access to space requires to define areas of interest for strategic technologies and means how to handle them. For several of the about 70 nations with national space agencies, a strategy like the Technology Nationalization Framework may be an interesting foundation to build and enhance local capacities. The TNF may be adapted, personalized or *nationalized* for their own cultural, political and economic environment.

12.9.2 Other Industrial Sectors

Looking at other industrial sectors than the space industry, the identification and handling of strategic technologies may be relevant to maintain and strengthen the economic power and sustainable growth of a country. Not only defense and security sectors and the high-tech industry, but also strategic sectors for a specific country may be included. Especially resource-rich developing and emerging countries may benefit from the ideas of the TNF to nationalize strategic technologies. The exploitation of these resources often requires technologies that are not available on the domestic market or subsequently exploitation is performed by multi-national corporations. Existing policies and requirements for local content may benefit by a systematic treatment of strategic technologies.

12.9.3 Use of TNF and Need for Adaptation in Other Countries/Industrial Sectors

The common ground and possible changes for the adaptation of the TNF in other countries or industrial sectors are pointed out in Table 12.3.

12.9.4 Limitations of the Research

The different steps of the TNF application were executed only by the authors on one pilot project. What if other persons would do it? Would they follow the same path and come to the same conclusions? Is the abstraction adequately accomplished? In some cases, the required control of the execution of activities could not be realized, e.g. the verification of the evaluation of competencies in the second step, or the verification of the TRL evaluation in the seventh step. Various case studies are required for validation and continuous improvement of the TNF and DfAutonomy. Is the level of abstraction of the TNF sufficient/general enough for future case studies in the aerospace sector or even in other industrial sectors and countries?

Table 12.3 Common ground and need for adaption of different steps of TNF

1st step—identification of strategic technologies to be nationalized
The first step is required in order to define the technologies necessary to obtain in order to assure non-dependency and feasibility of a Brazilian space program. The technologies can be segmented within the different space sectors of launch vehicles, space segment, applications, and ground segment. Stakeholders for the identification of strategic technologies (aerospace and other industrial sectors) have to be adapted to the country's political and economic environment
2nd step—evaluation of feasibility for nationalization
The second step is necessary to evaluate the feasibility to nationalize a certain technology, defined in the first step. This step is fundamental for the decision if the task of nationalization should continue or not. With this step, unnecessary waste of resources can be prevented in an early phase of nationalization. Furthermore, the transition from a desired technology to a specified product to be nationalized is achieved which includes the definition of a technology transferor. The idea of evaluation of competencies (technological capabilities, infrastructure and TRL) can be adopted or adapted. However, the methods and mechanisms of technology transfer may vary from case to case
3rd step—identification of critical elements
The identification and classification of critical elements and subsequent action lists to obtain national domain are essential procedures to quantify and qualify the technological dependence. Regardless of the sector or the technology, this step may be applied with little alterations to future frameworks and strategies
4th step—preparation of nationalization
The national industry and research sector are analyzed and the environment for a nationalization is prepared. This step is indispensable for successful reengineering, since an inappropriate environment may cause unnecessary delays and/or cost overruns. These preparations have to be considered on case-to-case basis
5th step—reverse engineering of original product
Independent of the industrial sector, reverse engineering is inherently case specific and has to be adapted to every single technology, product, system or subsystem
6th step—forward engineering of national product
The TNF is all about the development of national products and its transfer from a transferor. Consequently, the forward engineering of a product which can be designed, produced and operated within a country is the ultimate and common goal of any *Technology Nationalization Framework*
7th step—maturation of product
This step is specific for the Brazilian Space Sector, which has governmental research centers and an infant industry that predominantly supplies this public sector. Little export of space products exists in Brazil
8th step—transfer to national industry
This step is specific for the Brazilian Space Sector, which has governmental research centers and an infant industry that predominantly supplies this public sector. Little export of space products exists in Brazil

12.10 Conclusion and Outlook

This Chapter contributes to the dissemination of systems engineering best practices and Integrated Product Development approaches with the Brazilian perspective for technology transfer and nationalization of strategic technologies. Guidelines for developing this Technology Nationalization Framework brought the focus of attention on the right balance between functionality and operability which was accomplished by

1. a balance between pragmatism and formality,
2. a balance between accuracy and data requirement, and
3. a balance between focus and flexibility.

The Technology Nationalization Framework was created and compiled based upon a real-world problem. The pilot project of an aerospace actuation system verified the development of a theoretical approach for the nationalization and subsequent industrialization of high technology products for the Brazilian space sector. This novel framework embraces the whole cycle of technology nationalization, starting from its identification and ending with its industrialization and commercialization. It transfers a desired technology into a robust commercial product available in Brazil. The objective of this Technology Nationalization Framework is to assure that certain lacking strategic technologies only available abroad are converted into a national product that can be designed, produced and operated in Brazil for a defined period of time at a minimum risk of being dependent on export bans or unavailability of components. The TNF fosters innovation and competitiveness in Brazil and ensures non-dependence of strategic technologies by a balance between completely domestic development and blind implementation of technology transfer.

In the pilot project, a *right first time* could be accomplished so far. However, like any manufactured product, the TNF and its DfAutonomy are to be improved. Verification is not a step that can be skipped and the framework has to be tested on a number of case studies. The following questions should be addressed:

- What are the criteria for verification and validation?
- How to conduct the verification and validation?
- How to improve the framework and its tools?

With a wide spectrum of gained applications and expertise, TNF manuals and/or workbooks may be compiled and disseminated. The TNF and Design for Autonomy tool were developed with the focus on the space sector in Brazil. Certainly, the framework with its processes and tools may be adapted and adopted to other sectors/production industries in Brazil. Thinking in even broader terms, the framework could also serve as cause for though and example for other newly industrializing and developing countries.

References

1. Pessôa MVP, Trabasso LG (2017) The lean product design and development journey: a practical view. Springer, Berlin
2. BKCASE Editorial Board (2017) The guide to the systems engineering body of knowledge (SEBoK), v. 1.8. R.D. Adcock (EIC), Hoboken, NJ
3. International Electrotechnical Commission International Organization for Standardization, Institute of Electrical, and Electronic Engineers. Iso/iec/ieee 15288: Systems and software engineering—system life cycle processes. Technical report, 2015
4. International Council on Systems Engineering (2015) Systems engineering handbook v4. Technical report, 4th edn. Wiley, Hoboken
5. United States (2001) Systems engineering fundamentals. Technical report, United States Department of Defense
6. Mar BW (1997) Back to basics again: a scientific definition of systems engineering. INCOSE Int Symp 7(1):309–316
7. Wognum N, Bil C, Elgh F, Peruzzini M, Stjepandić J, Verhagen WJC (2019) Transdisciplinary systems engineering: implications, challenges and research agenda. Int J Agile Syst Manag 12(1):58–89
8. Sobolewski M (2017) Amorphous transdisciplinary service systems. Int J Agile Syst Manag 10(2):93–115
9. Stevens R, Brook P, Jackson K, Arnold S (1998) System Engineering coping with complexity, 1st edn. Prentice Hall
10. Biahmou A (2015) Systems engineering. In: Stjepandic J, Wognum M, Verhagen JC (eds) Concurrent engineering in the 21st century: foundations, developments and challenges. Springer International Publishing, Switzerland, pp 221–254
11. Chikofsky EJ, Cross JH (1990) Reverse engineering and design recovery: a taxonomy. Softw IEEE 7(1):13–17
12. Huang GQ (1996) Design for X—concurrent engineering imperatives. Chapman and Hall, Netherlands
13. Huang GQ, Mak KL (1997) The DFX shell: a generic framework for developing design for X tools. Robot Comput-Integr Manuf 13(3):271–280
14. Kuo T-C, Huang SH, Zhang H-C (2001) Design for manufacture and design for 'X': concepts, applications, and perspectives. Comput Ind Eng 41(3):241–260
15. Knowledge Based Systems. IDEF0 Function Modeling Method, 2015. Accessed 15 Oct 2015
16. European Space Agency (2015) European space technology master plan, 12th edn. Technical report, ESA-ESTEC, June 2015
17. Petroni G, Venturini K, Verbano C, Cantarello S (2009) Discovering the basic strategic orientation of big space agencies. Space Policy 25:45–62
18. World Intellectual Property Organization (2016) Accessed 18 Feb 2016
19. CGEE (2015) Inteligência Tecnológica: Observatório de Tecnologias Espaciais. Brasília, Brazil, December, 3rd 2015. Center for Strategic Studies and Management in Science, Technology and Innovation
20. Wekerle T, da Costa LEL, Trabasso LG (2016) Design for autonomy: an integrated product development tool for nationalization due to reengineering of complex products for the Brazilian space sector. Adv Transdiscipl Eng 4:632–641
21. Wekerle T, Trabasso LG, da Costa LEVL, Villela T, Brandão A, Leonardi R (2017) Design for autonomy: integrating technology transfer into product development process. J Ind Integr Manag 2(1):1750004
22. Tonnellier E, Terrien O (2012) PBS: a major enabler for systems engineering. INCOSE Int Symp 23(1):1–15
23. ANSI/EIA (2003) Processes for engineering a system. Technical report, American National Standards Institute (ANSI)/Electronic Industries Association (EIA), Philadelphia, PA, USA, 2003. ANSI/EIA 632-2003

24. Department of Defense (2009) Technology readiness assessment (TRA) deskbook. Technical report, Washington, DC, July 2009
25. CGEE (2014) unpublished. Center for Strategic Studies and Management in Science, Technology and Innovation
26. Council of the European Union (2009) Council regulation No. 428/2009 of 5 May 2009 setting up a community regime for the control of exports, transfer, brokering and transit of dual-use items. Off J Eur Union
27. European Space Agency (2015) Critical space technologies for European strategic non-dependence list of urgent actions for 2015/2017 V1.16. Technical report, ESA
28. Andreasen MM, Hein L (1987) Integrated product development. IFS-Springer, Berlin
29. Holland M, Nigischer C, Stjepandic J (2017) Copyright protection in additive manufacturing with blockchain approach. Adv Transdiscipl Eng 5(2017):914–921
30. International Organization for Standardization (2013) ISO 16290:2013
31. Triple Helix Research Group (2015) The triple helix concept, 2015. Accessed 20 Feb 2016
32. Biahmou A, Stjepandic J (2016) Towards agile enterprise rights management in engineering collaboration. Int J Agile Syst Manag 9:302–325
33. Presidência da República. LEI N° 12.598, 21 Mar 2012. Accessed 23 Feb 2016
34. Silveira V (2013) AEQ concentra produção em nova fábrica, 15 Oct 2013. Accessed 23 Feb 2016
35. Respondovesk W (2014) Resultados PSC Inova Aerodefesa. Reunião COMDEFESA—10/03/2014, FINEP
36. De Souza V, Borsato M (2016) Sustainable design and its interfaces: an overview. Int J Agile Syst Manag 9(3):183–211
37. Zhong RY, Ge W (2018) Internet of things enabled manufacturing: a review. Int J Agile Syst Manag 11(2):126–154
38. de Vasconcellosa RR, Amato Neto J (2012) Critical factors in technology transfer in the space sector: a case study of the partnership programs between the space agencies from Brazil ('AEB') and the USA ('NASA'). Produção 22(4):851–864
39. Kremic T (2003) Technology transfer: a contextual approach. J Technol Transf 28:149–158
40. Waswa PMB, Juma C (2012) Establishing a space sector for sustainable development in Kenya. Int J Technol Glob 6(1/2):152–169
41. Presidência da República. LEI N° 13.243, 11 Jan 2016. Accessed 23 Feb 2016
42. Presidência da República. LEI N° 8.666, 21 June 1993. Accessed 23 Feb 2016
43. Krishen K (2011) Multiple aspects of space technology transfer. IETE Tech Rev 28(3):195–206
44. Bach L, Cohendet P, Schenk E (2002) Technological transfers from the European space programs: a dynamic view and comparison with other R&D projects. J Technol Transf 27:321–338
45. Rauch E, Dallasega P, Matt DT (2017) Distributed manufacturing network models of smart and agile mini-factories. Int J Agile Syst Manag 10(3/4):185–205
46. Schmucker RH, Schiller M (2015) Raketenbedrohung 2.0—Technische und Politische Grundlagen. Mittler und Sohn, Hamburg
47. German Aerospace Center (2011) 40 Jahre deutsch-brasilianische Zusammenarbeit in der Luft- und Raumfahrt—40 anos de Cooperação teuto-brasileira no Transporte Aéreo e Espacial. Linder Höhe, 51147 Köln, Germany, Oct 2011
48. Wouters J, Hansen R (2015) Strategic autonomy in EU space policy: a conceptual and practical exploration. In: Al-Ekabi C (ed) European autonomy in space. Springer International Publishing Switzerland, pp 49–61

Chapter 13
Systems Engineering for Sustainable Mobility

Alain Biahmou

Abstract Nowadays, sustainability has established itself in the automotive industry and has evolved to an indispensable part of it. In contrast with its initial understanding as ecological improvement during development and production of vehicles, it has emerged to an advanced concept that considers much more, for instance the interaction of vehicles with the superordinate system they are included in. Therefore, not only the reduction of the pollution as well as of the resource consumption, but also the impact on the societal, economic and environmental development are of great importance. The current product development in many companies is still characterized by the fact that different disciplines create several partial models of the same product and provide many information only in documents. Periodic synchronizations of common parameters and models are performed. Information related to sustainability even when it exists is not consistent and not represented in models, which can be used for synchronization points. Therefore sustainability is often not really taken into account along the product life cycle. In order to master the complexity of smart products, which arises from customer behavior and requirements, but also from legal requirements related to sustainability, a proposal is made for Systems Engineering to integrate sustainability to a greater extent. Based on the main research directions over sustainability, such as innovative design concepts including alternative propulsions for less pollution, the safety and driver assistance for resource efficiency and life protection, the mastery of networked vehicles for instance to control the interaction of car with its superordinate system, adapted and even new methods as well as processes are needed in order to link Systems Engineering with Sustainability. This paper presents proposals of product development processes that take Systems Engineering methods into account as well as sustainable mobility. The prerequisites to realize such a product development process are described, whereby the whole product development cycle from the product concept down to disposal is taken into account.

Keywords Sustainability · Systems engineering · Model-based-enterprise · Digitalization

A. Biahmou (✉)
EDAG Engineering GmbH, Reesbergstr. 1, 36039 Fulda, Germany
e-mail: alain.biahmou@edag.de

J. Stjepandić et al. (eds.), *Systems Engineering in Research and Industrial Practice*,
https://doi.org/10.1007/978-3-030-33312-6_13

13.1 Introduction

The big cities are growing more and more and it is expected that the number of residents will continuously increase in those cities in the next years and decades. Thus, driving cars today in those cities means to be facing many traffic jams that in turn can increase power consumption as well as driving stress [1]. Even though public transport often is available, many people feel restricted due to the driving routes and departure times.

Furthermore, a high number of vehicles are available on roads, often just transporting a single passenger and contributing therefore to increasing delays due to congestions. Those congestions often can lead not only to a frustration of passengers but also to accidents [2, 3]. Furthermore, delays due to traffic congestions can lead also to a loss for economy as the people who are waiting in the traffic cannot realize their economic potential. Another frustration associated with car driving in big cities is the interminable search for parking and accidents due to human mistakes, in fact about 90% of all accidents [4]. Besides, the amount of time wasted in traffic is a loss of personal time that could be spent with family, with further education or even with leisure activities, just to name a few.

From an environmental point of view, the very high amount of vehicles that are available in dozens of cities has contributed to increase pollution, as many factories do. Even though cars are not the unique source of pollutants, it is important to note that pollution can significantly deteriorate the quality of life. Therefore, driving bans have been introduced in cities in Germany in order to reduce nitrogen oxide output [5]. In China for instance, people sometimes have to bear masks for protecting against smog [6].

All this calls for approaches for tackling the challenges related to sustainability, urbanization and autonomous driving.

This chapter presents technological approaches for enhancing sustainability, based on a holistic, transdisciplinary approach that considers the impact of transportation on environmental and socio-economic developments [7]. Important challenges of the automobile industry are presented in Sect. 13.2. In that section, the impact of urbanization is highlighted and some enablers of the Digitalization for the automotive are discussed: Principles and technologies for Driver Assistance Systems, connected vehicles, safety and security. The integration of those technologies to enhance sustainability is investigated. In fact, sustainability should be aligned with the important challenges of the automobile industry and considered as factor of equivalent worth during product creation. Moreover, proposals are made for linking Systems Engineering with sustainability. Those proposals take further parameters into account such as the entire life cycle of vehicles, Industry 4.0, Additive Manufacturing, Lifecycle Assessment, to name a few.

13.1.1 Sustainability

Sustainability can be defined as a set of environmental, economic and social conditions in which all society members have the capacity and opportunity to maintain and improve their quality of life indefinitely without degrading the quantity, quality or the availability of natural, economic and social resources [8]. While some approaches are giving the same priorities to society, environment and economy, some others assign the most importance to the planet, then to the society and last to the profit that is strived for by economy (see Fig. 13.1). From the point of view of the automotive industry, sustainability has become a purchase criterion for car buyers, who more and more have developed an ecological awareness. Furthermore, taking sustainability into account is also a criteria for investors who rely on sustainability rankings in order to decide whether they should invest in stock share of a company or not.

While the automotive industry has understood environment protection very well and considered it in the past, the direct preservation of life and health of persons and other living organisms is also a key factor for sustainability nowadays. Thus, the interaction of vehicles with the infrastructure as well as with all other actors that are involved in road traffic is one of the major topics related to sustainability in the next years.

To reach the objectives of a sustainable mobility such as zero emission resp. low carbon and nitrogen emission, additional efforts are needed in order to enhance the approaches that have been presented and discussed in the last years. In fact, sustainability consists of integrating environmental, social and economic considerations in business strategy.

Even though the early concepts of the automotive industry that where related to sustainability were focused just on the fuel consumption without taking the life cycle of fuel into account, more and more the vehicle is considered as a sub-system that should support sustainability regarding global issues of societal, economic and environmental development.

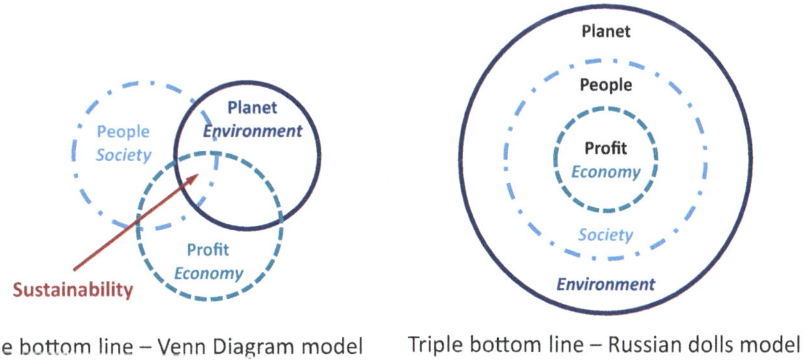

Fig. 13.1 Graphical representation of sustainability [8]

Therefore, enhancing sustainability ought to be apprehended as a holistic approach that involves the whole product lifecycle of vehicles and propellant, that is, from the product planning through manufacturing down to recycling.

In fact, there are some approaches targeting an improved fuel composition and efficiency in order to realize a better emissions balance, while further works target an optimization of engines in order to reduce fuel consumption. Approaches that bear most expectations deal with alternative drives. Especially the electric vehicle is pointing the way for the next years. However, challenges such as high charge acceptance rate for recharging, regenerative braking and the electric driving range are to be tackled, to name a few [9]. Thus, interim solutions are provided in the meantime, for instance hybrid and plug-in hybrid vehicles that combine the abilities of electric vehicles and combustion engines [10].

Furthermore, there are valuable concepts that consider the vehicle as part of a superordinate system and therefore take the interaction of the vehicle with its superordinate system (V2X) into account. For example, proposals have been provided to use the infrastructure for supporting the introduction and operation of electric vehicles, for instance through increasing the density of battery recharging stations, and through the continuous provisioning of vehicles with electric power based on induction. Moreover, the grid infrastructure can be involved from an economical point of view when it comes to managing electrical power for transportation. A use-case consists of automating the energy supply of electrical vehicles, whereby a car could purchase or sell energy depending on factors such as the actual electricity price, the driving profile of the car owners and the expected traffic flow based on predictive analyses [11]. Such a scenario requires the ability for cars to be connected to different networks. Conversely, the relevant networks (e.g. electricity grid) have to be adapted in order to align with those dynamic loads, which increase when the demand is very high.

To sum up, sustainability includes energy efficiency of vehicles and other transportation systems as well as life preservation of living beings existing in their superordinate system.

In order to consider sustainability to a greater extend during car development, it is important to align it with the challenges, which the automobile industry is facing. The main influencing factors that will be considered in this paper to boost sustainability in automotive are the use of alternative or improved propulsion, the use of driver assistance systems, the exploitation of the potential of connected vehicle and the integration of all these concepts afore-mentioned into a holistic approach. Besides, the consideration of sustainability implications in all processes along the product life cycle (e.g. operations optimization) will also be considered as influencing factor.

13.2 Factors Increasing the Complexity in Automobile Industry

13.2.1 Urbanization

Urbanization calls for concepts for shared mobility, even though individual mobility will still be necessary. Individual mobility certainly will be required by people who need strong personalized features or who are not willing to use shared vehicles.

The concept of car sharing is gaining more and more importance and many research works performed in the industry aims to connect cars with people and things. Some use cases consist of connecting cars to many people for planning purposes, for instance in case of a collective transport or when it comes to notifying their availability to a control instance (e.g. booking server). Doing this can help reducing many individual trips. In fact, many cars carrying only the driver often shares the same route and remain unused many hours a day, for instance after the driver is at work. Connecting cars with people can help joining for a lift as well as reducing individual car ownership, which yields resource saving and contribute therefore to enhance the sustainability.

In fact, giving up personal cars thanks to car sharing can lead to less cars in cities and therefore less parking spaces. Thus, no longer required parking surfaces can become urban green-spaces. Besides, less cars in cities can significantly reduce the amount of traffic jams, whereby passengers can gain more time to do their preferred activities.

The above-mentioned calls for new vehicle solutions for car sharing and associated services, especially Transportation on-demand resp. Transportation-as-a-Service (TaaS) and Mobility-as-a-Service. Suitable cars for TaaS ought to be conceptualized and manufactured for a very large customer spectrum. In Analogy to a smartphone that can be used nowadays even by very small children without having to read user instructions in advance, a relevant user experience is required as users should get in and start the journey intuitively. For this purpose, user-centric as well as design-for-sharing vehicles are needed.

The vehicles ought to be robust and to allow users plugging their personal settings that might be provided either by a device (for instance smartphone, token) or by an infrastructure such as a cloud, on which personal identities (IDs) are stored.

To sum up, shared mobility is a valuable approach to tackle the challenges of urbanization. However, user-centric and design-for-sharing vehicles, which can connect to people as well as to other things (e.g. infrastructure, other cars [12]) are needed.

13.2.2 Digitization in the Automotive

An important area that can provide appropriate methods and technology for enhancing sustainability is Digitalization. The digitalization provides means for bridging the analogue gaps between service providers and customers, but also between industrial processes, software and hardware. Thus, customers can for instance order, consume and pay for services using a digital interface [13]. Moreover, the interaction with consumer products (e.g. vehicles) but also with industrial machines and plant as well as development and manufacturing processes can be organized much better than in the past.

An obvious example of digitization in mobility is autonomous driving. Even though autonomous driving cars, which have been presented so far are still prototypes and yet not ready for a widespread use, autonomous driving is very promising for the future. Self-driving cars could take passengers to destination and drive afterwards to the next ones who have request transportation service. Such cars could autonomously book a free parking surface, get there and park and therefore free the passengers from long searching of parking surfaces, which can be tedious in a big city.

Autonomous driving offers new possibilities to passengers as the control is delegated to the car. Thus, passengers could lay in the car during a journey, read books, perform some office tasks or relax. Besides, autonomous driving will increase social mobility by giving seniors, injured-peoples and handicapped peoples the opportunity to move autonomously.

This calls for new design concepts, especially regarding vehicle architecture, body and interior. Due to the fact that no combustion engine will be in use, the front of cars can be shortened, paving the way for new designs. Therefore, more space is available and can be used to enlarge the car interior. Doing this provides additional space for introducing more features than nowadays, for instance large displays for surfing the internet or watching movies during the journey. This can reduce the stress that often is associated with driving in dozen of big cities world-wide.

On the other side, the new user experience (for instance laying and relaxing in a driving car, performing office tasks, etc.) requests new and appropriate safety and security concepts.

From an operation point of view, one of the main enablers for product digitization in the automotive are self-driving systems, also known as automated driving systems (ADS), which enable autonomous driving. ADS include many individual packages, which interact together in order to identify and avoid obstacles, to steer the car, to communicate, and so on. In fact, ADS help realizing the functions, which in case of classical cars are performed by a driver, as well as additional functions. Thus, ADS manage self-driving ability of cars (perception, decision and control as well as navigation planning and vehicle manipulation). All components that contribute to autonomous driving build an ADS together. Examples are assistance systems, sensors and actuators, cameras, connectivity devices, artificial intelligence, electronic control units and supporting services such as safety (including redundancy) and IT-security, to name a few [14].

13.2.2.1 Safety and Driver Assistance

Driver assistance systems (DAS) are enablers for autonomous driving and thus a key factor for implementing sustainable mobility goals. Therefore, they are gaining more and more importance in new vehicles. DAS can help avoiding critical situations or support drivers who already has got in these situations [15]. Moreover, they help increasing the driving comfort and support also autonomous tasks such as parking and driving. There are different types of driver assistance systems, from autonomous systems that provide information only to other internal car components over assistance systems that provide information to telematics services up to assistance systems communicating with the infrastructure [16].

Principally, driver assistance systems can focus on vehicle driving performance and behaviour, driver behaviour [17], input from external sources (e.g. traffic light information [18], digital terrain-information or satellite information) or on a combination of previous factors for decision making [19].

The driver remains the focal point of the vehicle system and therefore it is important to ascertain that s/he is able to command the vehicle at all time. Thus, driver behaviour such as fatigue can be tracked through assessment of facial features such as eye movement or alternatively by head pose estimation. Especially fatigue estimation points out the challenge of driver assistance systems to recognize in which situation a given decision should be taken. Head pose variations, changes due to illumination variations, hard shadow and occlusion are only a few amongst the parameter to be tracked. Moreover, the accuracy and robustness of a head pose estimation system against individual settings is to be ensured [20].

One of the pre-requisites for driver assistance systems to be adequately used in a vehicle is the real-time communication with other components such as the ECU (Electronic Control Unit) and the available bandwidth for signal transmission. Although automotive Ethernet is bearing a very high potential to help tackling this challenge, consequences regarding energy efficiency may arise depending on its implementations. Thus, to reach an optimized energy efficiency, it may be necessary to combine multiple approaches since each energy saving is relevant [21].

Furthermore, integrating driver assistance systems of different vendors in the vehicle requires standard protocols and interfaces, which for instance are taken into account by ADASIS-Standard [22].

Driver assistance systems (e.g. adaptive cruise control) can help reach a better fuel efficiency, which is a goal of sustainable mobility. Furthermore, they can save live through look ahead functions and therefore increase transport safety. However, the adequate recognition and assessment of the driving situation even when sensor data is not completely reliable [23] remains a challenge to be tackled by driver assistance systems. Especially when it comes to taking control of the vehicle or returning control to the driver, decision-making is crucial. In this context, controversial discussions are being held regarding the responsibility in case of accidents.

To sum up, a combination of driver assistance systems with relevant sensors, actuators, cameras as well as connectivity devices are essential in order to increase safety [24] and therefore to enhance sustainability.

Furthermore, driver assistance systems can help tackling the challenge of energy efficiency by improving driving style, which as is known led to a reduced consumption of fuel resp. battery power. For instance, acceleration as well as braking actions can be supported by speed regulating systems or by an intelligent speed adoption. The ability of cars to communicate with the infrastructure as well as with other road users serves as basis to realize the function afore-mentioned. Further advantages of driver assistance systems can help also achieving significant results in the development of autonomous driving vehicles [25, 26].

13.2.2.2 Connected Car

As mentioned above, Sustainability can be enhanced through resource saving by enabling car communication in order to align resource planning and consumption. For instance, a Vehicle-to-Infrastructure communication provides the ability to sending requests regarding the repartition of recharging stations along the planned route as well as the ability to communicate with these stations. Therefore, the route can be continuously analysed and updated depending for instance on the maximum reachable distance of a car, the battery status, the operability of the relevant charging stations and so on.

These needs for connectivity calls for solutions for connecting cars not only with the infrastructure, but also with all actors of the ecosystem they are running in. Therefore, Vehicle-to-X (e.g. car-to-car connection, car connection to other transportation systems such as tramway, car to Business connection, car-to-service and other necessary connections) have the potential to contribute to more Sustainability.

The applications of connected vehicles can be divided in the categories hard safety, soft safety, mobility and convenience [27]. Hard safety applications are critical and are destined to avoid accidents or reduce their consequences. This bears high requirement from the involved communication systems and infrastructure.

Soft safety systems are destined to increase safety by informing drivers about driving conditions that require care (e.g. icy roads), but no immediate action is required from the driver.

The main goal of mobility vehicle applications consists of facilitating the traffic flow, which can be realized for instance with traffic information service or navigation.

Vehicle convenience applications focus on enjoying the time spent on-board vehicles as well as driving comfort by providing additional services to customers, especially when using an autonomous driving vehicle. Among others, video gaming and web browsing can be provided to passengers for enjoying the trip [27].

In fact, Car-to-X facilities enable vehicles to send broadcast messages to other vehicles that are located in their communication range. Thus, vehicles can use that information in order to capture the behaviour (e.g. velocity, position, emergency electronic brake light messages, etc.) of neighbour vehicles and therefore avoid collisions. Besides, vehicle-to-vehicle communication can be realized with multi-hop message dissemination, whereby information sent by an arbitrary vehicle are relayed to further vehicles located in their own communication range. Assuming a very low number of hops, V2V multihop [28] message dissemination can be applied for hard safety applications, however, it is suitable for soft safety applications too. An important challenge in this case remains managing the performance especially when a high number of vehicles are involved and thus produce an important information overhead.

Another communication approach consists of forwarding the messages of sending vehicles to an infrastructure dissemination server or to an infrastructure application server that in turn would disseminate the information to neighbour vehicles [27].

In fact, many technologies are suitable for realizing vehicle communication. Depending among others on the type of application and the performance required, short-range radios including Bluetooth, WiFi, and dedicated short-range communications (DSRC) can be appropriate, but also Long-range radios including cellular networks, satellite services, and digital radio broadcast networks [27].

The V2V local communication as well as the I2V local broadcast can be implemented with short-range radio receivers. Additionally, I2V can be realized with cellular, satellite, or digital radio broadcast services in order to increase the range of reachable vehicles [27].

Many vendors are proposing solutions for connecting cars to other systems; however, most available systems are proprietary and not open for communication with systems of competitors. The car 2 car communication consortium has been created in order to pave the way for open cooperative and intelligent systems that support vehicle-to-vehicle as well as vehicle-to-infrastructure communication [Car2Car]. Based on that, driver assistance systems of a specific vendor could easily request information for instance from navigation systems of other vendors and combine that with inputs from the infrastructure (other provider or vendor) in order to control vehicle behaviour.

To sum up, vehicle networking impacts sustainability since it can help avoiding accidents with collision detection and avoidance mechanisms. Vehicle networking can help reducing congestion through traffic regulation and help improving the environmental impact (carbon footprint) of vehicles [29, 30] for instance through optimized routes and eco-friendly driving behaviour.

13.2.2.3 Security

The openness of vehicle architectures in order to integrate different components such as driver assistance systems as well as the car connectivity bear the risk of vehicles being hacked or manipulated (e.g. replacement of original components with counterfeits) by non-authorized persons. Therefore, the protection of vehicle systems requires holistic approaches that consider all relevant factors.

In fact, individuals, criminal organizations, government agencies as well as foreign governments can perform security and privacy attacks on connected vehicles, damage them and therefore lever their contribution to sustainability.

Attacks can consist of manipulating the driving parameters such as velocity through compromising of components (e.g. Electronic Control Unit, input values to CAN-Bus), but also manipulating the display as well as vehicle behaviour such as steering direction. Besides, attacks can be performed in real time in order to take control of vehicles. Furthermore, attacks can focus on a system breakdown or at least on a considerable degradation of system performance (e.g. a denial-of-service-attack, which is well known in conventional computer networks).

Apart from attacks that consist of influencing another vehicles and their passengers, privacy attacks are an important threat, for instance, a vehicle could act with the credential of another one and therefore create damages that could be assigned to innocents. For instance, a vehicle could make the so-called Sybil attack that consists of sending fake messages to other vehicles in order to let them behave as if there were more vehicles involved in the traffic than they actually are [27].

Further privacy attacks can consist of injecting spyware or physical components in order to collect driver related information without any permission. Not only the connection with internet, but also the mobile devices of passengers as well as the services and in-vehicle wireless devices they are using can provide an opened door for attacks. An example of a personal information is the movement profile that provides enough information to observe people illegally.

In order to enforce the security of mobility systems within a large vehicle network, it is important to implement a reliable authentication system using for instance signatures, digital certificates and a Private Key Infrastructure (PKI). Besides, identifying misbehaving vehicles and their revocation can help preventing major harms, especially in the cases when a very huge amount of vehicles would have been manipulated [27]. Data being exchanged can be checked regarding integrity, but it remains important to ensure data confidentiality [31]. Thus, cryptography can be used for protecting connections and integrated components [32].

13.3 Systems Engineering

The integration of the technologies, which have been mentioned above in order to tackle the challenges of the automobile industry and enhance sustainability will lead to an increase of complexity (e.g. functions, components, interfaces for communication with infrastructure and other vehicles, etc.). Some mechanical components that are available in vehicles nowadays will have to evolve to smart parts that will be necessary for implementing intelligent and dynamic vehicle behavior. Introducing new technologies, for instance regarding networked vehicles, is an additional factor of complexity since it bears also the risks mentioned above. There are some more examples that could be used for pointing out why and how the vehicles are getting more and more complex. Therefore, complexity management is a key factor for the development of future, sustainable vehicles.

In contrast, the development processes in most enterprises are not yet adapted to this new challenge. The development often is strongly driven by mechanics. Many companies are breaking down the work packages into different categories according to the engineering disciplines involved in the product development: mechanical, electric/electronics and software. Unfortunately, the communication between the different disciplines is error-prone as there is no common language for all disciplines in order to understand the virtual product. The use of SysML that can help tackling this issue by providing a common understanding of models, is rare [33]. This is exacerbated with the fact that different product models with different representations as well as semantics are used by the authoring disciplines (e.g., design, simulation) [34].

The dependencies between partial product models (e.g., geometrical model) are often not automatically tracked, analyzed or used, for instance, for model update purposes or for spontaneous investigations as well as decision making. Thus, in many cases, components and sections are not fitting with connecting elements [33]. A considerable effort is needed each time an overview of the whole vehicle product (including all partial models) is needed, for instance, for releasing or for prototype building. However, in order to obtain an efficient collaborative work, not only consistent data is relevant, but also processes, methods and tools as well as the organization are to be aligned [33].

A further drawback of conventional work breakdown is the missing of a systematic and well-organized knowledge management. Often, the knowledge is not gathered, formalized and represented for intelligent use, which can improve collaborative work. Thus, information related to products is kept in a large number of documents (see Fig. 13.2) and databases. Sometimes, the knowledge about the dependencies and the latest works is known only by few people. Therefore, it becomes very tedious to realize appropriate workflows for an automatic control of the product development, for instance, when the impact of a requirement should be quickly traced for decision making [33].

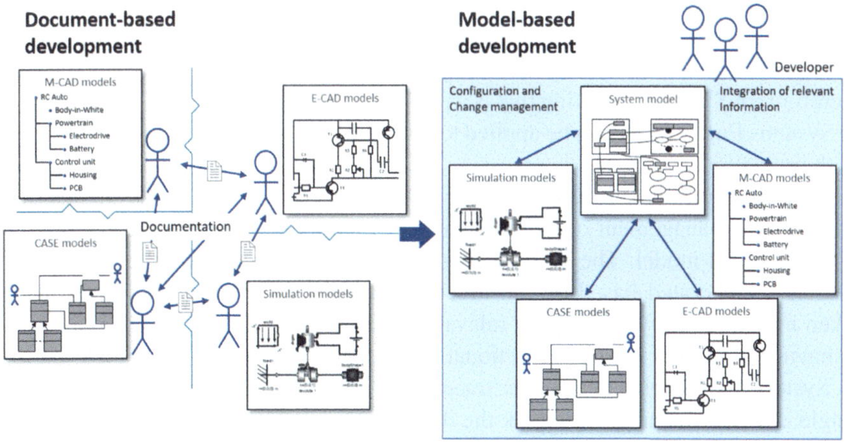

Fig. 13.2 Model-based-systems engineering [43]

This calls for approaches regarding an efficient way of working, out of which the company organization, processes and Product Lifecycle Management-Architecture (PLM) to be derived.

Considering the fact that vehicles nowadays are to be apprehended as a system of systems and therefore the evidence of the role that systems thinking [35–37] should play in the product development, Systems Engineering (SE) is an adequate approach as it emphasizes system thinking. SE as a complexity management approach [38, 39] is also appropriate in regard of the higher complexity to be expected from the vehicles that should enhance sustainability in the future [33]. In order not to blast the length of this chapter, reference is made to valuable scientific articles to Systems Engineering [11, 33, 37, 40–42].

Although SE provides a toolbox that can improve the development, manufacturing, distribution, use and disposal of future transport systems, not only the application of Systems Engineering is sufficient to reach the objectives of sustainable mobility.

Instead, multidisciplinary work is needed within but also across enterprises with the social, economic and also the public sectors. Thus, Systems Engineering for Sustainable Mobility consists of integrating all these systems into a global model and to understand their interrelationships in order to design appropriate solutions with a focus on the entire product lifecycle. That is, from the first product idea through the development, manufacturing, maintenance and down to disassembly and recycling.

13.4 Proposals to Link Systems Engineering to Sustainability

Sustainability consists of considering social, environmental and economic implications in company's strategies, processes and operations. It is related to the global word, which entails several systems (e.g., ecological) that are inter-related, calling therefore for Systems Thinking that is a pillar of Systems Engineering. The methods of Systems Engineering can be applied to model the systems aforementioned as well as their relations.

As defined in source [33], developing products according to systems engineering includes the management of requirements, which are the inputs for the elaboration of a functional model. The systems model, which is an architectural and behavior model, is elaborated based on the functional model. Thus, sustainability is to be taken into account by elaborating relevant requirements, which should be regarded as having equivalent worth to functional requirements.

Systems engineering enables the traceability of product requirements down to the single components and helps track the dependencies between the different product parameters. This ability is taken into account often when it comes to knowing for instance, which impact the modification of a specific requirement would have on further factors such as the product quality or product price. Thus, one further approach to consider sustainability when applying systems engineering consists of aligning the classical product models with social, environmental and economic models (see Fig. 13.3). Doing so, the impact of modifications on one single system model (e.g. ecological) can be predicted and the appropriate measures can be taken during product development.

However, it is necessary to identify the artifacts of social, environmental and economic models, which are appropriate to be considered in the frame of systems engineering. It is also necessary to identify the level of granularity of those models because several level of details can be considered when it comes to realizing an environmental model. Furthermore, appropriate methods for simulating the interactions of the different models represented on Fig. 13.3 are needed in order to be able to predict the impact of product properties on sustainability and vice versa.

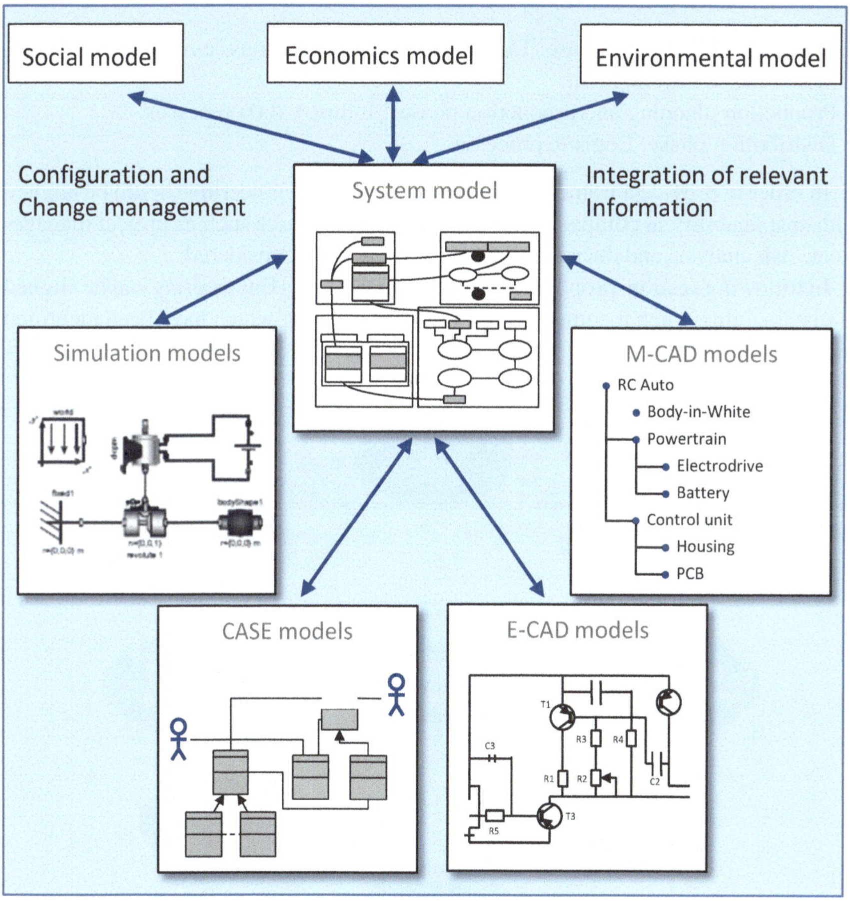

Fig. 13.3 Integrating sustainability with environmental, social, economic models. Figure adapted from Fig. 13.2

The objective of product development should be to anticipate in order to avoid later conflicts related to sustainability. Thus, findings that have been gain through simulation or impact analyses are to be formalized and reused during product development.

Applying Systems Engineering for sustainable mobility challenges consists also of enhancing standard SE-processes according to sustainability requirements, visions and goals, all considered under long-term prospects. From a point of view of the automotive, Design-to-Sustainability and Manufacture-to-Sustainability are to be realized and emphasized as they are able to impact development processes as well as sustainability to a great extent. For instance, design decisions can impact the manufacturing costs up to 80% [44, 45].

Base on the phases of the product lifecycle (see Fig. 13.4), some processes that can help enhancing sustainability are identified:

- Product development phase: Design of product functions, embodiment design, model-based development
- Production planning and production phases: Industry 4.0 processes
- Distribution phase: Logistic processes.

In order to provide a framework in which Systems Engineering should be aligned with sustainability in companies, cross-sectional processes such as project management, risk analysis and decision-making processes are considered.

In following section, proposals are made how Systems Engineering can be aligned to sustainability when it comes to perform the processes, which have been identified and mentioned above.

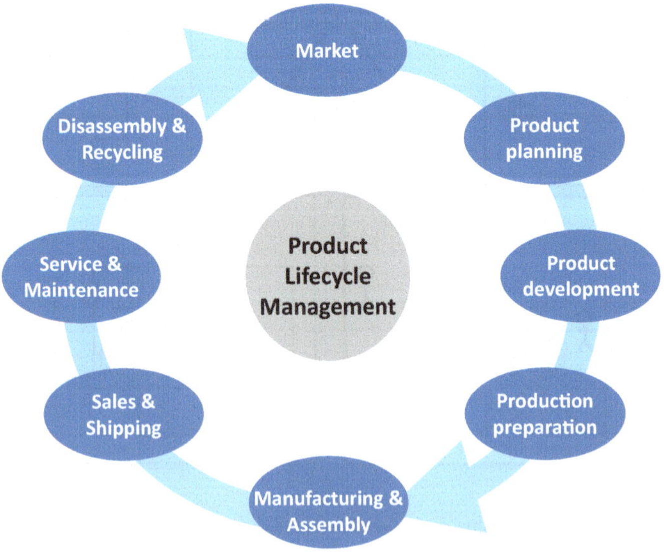

Fig. 13.4 Integrating sustainability with PLM [46]

13.4.1 Linking Functional Development to Sustainability

In practice, functional development is not performed explicitly in many companies. Even when a functional model is considered as a full phase of Systems Engineering, the focus is mostly on representing the function structure in order to identify solutions to some individual functions. Instead, functional design should be in line with Sustainability in order to affect the subsequent phases of product development such as logical design, detail design, verification and validation.

Implementing this consists of creating supporting functions that enhance the conventional functional model with parameters related to durability, re-use and recycling, among others.

Furthermore, the relationships between sustainability-oriented requirements and the functional model are to be managed, therefore first sustainability assessments should be performed with the functional and the logical model of vehicles. Doing so can help comparing different vehicle alternatives and adjusting functional models very early in order to avoid the costs of later changes. Besides, taking sustainability related requirements into account during functional design may provide a very interesting input for the elaboration of the subsequent systems models that impacts the product development in single disciplines.

13.4.2 Design to Sustainability

From the point of view of design, many aspects can be considered when it comes to aligning to sustainable mobility goals. First of all, the product characteristics (e.g. design [47]) as well as the process perspectives are to be distinguished, even though the company vision can impact the thinking of the staff and therefore is of great importance.

In fact, a code of ethic regarding sustainable development is to be adopted in order to orientate the work of engineers for enhancing sustainability. Such a code has been defined by the American Society of Civil Engineers (ASCE) [8].

One of the initial points where to impact sustainability is the vehicle architecture. In fact, the power supply and materials as well as the joining technologies that are chosen as well as the related services are of great importance.

A sustainable vehicle architecture can be achieved by integrating clean or greener energy sources in vehicles and transport systems in order to reduce the environmental impact. Besides, the energy factor ought to be considered not only regarding the end product, but its creation also. Thus, using renewable energies in the product creation processes is a logical consequence.

In fact, recycled and recyclable materials as well as sustainably-produced materials are to be used for components design and in case of existing products, some components can be replaced with materials that help obtaining a better environmental balance. The focus regarding sustainability is to be set not only on the material to be used for creating components, but also on the entire material lifecycle. Furthermore, not only environmental compatibility but also the impact of products on the users' health is to be highlighted. Thus, materials that are non-toxic for the humans, but also for other living actors of the ecosystem are more appropriate in order to realize a more sustainable product.

On the other hand, the processes related to product creation as well as associated services have to take the goals of sustainable mobility into account. For instance, the design department can apply a modular design methodology in order to ease the replacement of single components instead of the replacement of whole modules containing many components, when a function is impaired. Doing so can help also remanufacturing products in order to reach more sustainability.

Beside design for re-use is to be applied for instance through standardization of components, whereby parts are used in different product lines. This can provide not only an economical advantage but also an environmental one since in case a product of a specific line (e.g. car) is damaged, the probability to re-use the most of its parts in several other products is high.

Furthermore, the design methodology has to take processes such as disassembly and recycling into account. For this purpose, the tools and methods to be used during product disassembly are to be considered in the product development phase (design-to-recycling). For instance, when creating assemblies for a product, a joining technology can be chosen according to the targeted disassembly process. Besides, precise and efficient installation investigations are to be performed in order to support and ease the disassembly process. Taking disassembly into account can be realized also by assessing the energy consumption and by implementing rules for energy saving.

Generally, all types of information related to the disassembly process of products should be of equal value to functional requirements. Thus, that information is to be gathered, formalized and provided to the requirement model, the functional and the systems model. According to the RFLP-process [33] for mechatronics product creation, the system model should pilot the engineering disciplines (e.g. mechanical design, Electric/Electronics, Software).

In the case of mechanical design, the outputs of the systems model can lead to creating geometrical guidance to be provided to designers. This guidance can be realized as CAD templates or CAD macros [48]. Thus, providing guidance to mechanical designers can enforce the observance of sustainability-based specifications and therefore improve process quality.

A further design methodology for enhancing sustainability consists of aligning component modularization not only with manufacturing processes but also with disposal. Thus, products can be organized in small entities, which are made by parts that have the same material. Therefore, these small entities can be recycled together without having to be decomposes down to the single constituent.

From a general point of view, development and manufacturing processes are to be designed to reduce waste that is generated during product creation, but also during product usage. Waste can consist of emissions but also of material.

A further approach to enhance sustainability consists of customizing or re-designing existing products in order to prevent emissions from accessing the atmosphere. For instance, appropriate diesel filters can be used to process gas emission. In the case of factories, emissions can be reduced following the same principle (waste filtering) for gaseous waste or liquids waste substances. In the case material waste is being produced during manufacturing, solutions are to be investigated for re-using it as raw material for new products, even though recycling might be necessary for certain cases.

13.4.3 Linking Decision and Risk Analysis to Sustainability

Although sustainability can be identified as corporate objective, there are still obstacles when it comes to implementing its concepts in daily work. Therefore, the question of how to start and in which company areas to start are often raised. In fact, aligning product creation to sustainability is in most cases a change that should be considered as such. First of all, a sustainability strategy is to be defined, which considers sustainability as a full part of company culture, loss aversion of the staff, safeguards to ensure the transparency and accountability of the decision-making processes and the commitment of all disciplines. For this purpose, guidance for sustainable decision making is to be provided along the product life cycle.

The elaboration and publication of a sustainability road-map is a key factor to ease the communication. Based on such a road-map, suitable sustainability metrics are to be defined and monitored, whereby the environmental, societal and economic impacts of products and processes are tracked. This procedure is necessary as decisions related to sustainability can be subject to biases, due to the complexity of all systems involved. According to several studies, biases influence most of human decisions [49, 50]. Since it is about a multi criteria decision making (MCDM) process, it is important to ensure that the values of the company are more privileged against personal conviction, assumptions and values. Especially, it is important to highlight the fact that long-term-benefits are to be traded-off with short-term benefits in a context of values that have been elaborated and shared within the company.

The decision and risk analysis needs a shift of focus from solely the economical perspective to a sustainable one. The social and environmental implications of the decisions made by employees are to be highlighted in order to emphasize individual responsibility. However, these measures are to be accompanied with employee empowerment as well as methods and tools for helping analyzing and visualizing the consequences of decisions (e.g. design decisions) thoroughly [51]. For instance, when a program manager decides to apply new technologies such as hydrogen fuel cell or alternative materials in vehicle lines, the impact of these changes is to be measured not only regarding the final product (e.g. a car) or service (e.g. car sharing), but along the whole lifecycle that includes logistics, manufacturing processes and product usage, to name a few.

Although the approach of providing tailored tools that help linking decision making with sustainability might seem to be complex, solutions can be developed and provided for communities. Community members and relevant stakeholders can be given the opportunity to continuously feed a community database or tool, which helps to assess sustainability. Moreover, secondary decision impacts that might not be seen at first glance are to be determined during product creation and the associated risks to be assessed (e.g. with state-of-the-art sustainability supporting tools).

Although sustainability metrics have been provided by different research work [52], enhancing sustainability remains a quite complex challenge as many local and global systems and stake holders are involved, the responsibilities sometimes are overlapping and the interrelationships difficult to master.

Decision making however, is not limited to development processes but takes also the product behavior into account. More and more, mobility systems are given the ability to make decision autonomously. This is the case for driverless cars that are not equipped with steering wheel and pedals, but can decide autonomously how to behave depending on a real-time situation. This ability (in fact artificial intelligence) is realizable in the manufacturing with self-organizing systems capable to make decisions [53–55]. These characteristics offer an interesting opportunity to meet sustainability goals (e.g. energy efficiency through intelligent decision-making), but much more, safety and security are crucial.

Especially in the era of self-optimizing and self-driving systems, the impact of wrong decisions on the market can be disastrous, especially when the behavior of products is not compliant with general expectations of customers. For instance, the safety and security algorithms to be involved in autonomous cars are subjects of very controversial discussions as it is not clear for the moment whether such a car should prioritize some lives (e.g. children's lives) at the expense of other lives in case of emergency. This ethical dilemma emphasizes the need to strive for community consensus and to integrate global, political and socio-economic aspects when it comes to making decision for a sustainable mobility.

In order to broadening the linking of decision-making and risk analysis with sustainability, a very interesting point where to start are curriculums. Students should get acquainted with multi-objective decisions making as well as risk management considering primary and secondary impacts on society, environment and economy. Sustainability is to be considered in degree programs in order to impart the advantages that can be provided by innovative technologies such as Internet of Things (IoT) [56] or Additive Manufacturing [57, 58] for enhancing sustainability. Besides the roles of concepts such as design-to-sustainability and production-to-sustainability among others, are to be emphasized.

Taking not only economical criterions but also the social and public sectors into account increases the process complexity during product creation and therefore the project management. Thus, new concepts for engaging stakeholders and therefore new processes and methods that enable an interaction with relevant institutions are required. The staff and the stakeholders ought to share sustainable goals. Nowadays it is unavoidable for engineers to understand sustainability issues and subsequently, to make suitable decisions.

The aforementioned requirements emphasize the very high complexity that characterizes the development of mechatronics products. That is where systems engineering comes into play, as approach for mastering complexity.

13.4.4 Linking Project Management to Sustainability

Managing projects in the context of sustainable mobility cannot be focused only on technical aspects of products and services, such as the implementation of required functions. More and more, customers are interested in the social and environmental impact of products and services and consumer associations provide benchmarks that present sustainability tests and winner companies. For instance, using carcinogenic materials is an important reason from a customer point of view for not buying a product.

Therefore, the focus of engineering management has to be redefined in order to satisfy all stakeholders, that is, also social and environmental interest groups. Project leaders have to be aware of their global responsibility in order to consider not only the legal, but also the ethical issues that are related to their products and to the services related to those products. For this purpose, project management is to be geared towards sustainability.

Especially, project leaders ought to align their projects to the company sustainability goals and ensure that sustainability metrics are in range. For this purpose, conventional project objectives are to be enhanced with sustainability related expectations. In order to reach that objective, project leaders can be supported by sustainability representatives. Later can additionally perform internal audits in order to monitor the metrics of single projects and report the results to the management in order to get the company-wide status tracked. Based on such inputs, the management is able to review the global metrics continuously and therefore react early in order to get sustainability targets reached. For instance, measures can be taken early to ensure that requirements for specific certifications will remain fulfilled.

13.4.5 Model-Based Enterprise for Sustainability

Aligning the strategy and therefore the processes of a single company to sustainability goals is not sufficient to ensure that an equipment such as a vehicle has been produced to sustainability. In fact, companies and teams that are geographically distributed around the globe are involved in car creation. Besides those involved development teams belong to OEMs as well as to suppliers and are assigned with different tasks.

Moreover, the different companies involved use software tools of different vendors, which often make use of proprietary formats and therefore complicate an information exchange with each other. Thus, communicating the design intent from the product development to downstream processes such as manufacturing is characterized by information gaps. 2D-drawings, which are a very limited representation of the technical product documentation, are often used for communication. Even though 3D product data is exchanged, important information often is missing because it is available in company internal databases or documents. This leads to misunderstandings that can cause not only errors and subsequently manufacturing rework, but also wrong decisions from a sustainability point of view.

Eliminating errors that can lead to material wastage and providing all stakeholders with necessary up-to-date information in order to optimize processes, lead times and decision making are a sustainability matter.

Model-Based-Definition (MBD) is an approach to tackle that issue as it is a product model that provides relevant data sets (e.g. geometrical information, Product Manufacturing Information—PMI) [59] and that can be shared across companies in order to support downstream processes. Thus, information related to sustainability can be federated to MBD-models in order to reduce system-specific interfaces and to improve the interoperability of systems related to sustainability throughout the entire product lifecycle (see Fig. 13.5). In the praxis, not only one sustainability tool would suffice for elaborating and implementing a sustainability strategy. Instead, a sequence of tools that share information even across company boarders is required.

For this purpose, a standard model for sustainability information related to parts and assemblies as well as methods for creating, reading, updating, deleting and representing model information are needed.

The principle of MBD consists of providing a single source of truth for all parties involved in the product lifecycle. The advantages consist among others of improving collaboration and avoiding errors. Thus, feedbacks from manufacturing can be

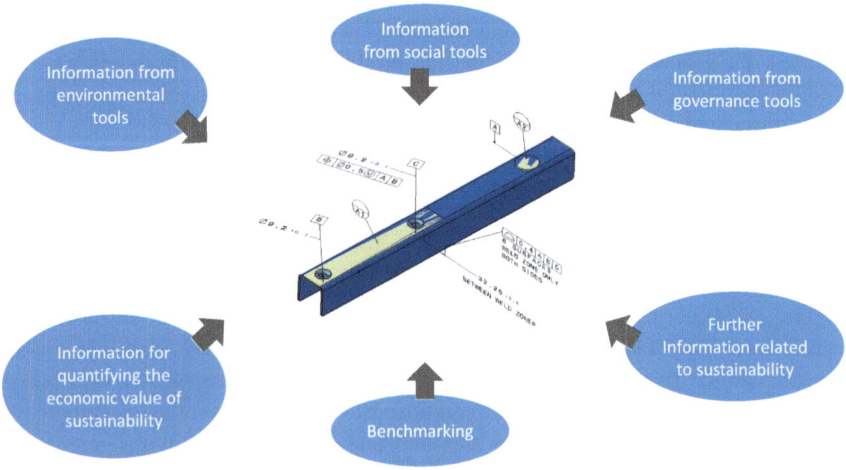

Fig. 13.5 Enhancing a MBD-model provided by [60] with sustainability related data sets

considered immediately in the product development, for instance to perform improvements related to sustainability.

In order to take the benefits of MBD along the value-added chain, a collaborative environment is needed, which helps bridging organizational and functional barriers as well as the obstacles due to proprietary formats. This is where Model-Based Enterprise (MBE) comes into play.

Since MBE helps following the processes from design concept through manufacturing down to disposal, it represents an interesting enabler for sustainability. In fact, MBE can help determining the impact of design decisions (e.g. material choice, vehicle architecture) throughout the product lifecycle. Based on this, real societal, environmental and economic implications can be tracked. These implications are to be considered during the creation of next products in order to ensure that created designs are aligned with sustainability requirements.

13.4.6 Industry 4.0 for Sustainability

One of the well-known principles of SE is Model-Based-Systems Engineering (MBSE) that consists of creating and using models representing specific perspectives of the virtual product (partial models) instead of using documents (see Fig. 13.2).

In fact, product models can be combined with process models and factory models in order to optimize all processes related to the product lifecycle [61]. This offers a basis for monitoring and enhancing sustainability since the information provided by created models can be used to improve manufacturing in general (e.g. material flow, lead time, equipment layout) and from a sustainability point of view (e.g. improve safety and operations, predict wastes) [62].

Moreover, MBSE is an interesting basis for Industry 4.0. Industry 4.0 consists of realizing a self-organizing manufacturing in which people, plants, machine tools and logistic systems can communicate and interact autonomously. For this purpose, a horizontal integration (cross-company oriented) is combine with a vertical integration (oriented to the digital factory) along the product lifecycle [63, 64].

According to Industry 4.0, the process model can interact with models of the equipment in order to define manufacturing sequence and therefore select appropriate machine tools for a specific product variant. On the other hand, the models of the equipment (e.g. machine tools of the shop floor) can interact with the product model in order to optimize the processing of a work piece. This ability to support manufacturing with a high level of variance, adaptive machining and self-organization as well as self-optimization are enablers for sustainability. The distance travelled by production workers, lead time and costs as well as waste can be reduced, contributing thereby for resources' saving. Besides, sustainable decisions can be dictated by software during product creation.

Furthermore, the digital twins [65] of the manufacturing equipment can be federated with data analytics models, whereby a connection with the Internet of Things

(IoT) helps collecting relevant data and performing real time analysis in order to support predictive analyses, process monitoring and control [66, 67]. Self-optimization can be realized by considering artificial intelligence (e.g. neural networks [68]) in the set up.

Thus, based on prediction, tools can be ordered at the right time to avoid delays in the production. Moreover, product quality can be ensured by real-time analysis of tool-related data, identification of defective tools and dictating tool replacement before performing scheduled manufacturing operations. Doing so helps reducing defective parts, which in many cases represent a material waste. Additionally, the set-up time of machine tools can generally be optimized and their associated data (operating related data, disassembly related data, energy saving data, communication with other systems) can be used to improve sustainability.

Industry 4.0 offers the ability to configure a flexible manufacturing that is in each case adapted to deliver an optimal result regarding product quality and sustainability as well as cost of ownership. Thus Industry 4.0 provides the pre-requisites for delivering new services using software agents, whereby value can be delivered with minimal resources [69]. For instance, features can be added to manufacturing systems and process optimization can be realized without the intervention of experts who would tie up resources.

In conclusion, in contrary to the past decades when the focus of manufacturing was set on productivity, manufacturing has evolved in many places to a discipline that emphasizes sustainability. Ensuring the safety of the staff, the separation as well as the reduction of wastes, converting waste into secondary products, the use of local materials as well as enhancing energy efficiency are just some principles of manufacturing-for-sustainability [70]. Through the realization of intelligent, self-organizing and high-automated manufacturing systems, Industry 4.0 is to be considered as an enabler for sustainability [71].

13.4.7 Logistics and Supply Chain for Sustainability

The organization and the processes of the conventional logistics can be modified in order to reduce emissions and therefore enhance sustainability. In fact, the product creation, especially in the automotive is characterized by networks of companies that perform the design and the manufacturing on different sites and continents. Components also are manufactured on several international sites and modules as well as whole vehicles are assembled and distributed around the world. Thus the manufacturing and distribution of the components involve many hundreds of trucks, which drive hundreds of thousands kilometers for distribution. Furthermore, the material waste and the emissions that arise from all the participating industrial processes are to be considered when it comes to enhance sustainability.

A solution approach consists of creating intelligent logistic models in order to optimize the travelling distance between involved factories (e.g. part supplier to module supplier, module-supplier to OEM). For this purpose, a connectivity to a cloud of services paves the way to functions for controlling and monitoring logistic processes, navigation to destination points. Additional advantages are function for regulating energy efficiency of involved systems as well as for ensuring safety of transported goods and safety of the surrounding environment.

The concept of manufacture-it-yourself is a very interesting approach to reduce the pollution and the material waste aforementioned as well as further negative environment impacts.

In fact, additive manufacturing is a process in which physical parts are created directly from CAD data and without the use of tools. Material layers are joined under computer control in order to create physical components [57, 58]. Even though support structures are created during the component creation process, additive manufacturing bears a high potential for savings as geometrical models can be manufactured without material surcharge that often is necessary in order to ensure manufacturability by conventional tools (see Fig. 13.6).

Therefore, designers are provided with the freedom to create load-optimized, multifunctional and topology-optimized components. That is, design can be focused on the needed functions, safety and stiffness instead of taking limitations of classic manufacturing processes into account.

A further advantage of additive manufacturing regarding sustainability is the saving of machine tools including the inputs and the energy they need. Besides the weight reduction of components and the possibility to use visionary solutions, which surely would have been unconceivable with traditional approaches are valuable advantages.

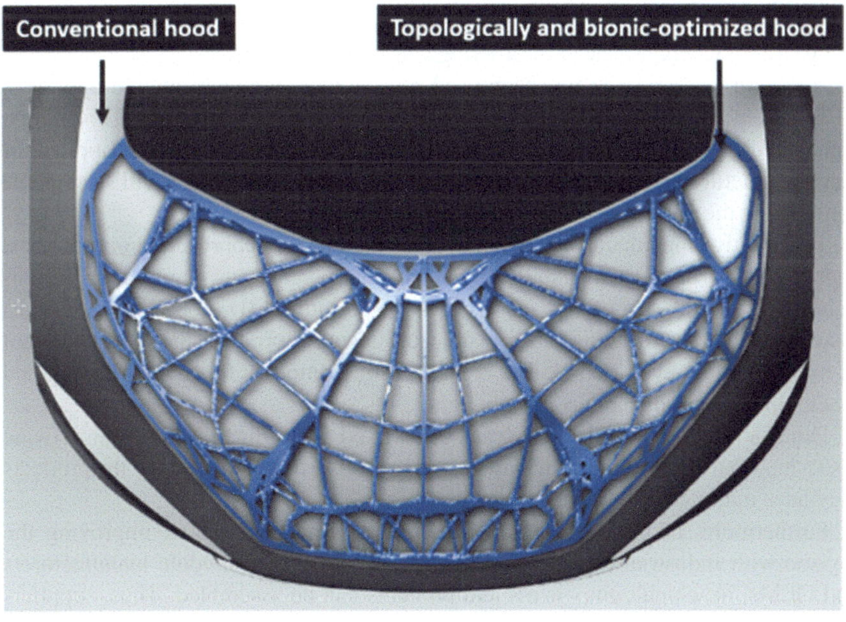

Fig. 13.6 Material saving with additive manufacturing

Fig. 13.7 EDAG Light Cocoon [EDAG]

An example illustrating this advantage is the EDAG Light Cocoon (see Fig. 13.7) that is an ultra-light vehicle concept developed at EDAG. The concept has been oriented to the approach that structures with a basic design inspired by nature are more stable and weigh less. Therefore, bionic principles have been applied in order to create a vehicle structure in analogy to leaves and bats' wings, which is efficient and waste-optimized.

In fact, the car body is not a closed surface. It is covered by a durable, weather-resistant cloth that enhances sustainability through weight reduction. Furthermore, neither paintwork nor costly repairs in case of damages are necessary. The personalization can be achieved using LED-lights that can provide different colors to EDAG-Light Cocoon instead of ordering additional accessories.

Although less material has been used compared to the conventional approach, all requirements regarding structurally relevant components are met. This can be considered as a proof that on one hand, bionic design can be successfully applied for vehicle structure and on the other hand, additive manufacturing is a real sustainable solution for the automotive.

The additive manufacturing provides a very interesting approach to optimize the supply chain from a sustainability point of view when using the manufacture-it-yourself principle. This approach consists of providing 3D-models to customers instead of hardware. For this to be possible, the design of products has to be rethought in order to allow even non-skill persons to use 3D-models for manufacturing and assembling the products they have bought.

Doing so brings not only the advantages related to sustainability, but can provide also a much stronger commitment of customers to the companies and to the products they have manufactured themselves.

Furthermore, additive manufacturing provides the potential of improving the logistic with industrial manufacturing centers located near module manufacturers and OEMs, in which components and products can be assembled. Thus, suppliers will send their models to module manufacturers or directly to OEM, who will use the capacity of those manufacturing centers in order to get their products manufactured by 3D-printers. Doing so helps shortening the transportation of parts around the world and therefore enhancing sustainability.

An adapted approach of manufacture-it-yourself-paradigm consists of involving local manufacturing stores (makers hubs) in order to manufacture products out of complex 3D-models and provide therefore support for private customers, who would not like to afford or operate 3D printers at home. This approach can reduce the number of home 3D printers, contributing therefore to resource saving and thus to sustainability.

At the same time, it cannot be denied that sending 3D models to customers or eventually to competitors bears the risk of piracy or intellectual property being theft. However, analyzing this challenge would go beyond the scope of this chapter. Instead, we refer to appropriate literature for intellectual property protection [59], especially in the case of additive manufacturing [72].

To sum up, additive manufacturing can be apprehended as a promising sustainability strategy that can certainly be expected to gain momentum in coming years.

Moreover, SE for sustainability opens new ways for engineering collaboration and trans-disciplinary engineering to develop new products (e.g. vehicles), new services or integral mobility concepts. However, an integrated process definition and implementation is needed for development activities starting from product design to the manufacturing. The adjustment between product design and quality assurance is required also. The processes afore-mentioned are necessary not only within a company, but also beyond companies borders.

13.5 Holistic Life Cycle Assessment as Essential Part of Systems Engineering

Life Cycle Assessment (LCA) is an approach for systematic analysis of the impacts that are related with all phases of the life cycle of a specific product or service. Thus, not only the product development phase is considered, but also the origin and development of raw material, the production and following stages up to disposal [73, 74].

An LCA encompasses a goal and scope definition, an inventory that delivers a matching of inputs (e.g. energy, material) and outputs (e.g. waste, emissions) for the lifecycle, and an assessment [73, 74]. The inventory consists of identifying and classifying environmentally harmful substances that issued, identifying the energy consumption and the materials that are needed for a product or service from concept down to disposal. The assessment consists of identifying relevant indicators (e.g. resource depletion) depending on initially defined goals. Assigning a weighting coefficient to indicators and therefore interpret the results and formulate recommendations are further assessment tasks. Thus, the assessment can be subjective as the weighting is arbitrary and the data quality can be poor.

Unlike many approaches that are focusing only on environmental impacts, also socio-economic consideration should play an important role for LCA, from a point of view of a sustainable mobility.

Generally, the designers do not receive a direct feedback of their work regarding sustainability. Many engineering partners perform design often and manufacturing is organized and performed by manufacturers, in the most cases different companies.

A high amount of components are provided as built since detail information for instance about geometry are part of intellectual property and therefore to be protected. Besides, manufacturing tools also are provided by several tool makers. Therefore, conducting LCA along the value-added chain is very difficult due to organizational barriers and confidentiality agreements, to name a few. Even though LCA can be conducted by OEMs, doing that would occur in the practice too late in the product creation phase if it were to be achieved at reasonable expenses. On contrary, conducting LCA can provide advantages already during the first phases of product development such as concept design and detailed design.

In the early phases of vehicle creation, the concepts of many alternative solutions (e.g. variants of parts) can be compared regarding also LCA results. Besides, new created parts can be compared with old ones in order to improve the LCA. Thus, LCA can be used as basis for decision making during development in order to anticipate customer requirements and expectations. Furthermore, it is important to consider that the design highly affects the manufacturing processes and therefore, making the right decision in early development phases has the potential to save later costs and rework.

In order to ease the realization of LCA during the whole product life cycle, LCA-related data sets can be assigned or added to 3D models and provided beyond companies borders. Thus, all involved development partners can input relevant data and therefore feed the party responsible for components with up-to-date information that will enable quick decision at any time in order to reach LCA objectives.

An additional approach for supporting the objectives afford-mentioned can consist of agreeing on a standard LCA-process for the automotive. This is important in order to obtain comparable results as many customers can rely on life cycle assessment results to support buying decision. Such a framework can help OEMs to perform reliable self-assessments.

13.6 Discussion

The major fields on which the automotive industry is working in order to tackle the challenges presented in this paper are finding solutions for sustainability, urbanization and digitalization. Regarding sustainability, even though governments have elaborated many standards, measures to enable customers to make a deep sustainability-related comparison of diverse vehicles are still missing. Such a sustainability index should take not only the resource consumption of a vehicle into account, but also its life cycle and the life cycle of the propellant.

For instance, an integrated assessment of the costs and impacts of extracting, processing, and delivering a fuel or energy source to automobiles ("well to tank") and of using that fuel or energy source and generating emissions ("tank to wheels") [4] can be considered in addition to sustainability implications along the vehicles life cycles. Thus, the approach of linking Systems Engineering with sustainability as presented in this paper is a step towards the ideal state afore-mentioned. For this concept to be implemented in the industry, existing processes and legacy tools are to be rethought. For companies, it is important to determine the individual appropriate steps to introduce such a way of working, but also to implement detailed processes and monitoring mechanisms for quality assurance. Generally speaking, new vehicle concepts from a point of view of architecture, propellant, sharing and disposal can help enhancing Sustainability.

Furthermore, Sustainability should play a more important role in education in order to ensure that future engineers already are used to systems thinking in a sustainability context.

Not only the automotive community, but also authorities ought to find suitable incentives for boosting the proliferation of electrical vehicles. Many proposals already have been published on this topic, even though the number of electric vehicles models remains marginal compared to vehicles with combustive engines.

The switch from personal owned cars to Shared Mobility also can bring advantages regarding sustainability as customers can pay only for use, without taking the risks such as vehicles loss in value, which are associated with car ownership. Using TaaS-vehicles (Transportation as a Service) can lead to reducing the number of cars available in the most urban areas and therefore the number of parking areas. Those areas could be used for instance to create new green zones. For this purpose, shared mobility should not be limited only on providing additional vehicles for sharing. Instead, zero-emissions vehicles are needed, which are distributed and aligned to the customers' mobility needs.

For a better service quality, connection is needed between cars and the car-sharing customers, but also between cars and other involved parties (e.g. infrastructure, pedestrian). In this context, real-time road data can help saving millions of lives as about 90% of accidents are caused by human errors [4]. Besides, networked vehicles combined with self-driving ability will open the door of social mobility to everyone such as blind persons, seniors and any disabled persons without driving license.

Furthermore, the information connected by sensors and assistance systems as well as connectivity enablers could be used for the development of new vehicles, but also for manufacturing, maintenance, sales, marketing, to name a few. From a customer point of view, self-driving systems has the potential of helping a driver feeling the dynamic capabilities of his car, which he likely would not be able to experience if he were to drive his vehicle at limit.

However, the connecting ability of cars could offer an opportunity to third parties to manipulate vehicles and therefore to attack persons driving connected vehicles. Additionally, privacy protection of persons using connected cars could be levered, as a high amount of information can be collected (e.g. driving style, motion profile).

The bottom line despite all the potential risks is that when thinking of sustainability as an integrator of environmental, social and economic factors, the advantages far outweigh: A self-driving networked electric vehicle produce zero emission, optimizes the routes, offers mobility to all, saves lives and therefore contributes to a high extent to sustainability.

Due to sustainability requirements, it can be expected that the angular stone of upcoming vehicles will not be the combustion engine anymore; instead, it will be the Automated Driving System (ADS), which enables autonomous driving (e.g. perception, decision and control as well as vehicle manipulation). The digitalization approaches (incl. Internet-of-Thing, Industry 4.0, additive Manufacturing, etc.) provide not only ways for saving resources but also for creating new services and business opportunities, which will contribute to enhance sustainability.

13.7 Conclusions and Outlook

Driving a car in big cities is often concomitant with traffic jams that in turn can increase not only the power consumption but also the stress of passengers. Thus, car owners spend a considerable time for unnecessary waiting in traffic and searching for parking, a time that could be invested for productivity.

Besides, the effects of environmental pollution have led to many cities being affected by brown haze, gridlock as well a shortage of parking. This trend is expected to be increasing in the next years because of the increasing agglomeration in such cities. As a consequence, some municipal administrations have taken measure for reducing pollution. For instance, Mexico City's *Hoy No Circula* ("no-drive days") program uses the license-plate numbers of vehicles to ration the number of days when they may be used. Further cities in Europe are following a similar approach with so-called green zones (low-emission zones) that are restricted to vehicles that fulfill specific standards and low-emissions zones to restrict vehicles with internal-combustion engines [4, 75, 76]. All this calls for approaches to enhance sustainability.

Although many approaches are being followed individually, this paper proposes a holistic approach that consists of linking Systems Engineering with sustainability in order to take sustainability implications into account along the whole product life cycle of vehicles. Especially the use of innovative approaches of the Digitalization such as Additive Manufacturing, Industry 4.0 and Internet of Things as well as Life Cycle Analysis for enhancing sustainability are highlighted.

In perspective, elaborating detailed scenarios and integrated processes as well as identifying the tools necessary for implementing the presented concepts are needed, as many gaps are still available from the first product idea through design down to manufacturing and recycling.

References

1. Henessy DA, Wiesenthal DL (1997) The relationship between traffic congestion, driver stress and direct versus indirect coping behaviours. Ergonomics 40(3):341–348
2. Dias C, Miska M, Kuwahara M, Warita H (2009) Relationship between congestion and traffic accidents on expressways: an investigation with bayesian belief networks. In: Proceedings of 40th annual meeting of infrastructure planning (JSCE)
3. Marchesini P, Weijermars W (2010) The relationship between road safety and congestion on motorways. Research report (https://www.google.de/url?sa=t&rct=j&q=&esrc=s&source=web&cd=8&cad=rja&uact=8&ved=2ahUKEwjU9_ixovzeAhWKaFAKHdBZDH4QFjAHegQIAhAC&url=https%3A%2F%2Fwww.swov.nl%2Fsites%2Fdefault%2Ffiles%2Fpublicaties%2Frapport%2Fr-2010-12.pdf&usg=AOvVaw2l395JNB7ZwWAadkyfh58S). Retrieved on 30.11.2018
4. Gao P, Hensley R, Zielke A (2014) A road map to the future for the auto industry. McKinsey study. Retrieved on 20161118 on http://www.mckinsey.com/industries/automotive-and-assembly/our-insights/a-road-map-to-the-future-for-the-auto-industry
5. Spiegel Online (2018) Dieselfahrverbot in Stuttgart—Was Autofahrer jetzt wissen sollten. http://www.spiegel.de/auto/aktuell/stuttgart-die-wichtigsten-fragen-zum-diesel-fahrverbot-a-1217930.html. Retrieved on 11.07.2018
6. Standway D (2018) China cuts smog but health damage already done: study. https://www.reuters.com/article/us-china-pollution-health/china-cuts-smog-but-health-damage-already-done-study-idUSKBN1HO0C4. Retrieved on 16.4.2018
7. Wognum N, Bil C, Elgh F, Peruzzini M, Stjepandić J, Verhagen WJC (2019) Transdisciplinary systems engineering: implications, challenges and research agenda. Int J Agile Syst Manag 12(1):58–89
8. Khisty CJ, Mohammadi J, Amekudzi AA (2012) Systems Engineering with Economics, Probability and Statistics, 2nd edn. J. Ross Publishing, Fort Lauderdale
9. Spath D, Pischetsrieder B (2010) Elektromobilität: Eine Technologie mit Historie und Zukunft. In: Elektromobilität—Potenziale und wissenschaftlich-technische Herausforderungen. Springer, Berlin
10. Husain I (2003) Electric and hybrid vehicles—design fundamentals. CRC Press, Boca Raton
11. Biahmou A (2015) Sustainable mobility. In: Stjepandic J, Wognum M, Verhagen JC (eds) Concurrent engineering in the 21st century: foundations, developments and challenges. Springer International Publishing, pp 779–803
12. Eichler S, Schroth C, Eberspächer J (2006) Car-to-Car Communication. VDE-Kongress—Innovations for Europe (Aachen), https://www.alexandria.unisg.ch/publications/30950. Retrieved 06.12.2018
13. Elfirdoussi S, Jarir Z (2019) An integrated approach towards service composition life cycle: a transportation process case study. J Ind Inf Integr (in press)
14. Sobolewski M (2017) Amorphous transdisciplinary service systems. Int J Agile Syst Manag 10(2):93–115
15. Knake-Langhorst S, Löper C, Schebitz N, Köster N (2014) Fahrerunterstützung beim Ein- und Ausfädeln. In: Siebenpfeiffer W (eds) Vernetztes Automobil, ATZ/MTZ-Fachbuch. Springer Fachmedien Wiesbaden
16. Li L, Wang F (2007) Advanced motion control and sensing for intelligent vehicles. Springer Science + Business Media, LLC, New York
17. Wang J, Li K, Lu X-Y (2014) Effects of human factors on driver behavior. In: Wang F (ed) Advances in intelligent vehicles. Elsevier, Waltham
18. Pu H (2016) Dynamic eHorizon with traffic light information for efficient urban traffic. In: Schulze T et al (eds) Advanced microsystems for automotive applications 2015-smart systems for green and automated driving. Springer International Publishing, Switzerland
19. Christen F, Eckstein L, Katriniok A, Abel D (2014) Sattelitenbasiertes Kollisionsvermeidungssystem. In: Siebenpfeiffer W (ed) Vernetztes Automobil, ATZ/MTZ-Fachbuch. Springer Fachmedien Wiesbaden

20. Tessema Y, Höffken M, Kresel U (2016) Driver head pose estimation by regression. In: Schulze T et al (eds) Advanced microsystems for automotive applications 2015-smart systems for green and automated driving. Springer International Publishing, Switzerland
21. Kunze S, Pöschl R, Grzemba A (2016) Comparison of energy optimization methods for automotive ethernet using idealized analytical models. In: Schulze T et al (eds) Advanced microsystems for automotive applications 2015-smart systems for green and automated driving. Springer International Publishing, Switzerland
22. Adasis-Forum (2015) https://adasis.org. Retrieved 7.12.2018
23. Schubert R (2014) Automatische Manöverentscheidungen auf Basis unsicherer Sensordaten. In: Siebenpfeiffer W (ed) Vernetztes Automobil, ATZ/MTZ-Fachbuch. Springer Fachmedien Wiesbaden
24. Yoshida Y, Pouke M, Tada M, Noma H, Noda M (2014) Evaluation method for safe driving skill based on driving behavior analysis and situational information at intersections. In: Schmidt G et al (eds) Smart mobile in-vehicle systems: next generation advancements. Springer Science + Business Media, New York
25. Divakarla KP, Emadi A, Razavi SN (2018) A cognitive advanced driver assistance systems (ADAS) architecture for autonomous-capable electrified vehicles. IEEE Trans Transp Electrif 5(1):48–58
26. Behere S (2016) Reference architecture for highly automated driving. Doctoral thesis in machine design, KTH Royal Institute of Technology, Stockholm, Sweden
27. Delgrossi L, Zhang T (2012) Vehicle safety communication—protocols, security, and privacy. Wiley, Hoboken
28. Bako B, Weber M (2011) Efficient information dissemination in VANETs. In: Almeida M (ed) Advances in vehicular networking technologies. InTech, Croatia
29. Naja R (2013) A survey of communications for intelligent transportation systems. In: Naja S (ed) Wireless vehicular networks for car collision avoidance. Springer Science + Business Media, New York
30. Vegni AM, Biagi M, Cusani R (2013) Smart vehicles, technologies and main applications in vehicular ad hoc networks. In: Giordano LG, Regiani L (eds) Vehicular technologies—deployment and applications. InTech, Croatia
31. Wekerle T, Pfouga A, Stjepandić J, Mai P (2018) Intellectual property protection in smart systems engineering on exchange of simulation models. In: Advances in transdisciplinary engineering, vol. 7. IOS Press, Amsterdam, pp 198–207. https://doi.org/10.3233/978-1-61499-898-3-198
32. Wolf M, Osterhues A (2014) Sichere Botschaften—Moderne Kryptographie zum Schutz von Steuergeräten. In: Siebenpfeiffer W (ed) Vernetztes Automobil, ATZ/MTZ-Fachbuch. Springer Fachmedien Wiesbaden
33. Biahmou A (2015) Systems engineering. In: Stjepandic J, Wognum M, Verhagen JC (eds) Concurrent engineering in the 21st century: foundations, developments and challenges. Springer International Publishing, Switzerland, pp 221–254
34. Sadlauer A, Hehenberger P (2017) Using design languages in model-based mechatronic system design processes. Int J Agile Syst Manag 10(1):73–91
35. Lamb CMT (2009) Collaborative systems thinking. An exploration of the mechanisms enabling systems thinking. Ph.D. thesis, Massachusetts Institute of Technology, Cambridge
36. Hitchins DK (2003) Advanced systems thinking, Engineering and management. Artech House, Boston
37. Tolk A, Adams KM, Keating CB (2011) Towards intelligence-based systems engineering and system of systems engineering. In: Tolk A, Jain LA (eds) Intelligence-based systems engineering. Springer, Berlin
38. Lindemann U, Maurer M, Braun T (2009) Structural complexity management. An approach for the field of product design. Springer, Berlin
39. Schuh G, Schwenk U (2001) Produktkomplexität managen. Hanser, München
40. Biahmou A, Fröhlich A, Stjepandić J (2010) Improving interoperability in mechatronic product development. In: Thoben KD et al (eds) Collaborative value creation throughout the whole lifecycle. Proceedings of PLM10 international conference, Inderscience, Geneve

41. Winzer P (2013) Generic systems engineering. Springer, Berlin
42. Haskins C (2011) International council on systems engineering (INCOSE) systems engineering handbook. A guide for system life cycle processes and activities, V. 3.2.1, INCOSE-TP-2003-002-03.2.1
43. Eigner M, Roubanov D (eds) (2014) Modellbasierte virtuelle Produktentwicklung. Springer, Berlin
44. Berthold A (2002) Der fertigungsorientierte Modellierer FERMOD als Erweiterung des Konstruktionssystem WISKON, S. 1-3, Kassel
45. Weber P (2011) Kostenbewusstes Entwickeln und Konstruieren, Grundlagen – Methoden –Beispiele, Expert Verlag, S. 1 ff, Renningen
46. Leber M, Ratte M (2018) Modelling company-specific PLM maturity levels for Industry 4.0. Prod Data J 1:44–47
47. Fraile del Pozo A, Pellicier EL, Arroyo Garcia JB, Torné A (2016) Influence of the design parameters of electric vehicles in the optimization of energy efficiency in urban routes. In: Schulze T et al (eds) Advanced microsystems for automotive applications 2015-smart systems for green and automated driving. Springer International Publishing, Switzerland
48. Kuhn O, Liese H, Stjepandic J (2011) Methodology for knowledge-based engineering template update, IFIP advances in information and communication technology, 355 AICT. Springer, Berlin, pp 178–191
49. Kahneman D (2013) Thinking, fast and slow. Farrar, Straus and Giroux
50. Dobelli R (2014) The art of thinking clearly. Farrar, Straus and Giroux
51. Emmer C, Fröhlich A, Stjepandic J (2013) Advanced engineering visualization with standardized 3D formats, IFIP advances in information and communication technology, vol 409. Springer, Berlin, pp 584–595
52. Schwarz JM, Beloff BR, Beaver E (2002) Use sustainability metrics to guide decision making. Chem Eng Prog 98:58–63
53. Leigh Reid R, Rogers KJ, Johnson ME, Liles DH (1996) Engineering the virtual enterprise. In: Automation & robotics research institute—conference '96, 1996, pp 485–490
54. Jiang P, Ding K, Leng J (2016) Towards a cyber-physical-social-connected and service-oriented manufacturing paradigm: social manufacturing. Manuf Lett 7:15–21
55. Koren Y (2013) The rapid responsiveness of RMS. Int J Prod Res 51(23–24):6817–6827
56. Zeinab KAM, Elmustafa SAA (2018) Internet of things applications, challenges and related future technologies. World Sci News 67(2):126–148
57. Gebhardt A, Kessler J (2016) Grundlagen und Anwendungen des Additiven Manufacturing (AM). Carl Hanser Verlag
58. Hagl R (2015) Das 3D-Druck-Kompendium – Leitfaden für Unternehmer, Berater und Innovationstreiber, 2. Auflage, Springer Gabler
59. Biahmou A, Stjepandic J (2016) Towards agile enterprise rights management in engineering collaboration. Int J Agile Syst Manag 9:302–325
60. Biahmou A, Emmer C, Pfouga A, Stjepandić J (2016) Digital master as an enabler for industry 4.0. In: Advances in transdisciplinary engineering, vol 4. IOS Press, Amsterdam, pp 672–681
61. Rauch E, Dallasega P, Matt DT (2017) Distributed manufacturing network models of smart and agile mini-factories. Int J Agile Syst Manag 10(3/4):185–205
62. Kim JH (2017) A review of cyber-physical system research relevant to the emerging IT trends: industry 4.0, IoT, big data, and cloud computing. J Ind Integr Manag 2(3):1750011. https://doi.org/10.1142/S2424862217500117
63. VDI/VDE-GMA: Statusreport Referenzarchitekturmodell Industrie 4.0. VDI/VDE-GMA, (2015)
64. Gausemeier J, Czaja A, Dülme C (2015) Innovationspotentiale auf dem Weg zu Industrie 4.0. In: Wissenschafts - und Industrieforum Intelligente Technische Systeme 2015, Heinz Nixdorf Institut
65. Qi Q, Tao F (2018) Digital twin and big data towards smart manufacturing and industry 4.0: 360 degree comparison. IEEE Access 1-1. https://doi.org/10.1109/access.2018.2793265

66. Wallis R, Stjepandic J, Rulhoff S, Stromberger F, Deuse J (2014) Intelligent utilization of digital manufacturing data in modern product emergence processes. Adv Transdiscipl Eng 1:261–270
67. Madni AM, Madni CM, Lucero SD (2019) Leveraging digital twin technology in model-based systems engineering. Systems 7:7. https://doi.org/10.3390/systems7010007
68. Chaudhry IA, Shami M, Khan A (2004) Manufacturing applications of artificial intelligence. J Eng Appl Sci 23:29–33
69. Tao F, Wang Y, Zuo Y, Yang H, Zhang M (2016) Internet of Things in product life-cycle energy management. J Ind Inf Integr 1:26–39
70. Sharif Ullah AMM (2019) Modeling and simulation of complex manufacturing phenomena using sensor signals from the perspective of Industry 4.0. Adv Eng Inform 39:1–13
71. Zhong RY, Ge W (2018) Internet of things enabled manufacturing: a review. Int J Agile Syst Manag 11(2):126–154
72. Holland M, Nigischer C, Stjepandic J (2017) Copyright protection in additive manufacturing with blockchain approach. In: Chen CH et al (eds) Transdisciplinary engineering: a paradigm shift. IO Press, Amsterdam, pp 914–921
73. Klöpffer W, Grahl B (2014) Life cycle assessment (LCA): a guide to best practice. Wiley-VCH Verlag, Weinheim
74. Hauschild MZ, Rosenbaum RK, Olsen SI (2017) Lifecycle assessment: theory and practice. Springer International Publishing, Switzerland
75. Obrecht M, Rosi B, Potrc T (2017) Review of low emission zones in Europe: case of London and German cities. Tehnicki Glasnik 11(1–2):55–62
76. Jiang W, Boltze M, Groer S, Scheuvens D (2016) Impacts of low emission zones in Germany on air pollution levels. Transp Res Procedia 25(2017):3370–3382

Part IV
Current Challenges

Chapter 14
Future Perspectives in Systems Engineering

Wim J. C. Verhagen, Josip Stjepandić and Nel Wognum

Abstract Systems Engineering (SE) is a well-established field of research and practice. Nevertheless, the theory underlying SE is experiencing significant development, directly and in association with advancements in closely associated research domains. In this final Chapter, a socio-technical perspective is applied to identify and describe major trends in SE, as well as identifying future challenges in theory and application of SE. In doing so, trends are identified for (1) strategic issues from a product and process lifecycle perspective; (2) stakeholder representation and involvement; (3) current and future technologies employed to enable SE; (4) knowledge and skills as contributed by people and teams; and (5) structures to enable transdisciplinary activities supporting a socio-technical system perspective in systems development. Challenges remain present regarding these dimensions; SE requires methods and tools that are suitable to support the dynamic and evolving nature of the systems that need to be developed including the development system itself. Besides, management of SE projects for solving complex societal problems requires people with vision and power to motivate and mobilize the necessary people and value their respective input in the overall task. Transdisciplinary Engineering is introduced as an approach in which Systems Thinking and System Approaches interoperate, taking into account the different levels of abstraction of the system of focus.

Keywords Systems engineering · Transdisciplinary engineering · Socio-technical systems · Systems thinking

W. J. C. Verhagen (✉) · N. Wognum
Technical University Delft (TU Delft), Delft, The Netherlands
e-mail: w.j.c.verhagen@tudelft.nl

N. Wognum
e-mail: p.m.wognum@tudelft.nl

J. Stjepandić
PROSTEP AG, Darmstadt, Germany
e-mail: josip.stjepandic@prostep.com

J. Stjepandić et al. (eds.), *Systems Engineering in Research and Industrial Practice*,
https://doi.org/10.1007/978-3-030-33312-6_14

14.1 Introduction

The foundations, methods and recent developments regarding Systems Engineering (SE) and its applications in several industries have been discussed at length in the preceding Chapters of this book. As part of the respective Chapters, authors have discussed specific research and practical challenges in isolation. The aim of this final Chapter is to integrate and present current trends and challenges with respect to SE and its associated research (sub) domains in a comprehensive overview. To perform this integration in a structured, methodical manner, the socio-technical dimensions of SE are explored in more detail.

The proposed dimensions are based upon prior research [1, 2] and are as follows:

- **Strategy/goals**: overall goal or vision of Systems Engineering and its particular application(s), which may include aspects such as profitability, sustainability, availability, efficiency, etc. Strategy and goals become visible in a product/project portfolio or ideas and requirements for new products and associated processes.
- **Stakeholders**: actors, functions, disciplines or roles that are involved in the processes that need to be executed in the organizational system. Not only directly involved stakeholders (internal organisation, supply chain partners) may be considered, but also external stakeholders that influence the process but have limited direct involvement (e.g., governmental institutions, technology providers).
- **Technology**: the technologies required to control and manage the processes associated with the development system. This includes hardware as well as software, including technology like internet. This aspect also includes product and process technology needed to develop the product system.
- **Knowledge/information**: the knowledge or information required to initiate and sustain the processes composing the development system. This comprises the tacit as well as explicit knowledge available in people, individually or collectively, and information systems. It may include internal and/or external information sources. This aspect includes the knowledge and information on the product system to be developed.
- **Organisation/structure**: the configuration of the processes required to run the development system. This may include aspects such as location, hierarchy, teams, communication lines, degree of outsourcing, organisational arrangements, comprising rules and norms, and so on. These aspects form the structure and culture of an organisation.

These elements make up an organizational system that performs processes to achieve the goals according to a particular strategy. These goals need to be aligned with the needs in the environment of the organizational system. These needs may consist of a problem that can be characterised as a, possibly complex, technical problem [3]. In this case, a hard SE approach might suffice to solve the problem. The needs may, however, also reflect a complex societal problem [3], like a sustainability problem. For example, the current problem with plastic waste might be referred to as a complex societal problem. Even when a waste treatment and destruction system is in place, plastic waste remains a problem. However, bringing plastic products into societies without such a system in place, creates huge problems. These problems are gaining more awareness, but are hard to solve. Such problems need at least a critical systems approach is needed to solve such problems, requiring the involvement of many different stakeholders with multiple views on the problem at hand.

In essence, the outcome of the processes is meant to satisfy needs in the environment and is to be used in and accepted by the environment of the system by clients, customers, citizens, etc., for which the outcome is meant. Some stakeholders, the external ones as mentioned above, are part of this environment. In Fig. 14.1, the system is depicted. Some elements cross the system borders, because they may also be part of the environment of the system.

The framework is an analytical system-based framework that helps to structure knowledge and information on a complex organizational system, like in Systems Engineering and other system approaches [3], Concurrent Engineering or Transdisciplinary Engineering. It provides a holistic view on a certain system of focus. The system of focus, including its processes of focus, need to be framed beforehand.

The framework is applicable and has been applied to a variety of contexts (see e.g., [2]). The various elements of the framework require more specific theories, method and tools when more in-depth investigation is needed, for example for analyzing specific problems that pop-up after the system and its elements have been studied holistically. This holistic study requires already a massive amount of information to be provided by people, e.g., through interviews, or by documents, like product, process, and specific project documents. In addition, not only engineering and design

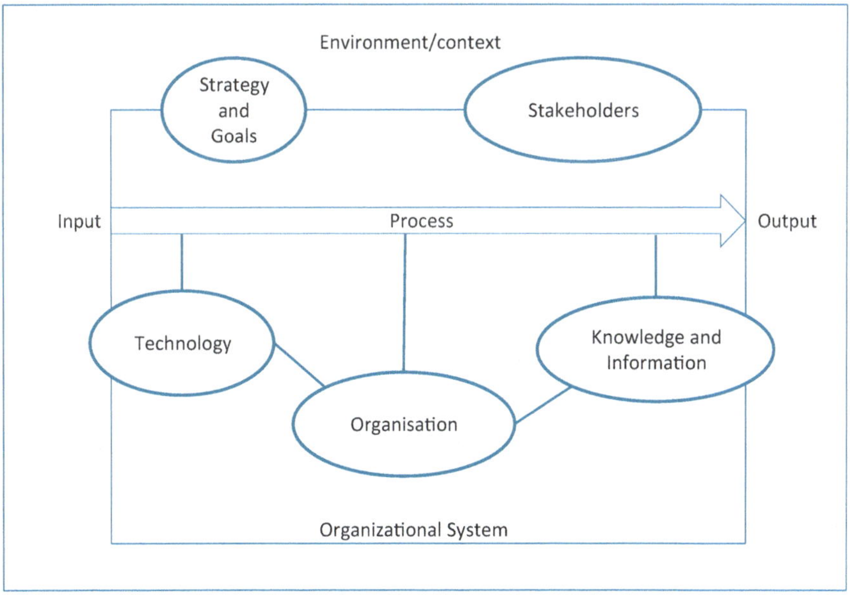

Fig. 14.1 Organizational system with its elements and environment

methods may be needed, but also methods and tools from other disciplines, depending on the nature of the subproblem to be tackled.

The outline of the chapter is as follows. In Sect. 14.2, the trends in SE are discussed, following the five dimensions given above. In a similar fashion, Sect. 14.3 highlights challenges in research and practice. Finally, a brief conclusion on the presented work is given in Sect. 14.4.

14.2 Trends in Systems Engineering

In this section, trends in research and practice for the (near) future are identified. This overview of trends covers different levels of system scope and technological uncertainty, as applied to the five dimensions identified in the introduction. Together, this allows for a more accurate expression of the five dimensions in terms of aggregation levels. The levels of system scope and technological uncertainty are a well-established principle which is in line with the two-dimensional taxonomy of Systems Engineering proposed by Shenhar and Bonen [4]. In particular, the following major domains and associated developments can be mapped along these axes:

- **System scope**: the additional scale, volume and complexity posed by new concepts and associated domains such as Cyber-Physical Systems and the Internet of Things (see Chap. 6) ensure that SE methods for modelling and simulating systems and their interactions must keep pace [5]. These developments increasingly demand an interdisciplinary or transdisciplinary approach to solve SE challenges, as the underlying constructs are themselves inter- or transdisciplinary. This realization is reinforced when considering different levels of aggregation, across system scope and time [6]. In particular, current research is focused on covering and deepening the understanding of the System-of-Systems paradigm. With respect to time, the shift of industry and academia towards an integrated perspective on products and services (see Chaps. 7 and 10) implies that a full lifecycle perspective on development is becoming prevalent [7]. This perspective is holistic: it is not constrained to development, but remains active beyond the beginning-of-life (BOL) phase of the product lifecycle, with for instance the construction of digital twins and condition/health monitoring allowing for more effective and sustainable operation, maintenance and disposal of products [8, 9].
- **Technological uncertainty**: Shenhar and Bonen [4] define four levels of technological uncertainty, ranging from low-tech to 'superhigh-tech' technologies. Different levels of technological uncertainty require different levels of integration; as Shenhar and Bonen [4] note, "in lower-tech systems, assembling, integrating, and testing all subsystems into one unit is usually not an issue, since all pieces readily fit together. Integration difficulties become prominent, however, in the higher uncertainty type system projects. In these projects, the successful production of the separate subunits is one thing, while integrating them into one working piece is another." In addition, "there are problems of configuration and risk management

[4]. Recent trends regarding digitization (see e.g. Chap. 12) as well as consistency and traceability of data, information and knowledge (Chap. 5) hold promise in addressing integration, configuration and risk aspects of high-tech product development.

Keeping the previous in mind, the next subsections consider trends in SE goals and strategies, stakeholders, technology, information and knowledge, and structures and organization respectively.

14.2.1 Trends in SE Goals and Strategies

Product development is an increasingly global activity, with co-creation and co-production spurring the interconnectedness of what used to be relatively monolithic activities [10]. This poses significant challenges on interoperability of methods, tools as well as organizations and users. In relation with this, the dynamic interactions between technical and social characteristics of new product (and service) development need to be taken into account in models, methods and tools. A transdisciplinary approach is gaining more prominence as it answers towards these demands: its focus on 'human-in-the-loop' modelling representations in which methodologies, tools and actors from multiple disciplines are combined is suitable for socio-technical systems engineering [11, 12].

Integration and interoperability are assumed to speed up development and lower costs, yet developing and implementing interoperable, integrated solutions can also be seen as a driver of complexity. From this perspective, the adoption of inter- or transdisciplinary approaches can be seen as both a viable strategy as well as another compounding factor in complexity, as transdisciplinarity is characterized by a 'catch-22': its potential and value have been shown in several cases, yet the characteristics of these cases are so particular that a generalized methodological approach is hard to distill [13–17]. Such challenging cases are however precisely where a transdisciplinary approach holds significant potential in comparison to competing approaches. Nevertheless, adopting transdisciplinarity will remain a complex and time-consuming endeavor in the (near) future, as there in essence is no standard map to guide practitioners, because of the different and particular characteristics of complex problems that require a sustainable, usable, and useful solution.

Another major strategic shift is servitization of manufacturing industries, i.e., the innovation of organization's capabilities and processes to shift from selling products to selling integrated products and services that deliver added value [18]. Servitization provides means for companies to move up the value chain and exploit higher value business activities, leading to product-service system (PSS) development. This shift in strategy and associated goals is increasingly being supported by research initiatives, as detailed in Chaps. 7 and 10.

14.2.2 Trends in SE Stakeholders

From one perspective, the stakeholders in SE can be considered to be stable: since the inception of SE, the consideration of all relevant stakeholders in a problem (e.g. designers, manufacturers, operators, legislators, end users, society, etc.) is a hallmark of available methods and tools. In particular, successful functional analysis and associated requirements engineering [18] is dependent on adequate consideration and involvement of stakeholders.

From another perspective, several trends in SE stakeholder considerations can be discerned. The first is related to the aforementioned growth in complexity and increasing integration and interoperability: with increasing complexity and integration, the size and complexity of the stakeholder environment are also expanding. In particular, the realization of complex systems usually requires the temporary collaboration of a multitude of stakeholders from different domains, such as hardware, software and services [18].

However, in recent years the consideration of stakeholders and their requirements has moved beyond the predominant consideration of technical requirements by technical stakeholders: inter- and transdisciplinary approaches aim to adequately represent and involve societal stakeholders using methodologies from a non-technical background (e.g., cognitive methodologies) [19]. Transdisciplinarity, however, implies that collaboration between stakeholders is essential—not merely at an academic or disciplinary collaboration level, but through systematic, repeatable collaboration with people affected by the stakeholders. In such a way, transdisciplinary collaboration becomes uniquely capable of engaging with different ways of knowing the world, generating new knowledge, and helping stakeholders understand and incorporate the results in a common, integral product or product-service system.

A second major trend relates to the ubiquity of data driven by condition-monitored technology and the Internet of Things. From a stakeholder perspective, this generates major concerns (and therefore requirements) regarding the ownership and security of data [20]. This intertwines with a third major trend regarding stakeholders: the shift towards Product-Service Systems and a resulting diffusion of characteristics between design, manufacturing, operator and maintenance organizations. Design for all is increasingly being replaced by design by all, as described in Chap. 6 of this book. In other words, organizations are increasingly focusing on value creation outside their boundaries, with value being created through interplay of customers, competitors, collaborators and the wider community. A major component in this is to leverage data, through for instance Big Data and Internet of Things applications [21]. Depending on the exploitation of the product (e.g., a passenger car), a vendor can build or have built a user profile of the customer and, based on this, offer additional services (route optimization, payment, infotainment, maintenance) using the recorded operating data [1]. However, this assumes that the end user and the various organisations along the value chain are not only able but also willing to share their data. This assumption is increasingly being challenged in today's society; for instance, consider the European Union's GDPR legislation.

14.2.3 Trends in SE Technology

SE technology trends are closely related to the strategic trends that have been identified.

Interoperability and integration lean heavily on the availability of data across products, systems and lifecycle phases [22]. Technologies such as digital twins can be leveraged in combination with manufacturing, operations and maintenance data to perform real-time analyses supporting predictive maintenance, process monitoring and control (Fig. 14.2) [23]. In addition, the advent of artificial intelligence (including shallow and deep learning approaches) increasingly replaces human intervention in model construction, learning and application [24].

The latter points runs counter to the increasing trend of applying transdisciplinarity in SE approaches. From a technological standpoint, this calls for robust tools to support the application of a variety of methodologies in product development. As of now, a major focus in research and development concerns multidisciplinary applications, in particular for (automated) multidisciplinary optimization approaches [25–27]. Transdisciplinary approaches are supported less thoroughly, though multi-agent systems (MAS) and associated tools hold considerable promise in supporting transdisciplinary modelling and analysis. Much current research aims to offer well-founded semantics and formalization techniques to support both qualitative and quantitative methodologies in MAS applications [28]. The involvement of a people from various disciplines requires, that tools are usable by the various disciplines.

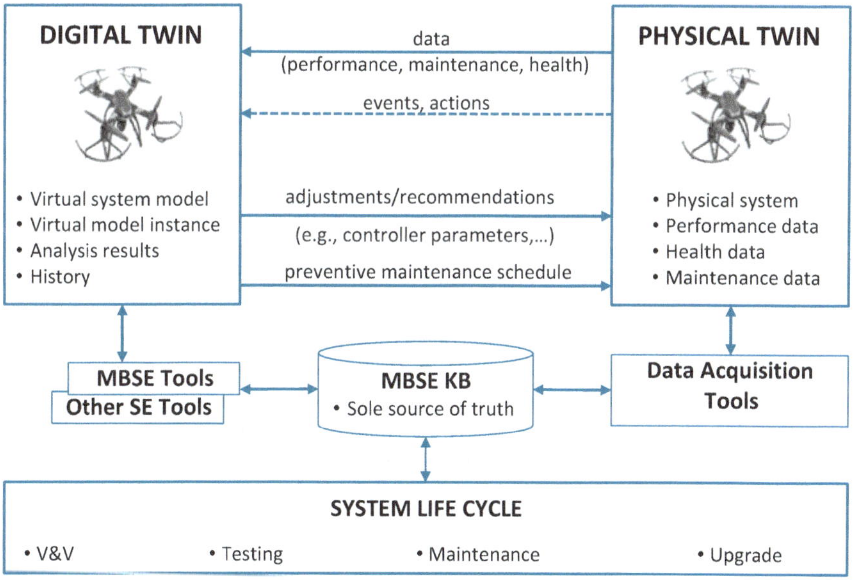

Fig. 14.2 Digital twin concept within model based systems engineering framework [23]

A final consideration with respect to SE technology trends is the increased uptake of methods for analyzing tacit and explicit knowledge. For these aspects, Natural Language Processing (NLP) has been validly adopted to enable formalization of tacit knowledge [29], according to a transdisciplinary approach, as pointed out in Chap. 7.

14.2.4 Trends in SE Information and Knowledge

The aforementioned trends towards integration and interoperability have an impact on the exchange of data, information and knowledge. Using technology means, sharing has become easier than ever. The amount of data being created, stored and used every day is growing exponentially. The way in which the knowledge is used in the design process is changing continuously. Some of the driving factors have been addressed in previous research [30–32] and include persistent storage of design information, and development of well-founded semantics for encoding design knowledge and facilitating interoperability. From a user perspective, natural user interfaces facilitate intuitive human–computer interaction with little user training or instruction, as evident in the ubiquity of touch-based personal computing devices like smartphones and tablets. The portable and ubiquitous nature of tablet computers make them ideal for collaborative design processes like the recording and progressive documentation of design discussions. Though this may prove to be an important factor towards mass collaboration and the democratizing of the design process, this hasn't transpired on a large scale as of yet.

Another research trend with respect to knowledge and information concerns human involvement in development [33]. Humans need to be 'in the loop', especially in the earlier phases of design. Emergent behavior cannot be explained sufficiently in deterministic approaches to design, because interaction between components and their behaviors is not well understood [34]. In addition, the impact of a design on its context needs to be anticipated as much as possible to achieve the necessary sustainability, although this impact evolves through the emerging characteristics of a product or service in use [35]. As highlighted before, socio-technical modelling approaches are necessary to model and evaluate this emergent behavior.

As mentioned before, a major trend in research and practice related to information and knowledge concerns intellectual property (IP) and its protection. Increasing cooperation between stakeholders necessitates intellectual property protection and enterprise rights management. Methods for patent infringement tracking as well as for IP protection [36, 37] in information and data flow must be developed to a further extent. In this light, the traceability of data, information and knowledge over a product's life, as well as the decisions being taken regarding its design, manufacturing, use and disposal, will take on an even greater importance, as traceability aspects can be used to monitor and control IP issues over time [38].

14.2.5 Trends in SE Structures and Organisation

Many modern products have achieved such a level of complexity that their behavior cannot be predicted in a sufficient way, followed by unexpected behavior or even failures. System thinking is a holistic analysis approach that focuses on the way constituent parts of a system interrelate and how systems work over time and within the context of larger systems [4]. This contrasts with traditional analysis, which studies systems by breaking them down into their separate elements. Systems thinking can be used in any area of research and science (medical, environmental, economic, etc.). Its application in an engineering context is known as systems engineering (SE)—the subject of this book. The following key trends drive process changes for system thinking [39], some of which have been highlighted in previous subsections:

1. Increasingly complex, global systems of systems: Products become more complex and connected, while they also interact with the context of use.
2. Emergent requirements: The most appropriate user interfaces and collaboration modes for a complex human-intensive system are not specifiable in advance, but emerge with usage.
3. Rapid change: Trying to stay competitive in a world of increasingly rapid changes requires new levels of agility, and shorter times between new releases of products and services.
4. High assurance of qualities: At the same time that systems engineering and development need to become more agile, the growing interdependence of systems and people requires systems to have higher assurance levels. It is even harder to get agreement among multiple system owners with widely disparate quality priorities.

A core aspect to address these key trends is found in transdisciplinarity. Transdisciplinarity covers the deep integration of various forms of research or expertise to create a holistic approach [16, 17]. As the prefix "trans" indicates, trans-disciplinarity concerns issues that are between the disciplines, across the different disciplines, and beyond each individual discipline simultaneously. When for instance contrary targets exist, transdisciplinarity can help determine the most relevant problems and solution approaches. It shares capabilities like interdependence with the network-centric world.

14.3 Challenges in Systems Engineering

In this section, challenges in research and practice for the (near) future are identified. This section builds upon the trends identified in Sect. 14.2—whereas current initiatives were described in detail there, this Section will focus on major assumptions, limitations and gaps in SE research and applications.

14.3.1 Challenges in SE Goals and Strategies

SE goals and strategies are characterised by expansion across time and space: globalization, increased integration, interoperability and servitization—all of which leverage increased availability of data and ubiquity in communication technology—are moving forward. In this light, two major challenges can be identified relative to technological scope and uncertainty.

First of all, the increasing scope and interconnectedness of systems and their development effort reinforce the observations by Nielsen [40] regarding alignment of development scope and uncertainty. In particular, whereas individual components or even systems can be addressed using traditional approaches, System-of-Systems (SoS) development requires improved methods for "adaptively characterizing and analysing (emergent) behaviour in SoS. This call for improved methods is mirrored in previous literature [40] to anticipate unintended behaviour in SoS operation. In Chap. 4, Mo and Beckett address the evolution over time of SoS, where no single controlling authority is present to specify SoS from a top-down set of requirements, and where individual systems composing the SoS are not stable over time, but change continuously over their lifecycle, because of changing needs in their context, which may also be the larger overall system. Notably, it is observed that "global as well as individual goals, which may be contradicting and change over time. Researchers have observed that whilst the actor, activity and resource elements relating to a particular situation may not change very fast, the connections between them can" (Chap. 4). In this light, new methods must be developed that support a "range of trade-offs between attributes such as adaptability and modularity that can be simulated in SoS architecture models" [40].

A second major challenge concerns the evaluation and performance assessment of systems (of systems). Especially at a higher level of aggregation (i.e., an SoS level), overall objectives can be vaguely identified if at all. As a result, SoS Key Performance Indicators (KPIs) are limitedly present in literature and practice [41]. Practical examples include a lack of integration between Product Lifecycle Management (PLM) and Service Lifecycle Management (SLM) metrics and associated feedback, as highlighted in Chaps. 7 and 10. In addition, KPIs for sustainability are not adequately taken into account in system development, as stated in Chap. 13. Methods should be developed that allow for a structured, systematic assessment of SoS performance, addressing questions regarding the aggregation of individual system performance while properly accounting for system interdependencies and emergent behavior. It is important, in addition, to identify ownership for such KPIs, because measurements require follow-up for improvement.

14.3.2 Challenges in SE Stakeholders

Trends in SE stakeholder perspectives center on the adoption of novel methodologies for addressing emergent behaviour in SoS, as well as handling and leveraging the ubiquity of data under the IoT/Industry 4.0 paradigms. The following challenges can be discerned in relation to these developments as well as the aforementioned challenges in SE goals and objectives:

- **Stakeholder representation, involvement and support**: stakeholders must be represented and involved in the development, production, use and disposal phases of systems and SoS. As mentioned in Sect. 14.2.2, a transdisciplinary approach is one of the ways in which this can be achieved: not only are stakeholders represented, but they need to provide and apply their methodologies, tools and metrics. However, a potential downside is the added effort in ensuring adequate representation and involvement of diverse stakeholders, with diverse methodologies, tools and techniques. In such complex environments, ownership and traceability of data, information, knowledge and ensuing decisions present huge challenges. These challenges are compounded by the element of time: how to ensure that essential information is stored over time and remains accessible to the appropriate stakeholders, with the right permissions?
- **Stakeholder involvement in system performance assessment**: Nielsen highlight previous research which identifies challenges in topics such as "orchestrating activities between diverse stakeholders, in measuring and evaluating the effectiveness of system design, and in creating adaptive system infrastructures" [42]. In particular, the assessment of evolving system compositions is highlighted, an item that has also been described in Chap. 2. Further challenges are related to knowledge sharing between geographically and organizationally dispersed stakeholders.

14.3.3 Challenges in SE Technology

From a technological perspective, several trends in SE have been highlighted in Sect. 14.2.3. In terms of technological challenges in SE, several points can be identified:

- **Supporting formal modelling**: as noted by several authors, modelling systems, and in particular systems of systems (SoS), requires well-founded formal modelling methods [43]. An associated challenge is to ensure that modelling languages have a well-founded semantic basis [42]. This may assist in the translation of conceptual, qualitative representations of SoS into formal, mathematical representations, which are able to be explored and analysed in both qualitative and quantitative modes. In SE, models must be dynamic because of the changing needs for using the various models.

- **Supporting analysis of large-scale systems**: with increasing computational power, digital twins and simulations of systems are increasingly suitable for scaling up towards SoS levels. Simulations of SoS may point out valuable lessons regarding behavior, evolution and performance of systems (of systems) [44]. In addition, the aspect of usability should be taken into consideration: with increasing complexity, technologies such as digital twins offer advanced means of visualization and interactivity with underlying systems, and support multiple viewpoints [45]. The latter point is particularly important in the light of allowing multiple stakeholders to interact with an artefact or system.

14.3.4 Challenges in SE Information and Knowledge

In terms of SE information and knowledge, two interrelated aspects pose significant challenges. The first aspect is the management of information and knowledge over time. Especially in the light of Product-Service System development, the increased complexity and higher quantity of data and knowledge exchanged during PSS design and development should be appropriately supported. While various researchers have discussed the basic tenets of knowledge management [46], the aspect of managing knowledge over long timespans and across dynamic product and process configurations has not received much attention [30, 47]. This is reinforced by observations in Chap. 5, where it is pointed out that "dependencies between partial product models (e.g., geometrical model) are often not automatically tracked, analyzed or used, for instance, for model update purposes or for spontaneous investigations as well as decision making".

A related challenge considers the traceability of information and knowledge over time, given dynamic evolution of system and SoS configurations [42]. As highlighted in Chap. 5, it is crucial to "enable traceability of decisions taken, tasks executed, knowledge used and artefacts developed throughout the whole lifecycle of an individual product". Still, the majority of SE methods and tools is aimed at supporting the development phase of the product lifecycle. There is relatively little attention towards the ability to maintain, expand or reuse information and knowledge during subsequent stages of the product or system lifecycle. More egregiously, as pointed out in Chap. 12, information related to sustainability is usually not represented in models, and is not consistent over time.

A final challenge relates to the lack of available formal methods for modelling, analysis and evaluation of systems (of systems), as pointed out earlier [48]. From the viewpoint of information and knowledge management over time, not only must these methods be developed, they should also be adequately supported through life. This includes consideration of verification methods and technologies, giving support to requirements analysis, design, testing, and code generation [49].

A further challenge is to capture stakeholder knowledge, which could be qualitative, though not all knowledge can be captured. The input of people remains paramount.

14.3.5 Challenges in SE Structures and Organisation

A major challenge concerning SE structures and organisation relates to system networks, in particular on the representation and analysis of SoS. As Maier indicates [50], the state of the art lacks methods and tools to describe and analyse the upper layers of systems of systems, where this refers to interactions among network elements [42]. In addition, suitable descriptive frameworks for properly describing and characterizing systems may be lacking.

This element is compounded when considering the dynamics of and within networks over time. As highlighted in Chap. 4, "individuals and organisations participate voluntarily in the networks. They can come and go at any time without warning". This implies that systems of systems and the associated networks are subject to change over time, requiring flexible, adaptable modelling, analysis and evaluation methods and tools [51]. The extension of networks to include IoT abstractions, with autonomous reconfiguration of systems based on context and need, as stated in Chap. 13 only emphasizes this necessity.

The advent of SoS, including cyber-physical systems and the Internet of Things, introduces organisational challenges regarding safeguarding the privacy of individuals, organisations and their data, while still supporting scalability and integration across networks and systems [52].

In addition, it is often difficult for people to understand their position and role within a SoS. People need to be aware of the importance of their task, of the impact of insufficient behaviour, as well as the need for their contribution. Regular communication between people involved in a system and also across systems is necessary [53]. A SoS in systems engineering also requires a visionary leader to manage the process(es) and to keep subsystems aligned and the overall system aligned with its context [18].

An extensive description framework would help to:

1. Characterise the system and its elements
2. Identify the interrelationships between elements, the hierarchical structure
3. Identify the context of the overall system, the needs to be satisfied, and the mutual relationships between context and system
4. Identify the stakeholders involved, overall and per subsystem
5. Identify the processes in the system and the people involved
6. Identify the necessary organisation and organisational rules, including authorities, responsibilities, and rights for the overall system and its subsystems
7. Give priorities for subsystem processes
8. Etc.

14.4 Conclusions

This final chapter has presented an overview of the main trends and challenges associated with Systems Engineering.

In this chapter, challenges in Systems Engineering have been identified. As has become evident, SE is an encompassing approach for tackling complex, real-world problems. SE is also an important approach in overarching approaches like Concurrent Engineering and Transdisciplinary Engineering. In essence, multiple disciplines, multiple functional roles, and multiple stakeholders need to collaborate in the processes making up the engineering systems. Moreover, a lifecycle perspective is essential in achieving a solution that is both useful and usable in the context in which the complex problem exists.

Throughout this chapter multiple aspects of Systems Engineering have been described based on the previous chapters of this book. For all the aspects making up a system, various challenges have been identified. Most importantly, in all aspects, the main challenge lies in developing methods and tools that are suitable to support the dynamic and evolving nature of the systems that need to be developed including the development system itself. Below, the various challenges will be listed.

The methods and tools that are desired need to:

1. Set up and manage performance measurements, preferably at a higher level of aggregation, including attention for sustainability [54]. Dependencies between subsystems need to be taken into account. In addition, ownership of performance measurement is required to guarantee follow-up.
2. Handle emergent behavior of stakeholders and handle data specifically for stakeholders. The methods and tools should incorporate suitable stakeholder representation [41]. Involvement of stakeholders in system performance measurement is needed to increase commitment and ownership.
3. Formally model and represent systems at multiple levels. Modelling should be both qualitative and quantitative, requiring suitable languages understandable for stakeholders [43, 48]. Moreover, dynamics should be supported.
4. Analyse large-scale systems [9].
5. Manage large quantities of information over time [45, 55].
6. Support traceability throughout the whole product-lifecycle, including sustainability, until now a reasonably weakly addressed area [39]. For this purpose formal modeling methods are needed including verification methods.
7. Capture stakeholder knowledge [31].
8. Represent and analyse SoSs, especially the upper layers. Again, representation languages and methods should be suited to support dynamics and allow adaptation and evaluation. The context of the Internet of Things makes such flexible modeling and evaluation extremely necessary and urgent [5].
9. Make people understand their role in the development system [56].

Although much effort has already been made, as has been shown in the previous chapters, much more work is still needed. Especially for the higher system levels more integrated approaches and efforts are desired, that value the different roles and disciplines of the people and stakeholders involved as well as the complex nature of systems with multiple levels and multiple aspects. As indicated before, managing an SE project for solving complex societal problems requires people with vision and power to motivate and mobilize the necessary people and value their respective input in the overall task.

In this respect, SE might not be sufficient for solving complex societal problems. Especially on the higher system levels, other approaches are needed. System approaches like SSM (Soft Systems Methodology) and CST (Critical System Thinking) are more suitable [3]. In these approaches stakeholders from various background need to be involved, who bring their own interpretation of the problem and different view on possible solutions. In addition, technical as well as social sciences need to be involved, which bring their own methods and tools for tackling (part of) the problem.

Transdisciplinary Engineering is an approach in which Systems Thinking and System Approaches make up an important part, taking into account the different levels of abstraction of the system of focus. There should, however, be openness to the input of other disciplines, including their methods and tools needed to deal with their aspect of the overall problem. Selecting methods, applying them, as well as further developing these methods in the context of complex societal problems, cannot be the task of one discipline alone [42].

References

1. Verhagen WJC, Stjepandić J, Wognum N (2015) Challenges of CE. In: Stjepandić J, Wognum N, Verhagen WJC (eds) Concurrent engineering in the 21st century. Springer International Publishing, pp 807–833
2. Wognum PM, Krabbendam JJ, Buhl H, Ma X, Kenett R (2004) Improving enterprise system support—a case-based approach. Adv Eng Inform 18:241–253
3. da Costa J Jr, Diehl JC, Snelders D (2019) A framework for a systems design approach to complex societal problems. Des Sci 5. https://doi.org/10.1017/dsj.2018.16
4. Shenhar AJ, Bonen Z (1997) The new taxonomy of systems: toward an adaptive systems engineering framework. IEEE Trans Syst Man Cybern Part A Syst Hum 27(2):137–145
5. Zhong R, Ge W (2018) Internet of things enabled manufacturing: a review. Int J Agile Syst Manage 11:126–154
6. Branger J, Pang Z (2015) From automated home to sustainable, healthy and manufacturing home: a new story enabled by the Internet-of-Things and Industry 4.0. J Manage Anal 2:314–332
7. Gumus B, Ertas A, Tate D, Cicek I (2008) The transdisciplinary product development lifecycle model. J Eng Des 19:185–200
8. Borsato M, Wognum N, Peruzzini M, Stjepandic J (2016) Transdisciplinary engineering: crossing boundaries. In: Proceedings of the 23rd ISPE inc. international conference on transdisciplinary engineering, 3–7 Oct 2016. IOS Press, Amsterdam
9. Kahlen F-J, Flumerfelt S, Alves A (2016) Transdisciplinary perspectives on complex systems: new findings and approaches. Springer International Publishing, Switzerland
10. Chesbrough H, Vanhaverbeke W, West J (2006) Open innovation: researching a new paradigm. Oxford University Press on Demand

11. Sobolewski M (2017) Amorphous transdisciplinary service systems. Int J Agile Syst Manage 10(2):93–115
12. Peruzzini M, Pellicciari M (2017) A framework to design a human-centred adaptive manufacturing system for aging workers. Adv Eng Inform 33:330–349
13. Fuqua J, Gress J, Harvey R, Phillips K, Baezconde-Garbanati L, Unger J, Palmer P, Clark MA, Colby SM, Morgan G, Trochim W (2003) Evaluating transdisciplinary science. Nicotine Tob Res 5(suppl 1):S21–S39
14. Wickson F, Carew AL, Russell AW (2006) Transdisciplinary research: characteristics, quandaries and quality. Futures 38(9):1046–1059
15. Stokols D (2006) Toward a science of transdisciplinary action research. Am J Community Psychol 38(1):63–77
16. Klein JT (2008) Evaluation of interdisciplinary and transdisciplinary research: a literature review. Am J Prev Med 35(suppl 2):S116–S123
17. Wognum N, Bil C, Elgh F, Peruzzini M, Stjepandić J, Verhagen WJC (2019) Transdisciplinary systems engineering: implications, challenges and research agenda. Int J Agile Syst Manage 12(1):58–89
18. Beckett RC, Vachhrajani H (2017) Transdisciplinary innovation: connecting ideas from professional and user networks. J Ind Integr Manage 02(04):1750016. https://doi.org/10.1142/S2424862217500166
19. Stjepandić J, Wognum N, Verhagen WJC (2015) Concurrent engineering in the 21st century: foundations, developments and challenges. Springer International Switzerland
20. Polk M (2015) Transdisciplinary co-production: designing and testing a transdisciplinary research framework for societal problem solving. Futures 65:110–122
21. Wekerle T, Pfouga A, Stjepandic J, Mai P (2018) Intellectual property protection in smart systems engineering on exchange of simulation models. Adv Transdisc Eng 7:198–207
22. Civerchia F, Bocchino S, Salvadori C, Rossi E, Maggiani L, Petracca M (2017) Industrial Internet of Things monitoring solution for advanced predictive maintenance applications. J Ind Inform Integr 7:4–12
23. Lu Y (2016) Industrial integration: a literature review. J Ind Integr Manage 1(2):1650007. https://doi.org/10.1142/s242486221650007x
24. Madni AM, Madni CC, Lucero SD (2019) Leveraging digital twin technology in model-based systems engineering. Systems 7(1):7. https://doi.org/10.3390/systems7010007
25. Cheng Y, Chen K, Sun H, Zhang Y, Tao F (2018) Data and knowledge mining with big data towards smart production. J Ind Inf Integr 9:1–13
26. La Rocca G, Van Tooren MJL (2007) Enabling distributed multi-disciplinary design of complex products: a knowledge based engineering approach. J Des Res 5(3):333–352
27. La Rocca G, van Tooren MJL (2009) Knowledge-based engineering approach to support aircraft multidisciplinary design and optimization. J Aircr 46(6):1875–1885
28. Van der Velden C, Bil C, Xu X (2012) Adaptable methodology for automation application development. Adv Eng Inform 26(2):231–250
29. Sharpanskykh A, Treur J (2010) A temporal trace language for formal modelling and analysis of agent systems. In: Dastani M, Hindriks KV, Meyer J-JC (eds) Specification and verification of multi-agent systems. Springer, US, pp 317–352
30. Carneiro HCC, Pedreira CE, França FMG, Lima PMV (2017) A universal multilingual weightless neural network tagger via quantitative linguistics. Neural Netw 91(1):85–101
31. Elgh F (2008) Supporting management and maintenance of manufacturing knowledge in design automation systems. Adv Eng Inform 22(4):445–456
32. Stjepandić J, Verhagen WJC, Liese H, Bermell-Garcia P (2015) Knowledge-based engineering. In: Stjepandić J, Wognum N, Verhagen WJC (eds) Concurrent engineering in the 21st century. Springer International Publishing, pp 255–286
33. Verhagen WJC, De Vrught B, Schut J, Curran R (2015) A method for identification of automation potential through modelling of engineering processes and quantification of information waste. Adv Eng Inform 29(3):307–321

34. Li S, Tang D, Yang J, Wang Q, Ullah I, Zhu H (2019) A novel approach for capturing and evaluating dynamic consumer requirements in open design. Adv Eng Inform 39:95–111
35. Zhang X, Hu F, Zhou K, Sato K (2017) Reflecting meaning of user experience: semiotics approach to product architecture design. In: Chen C-H et al (eds) Advances in transdisciplinary engineering, vol 5. IOS Press, Amsterdam, pp 737–744
36. Wang W, Hu F (2018) Service design of urban bloodmobile based on PSS design support tools. In: Peruzzini M et al (eds) Advances in transdisciplinary engineering, vol 7. IOS Press, Amsterdam, pp 917–926
37. Wognum N, Trappey A (2008) PLM challenges. Adv Eng Inform 22(4):419–420
38. Biahmou A, Stjepandic J (2016) Towards agile enterprise rights management in engineering collaboration. Int J Agile Syst Manage 9(4):302–325
39. Elgh F (2014) Automated engineer-to-order systems a task oriented approach to enable traceability of design rationale. Int J Agile Syst Manage 7(3–4):324–347
40. Nielsen CB et al (2015) Systems of systems engineering: basic concepts, model-based techniques, and research directions. J ACM Comput Surv 48(2):1–41
41. Ring J, Madni AM (2005) Key challenges and opportunities in 'system of systems' engineering. In: 2005 IEEE international conference on systems, man and cybernetics
42. Ertas A (2010) Understanding transdiscipline and transdisciplinary process. Transdisc J Eng Sci 1:55–73
43. Jain S, Hutchings CW, Lee Y-TT (2015) Building analytical support for homeland security. In: Rainey LB, Tolk A (eds) Modeling and simulation support for system of systems engineering applications. Wiley, Hoboken, pp 219–248
44. Fernandez R, Kienbaum GS, Neto ÁA, Ferreira MGV (2016) T-PROST: a transdisciplinary process oriented framework to support the product design phase in systems concurrent engineering. In: Borsato M, Wognum N, Peruzzini M, Stjepandic J (eds) Advances in transdisciplinary engineering, vol 4. IOS Press, Amsterdam, pp 758–767
45. Kienbaum GS, Fernandez R, Silva EKT, Maria R, Coicev M, Gartenkraut E, Rodrigues M, Neto AA, Ferreira M (2016) A transdisciplinary process oriented framework to support generic PLM implementation for use by small and medium enterprises. In: Borsato M, Wognum N, Peruzzini M, Stjepandic J (eds) Advances in transdisciplinary engineering, vol 4. IOS Press, Amsterdam, pp 808–817
46. Emmer C, Fröhlich A, Stjepandic J (2013) Advanced engineering visualization with standardized 3D formats. IFIP Adv Inf Commun Technol 409:584–595
47. Alavi M, Leidner DE (2001) Review: knowledge management and knowledge management systems: conceptual foundations and research issues. MIS Q 25(1):107–136
48. Bermell-Garcia P, Verhagen WJC, Astwood S, Krishnamurthy K, Johnson JL, Ruiz D, Scott G, Curran R (2012) A framework for management of knowledge-based engineering applications as software services: enabling personalization and codification. Adv Eng Inform 26(2):219–230
49. Stolt R, Johansson J, André S, Heikkinen T, Elgh F (2016) How to challenge fluctuating requirements – results from three companies. In: Borsato M, Wognum N, Peruzzini M, Stjepandic J (eds) Advances in transdisciplinary engineering, vol 4. IOS Press, Amsterdam, pp 1061–1070
50. Beisheim N, Kiesel M, Rudolph S (2018) Digital manufacturing and virtual commissioning of intelligent factories and Industry 4.0 systems using graph-based design languages. In: Peruzzini M et al (eds) Advances in transdisciplinary engineering, vol 7. IOS Press, Amsterdam, pp 93–102
51. Maier MW (2005) Research challenges for systems-of-systems. In: 2005 IEEE international conference on systems, man and cybernetics
52. Lu Y (2018) Cybersecurity research: a review of current research topics. J Ind Integr Manage 03(04):1850014. https://doi.org/10.1142/S2424862218500148
53. DeTombe D (2015) Human problem handling. In: Handling societal complexity. A study of the theory of the methodology of societal complexity and the COMPRAM methodology. Springer, Heidelberg, New York, Dordrecht, London, pp 81–154
54. Gaziulusoy AI, Ryan C, McGrail S, Chandler P, Twomey P (2016) Identifying and addressing challenges faced by transdisciplinary research teams in climate change research. J Clean Prod 123:55–64

55. Verhagen W, de Boer L, Curran R (2017) Component-based data-driven predictive maintenance to reduce unscheduled maintenance events. In: Chen C-H et al (eds) Advances in transdisciplinary engineering, vol 5. IOS Press, Amsterdam, pp 3–10
56. Orellana DW, Madni AM (2014) Human system integration ontology: enhancing model based systems engineering to evaluate human-system performance. Procedia Comput Sci 28:19–25

The manufacturer's authorised representative in the EU is Springer Nature Customer Service Centre GmbH, Europaplatz 3, 69115 Heidelberg, Germany. If you have any concerns regarding our products, please contact ProductSafety@springernature.com

Printed and bound by CPI Group (UK) Ltd, Croydon, CR0 4YY
30/09/2026
02225657-0012